Algebraische Modelle, Kategorien und Gruppoide

STUDIEN ZUR ALGEBRA UND IHRE ANWENDUNGEN

Im Auftrage des
Zentralinstituts für Mathematik und Mechanik
der Akademie der Wissenschaften der DDR
herausgegeben von L. Budach und H.-J. Hoehnke

BAND 7

ALGEBRAISCHE MODELLE, KATEGORIEN UND GRUPPOIDE

ALGEBRAISCHE MODELLE, KATEGORIEN UND GRUPPOIDE

Herausgegeben von Dr. HANS-J. HOEHNKE

AKADEMIE-VERLAG · BERLIN
1979

Zentralinstitut für Mathematik und Mechanik, DDR-108 Berlin, Mohrenstr. 39
Erschienen im Akademie-Verlag, DDR-108 Berlin, Leipziger Straße 3–4

Lizenznummer: 202 · 100/409/79
Gesamtherstellung: VEB Druckerei „Thomas Müntzer“, 582 Bad Langensalza
Bestellnummer: 7626422 (2174/7) · LSV 1025
Printed in GDR
DDR 36,– M

Inhalt

E. Realisierungstheorie

F. Modelltheorie

G. Kompliziertheitstheorie

Einleitung

Die Beiträge dieses Bandes gruppieren sich um folgende Themen:

A. Allgemeine Kategorientheorie
B. Spezielle Kategorien
C. Gruppen und Gruppoide
D. Algebren über Monaden
E. Realisierungstheorie
F. Modelltheorie
G. Kompliziertheitstheorie

A. Allgemeine Kategorientheorie

1. Nach LAWVERE und TIERNEY ist ein elementarer Topos eine Kategorie E mit (i) endlichen Limites, (ii) Exponenzierung und (iii) einem Teilobjekte-Klassifikator t: $1 \to \Omega$. Beispiele von Topoi (in jedem Falle als Kategorien variabler Mengen vorstellbar) sind: Mengen, endliche Mengen, Mengen in einem KRIPKE-Modell der Mengenlehre, allgemeiner jede Funktor-Kategorie $\mathsf{S}^{\mathfrak{C}}$, wo $\mathfrak{C}$ eine kleine Kategorie und S die Kategorie der Mengen ist. Die Kategorie der Garben über einem topologischen Raum ist ein Topos, und die triviale Kategorie 1 ist ein Topos. Wie die Logik in der Mengenlehre entwickelt werden kann, so kann sie auch innerhalb eines Topos entwickelt werden. Dabei spielt Ω die Rolle des Objekts der Wahrheitswerte. Da die Mengen variabel sind, ist die Logik nicht klassisch, sondern intuitionistisch. Der Artikel von COLE gibt mit einem Minimum an technischem Apparat einen Überblick über die grundlegenden Resultate der Theorie der klassifizierenden Topoi und beschränkt sich daher auf Topoi in der Kategorie S.

2. Bekanntlich (vgl. GABRIEL und ULMER, KAPHENGST und REICHEL) existieren zu jeder kleinen Kategorie C freie Limes-Erweiterungen und freie Colimes-Erweiterungen. Es erhebt sich die Frage, ob stets eine freie Limes-Colimes-Erweiterung existiert. Für kleine Limites und kleine Colimites existiert eine solche Erweiterung im allgemeinen nicht (vgl. M. DEHEN, 1969). KAPHENGST und REICHEL geben eine positive Antwort im Fall endlicher Limites und endlicher Colimites. Die Erweiterung wird mittels der Theorie der heterogenen gleichungspartiellen Algebren gebildet. Die Konstruktion ist für die Beschreibung algebraischer Eigenschaften von komplizierten partiellen Operationen in der Theorie der Programmiersprachen von Bedeutung.

B. Spezielle Kategorien

3. Mac Lane (1961) und Puppe (1962) begannen damit, Ideen und Methoden der Kategorientheorie zum Studium der Theorie binärer Relationen einzusetzen. In der Folgezeit wurden von verschiedenen Autoren Kategorien binärer Relationen über abelschen, exakten, regulären und anderen Klassen von Kategorien betrachtet. Calenko gibt hier eine Klassifizierung von Kategorien binärer Relationen im Rahmen der umfassenderen Klasse der Kategorien mit Involution (I-Kategorien) in Abhängigkeit von der Kategorie aller darin enthaltenen funktionalen Morphismen. Die Bedeutung dieser Resultate beruht darauf, daß die Theorie binärer Relationen in verschiedenartigen Gebieten Anwendung findet, wie zum Beispiel in der Funktionalanalysis, der homologischen Algebra, der mathematischen Linguistik und der Modellierung automatisierter Informationssysteme.

4. Bei verschiedenen Anwendungen treten Kategorien mit Epi-Mono-Faktorisierung ihrer Morphismen auf, so bei der abstrakten kategorientheoretischen Beschreibung der zu einer Varietät von universalen Algebren gehörigen Kategorie. Schwab betrachtet eine spezielle derartige Klasse von Kategorien, die Klasse der G-regulären Kategorien. Sie besteht aus endlich-vollständigen Kategorien, in denen Faserprodukte für G-reguläre Epimorphismen (das sind Coequalizer von Morphismenpaaren) existieren und jeder Morphismus f eine G-reguläre Faktorisierung $f = uh$ zuläßt (d. h. u Mono und h G-regulärer Epi). Von dieser Art sind von-Neumann-reguläre Kategorien R mit gewissen speziellen Eigenschaften. Wird vorausgesetzt, daß R Faserprodukte hat, so kann die Kawaharasche Konstruktion von Relationen benutzt und die I-Kategorie $\mathsf{Rel}\ \mathsf{R}$ der Relationen gebildet werden. Dies wird auf die Konstruktion von Kongruenzen und zur Gewinnung von Isomorphie-Theoremen in R angewandt.

5. Inverse Kategorien sind Kategorien mit eindeutig bestimmten verallgemeinerten Inversen und bilden gleichzeitig eine Verallgemeinerung der inversen Halbgruppen und der von-Neumann-regulären Kategorien.

Kastl gewinnt ein Darstellungstheorem für inverse Kategorien, das als Spezialfälle Verallgemeinerungen von Darstellungssätzen von Ljapin und Preston für inverse Halbgruppen als inverse Halbgruppen von injektiven partiellen Abbildungen einer Menge enthält. Dieses Theorem beruht auf zwei Feststellungen. Erstens folgt aus einem Kriterium von Isbell, daß jede inverse Kategorie konkret ist. Zweitens: Es sei X eine Kategorie mit Epi-Mono-Faktorisierung und $\mathfrak{M}$ eine Teilkategorie von X mit $\operatorname{Iso} \mathsf{X} \subseteqq \mathfrak{M} \subseteqq \operatorname{Mono} \mathsf{X}$. Die Kategorie X sei $\mathfrak{M}$-lokal-klein, habe $\mathfrak{M}$-Durchschnitte und genüge der Bedingung $xy \in \mathfrak{M} \Rightarrow x \in \mathfrak{M}$. Bildet man die Kategorien $\mathfrak{M}\text{-}\mathsf{PX} \supseteqq \mathfrak{M}\text{-}\mathsf{PBX}$ der partiellen bzw. partiell-bijektiven Morphismen $a \xrightarrow{(m,n)} b$ als Äquivalenzklassen (m, n) (wo $m\colon c \to a \in \mathfrak{M}$, $n\colon c \to b \in \mathsf{X}$ bzw. $m, n \in \mathfrak{M}$) bezüglich einer sich in natürlicher Weise ergebenden Äquivalenz, so ist jede inverse Teilkategorie von X isomorph zu einer Teilkategorie von $\mathfrak{M}\text{-}\mathsf{PBX}$, und ein entsprechender Satz folgt für inverse Teilkategorien von $\mathfrak{M}\text{-}\mathsf{PX}$.

C. Gruppen und Gruppoide

6. Die Menge $\mathfrak{A}$ aller Isomorphismen zwischen den Teilalgebren einer universalen Algebra A bildet bezüglich der Hintereinanderausführung dieser Funktionen und bezüglich $(\subseteqq) \cap (\mathfrak{A} \times \mathfrak{A})$ ein Beispiel eines induktiven Gruppoids im Sinne von EHRESMANN, das induktive Gruppoid $\mathcal{A}(\mathsf{A})$ der partiellen Automorphismen von A. MICHLER und SCHRECKENBERGER nennen ein induktives Gruppoid $\mathscr{C}$ algebraisch, wenn eine universale Algebra $\mathfrak{A}$ mit $\mathscr{C} \cong \mathcal{A}(\mathsf{A})$ existiert. Die vorliegende Arbeit gibt einen grundlegenden Beitrag zur Frage der Vollständigkeit algebraischer induktiver Gruppoide.

7. Es sei $A^{(n)}$ bzw. $B^{(n)}$ die Gruppe aller affinen Transformationen bzw. aller Bewegungen des R^n. Zwei Untergruppen G_1 und G_2 von $B^{(n)}$ heißen geometrisch äquivalent in einer Menge $U \subseteqq A^{(n)}$, wenn $G_1 = \alpha G_2 \alpha^{-1}$ für geeignetes $\alpha \in U$ ist. Dieser Äquivalenzbegriff spielt eine wichtige Rolle in der Kristallographie, wo man im Fall $n = 3$, $U = A^{(3)}$, die 230 Raumgruppen(-Klassen) erhält. Verfeinerte Fragestellungen führen FICHTNER auf einen allgemeineren Begriff der geometrischen Äquivalenz von Gruppoiden. Dabei ergeben sich Beziehungen zu Untersuchungen von WECHLER (1972) zur Symmetrie-Beschreibung physikalischer Systeme.

8. SACHS (1963) schlug die Untersuchung einer Korrespondenz vor zwischen Gruppen und Invariantensystemen aus orientierbaren Flächen F_i und Graphen G_i, welche diese Flächen triangulieren. VOSS untersucht diese interessante Korrespondenz in verschiedenen konkreten Fällen.

D. Algebren über Monaden

9. KURPIEL definiert ein Paar von Funktoren, Semantik und Struktur, zwischen der Kategorie der Monaden und der Kategorie der Funktoren, die rechtsseitige KAN-Erweiterungen entlang sich selbst besitzen. Damit kann eine algebraische Form des BECKschen Vergleichstheorems abgeleitet werden.

10. Wie A. DAY (1975) zeigte, lassen sich die stetigen Verbände im Sinne von SCOTT (1971 a) als Algebren über einer Monade definieren. JARZEMBSKI betrachtet hier den Fall der vollständigen Verbände, die zusammen mit den Supremum-erhaltenden Abbildungen eine Kategorie Lat bilden. Zunächst wird allgemein für einen Endofunktor $R\colon \mathsf{C} \to \mathsf{C}$ einer Kategorie C mit endlichen Coprodukten die Frage der Existenz freier R-Algebren und die zu R gehörige Monade UV untersucht. Der spezielle Endofunktor $S\colon \mathsf{Lat} \to \mathsf{Lat}$ mit $S(X) = B \times X^2 \sqcup X^2$ (B ein vollständiger Verband) ergibt dann für jeden vollständigen Verband X die freie S-Algebra über X, die mit dem Verband der Flußdiagramme über dem Alphabet X im Sinne von SCOTT (1971 b) identisch ist.

E. Realisierungstheorie

11. Die Kategorientheorie hat sich als ein geeignetes Mittel zur Einführung algebraischer Begriffe und Methoden in die Automatentheorie erwiesen. Dazu ist eine Fülle von Untersuchungen erschienen. ADÁMEK gibt eine systematische Übersicht über die grundlegendsten Ergebnisse dieser Richtung unter Hervorhebung der Theorie der minimalen Realisierung und der NERODE-Äquivalenzen. Durch Verwendung des Begriffs einer F-Algebra (d. h. einer funktoriellen Algebra im Sinne von ADÁMEK, KOUBEK und anderen) wird größtmögliche Klarheit der Begriffsbildungen erreicht.

12. EHRIG und KREOWSKI geben eine axiomatische Darstellung der Theorie der minimalen Realisierung in der Automaten- und Systemtheorie unter möglichst schwachen Voraussetzungen. Dabei wird die F-Morphismendarstellung (F ein Funktor der Form $\mathsf{G}_0 \times \mathsf{K} \rightarrow \mathsf{Set}$) im Sinne von EHRIG (1974) angewandt. Es ist zu erwarten, daß sich viele der von der Gruppe ADÁMEK-KOUBEK-REITERMANN-TRNKOVÁ aufgestellten Resultate über ARBIB-MANES-Maschinen auf den allgemeineren, verschiedene interessante Anwendungen gestattenden Kontext des vorliegenden Beitrags übertragen lassen.

F. Modelltheorie

13. VOLGER gibt eine Übersicht über aktuelle Untersuchungen zu den algebraischen Grundlagen der Modelltheorie. Dazu werden Garben von Algebren über BOOLEschen Räumen, insbesondere BOOLEsche Garben eingeführt und in Verbindung mit dem Begriff einer Diskriminator-Varietät die Rolle des ternären Diskriminators für diese Garben studiert. Sodann werden die Darstellungs- und Dualitätssätze von KEIMEL und WERNER (1975) und Beispiele für Diskriminator-Varietäten angegeben. Ein verallgemeinertes FEFERMAN-VAUGHT-Theorem, das Eigenschaften der Algebra der globalen Schnitte mit solchen der Halme verbindet, gestattet es, Transfer-Theoreme für modelltheoretische Eigenschaften zu gewinnen. Damit können die Entscheidbarkeitsresultate von BURRIS und WERNER (1975) für residual endliche Diskriminator-Varietäten bewiesen werden. Schließlich werden Anwendungen auf die Frage der ω-Kategorizität und andere modelltheoretische Eigenschaften beschrieben.

G. Kompliziertheitstheorie

14. TRACHTENBROT gibt in dem vorliegenden Beitrag hauptsächlich eine Übersicht über den Stand der Untersuchungen des Problems, wie die Wahl des Parameters p des einer TURING-Maschine beigefügeten BERNOULLI-Generators D_p (der in jedem Takt zu den Bandsymbolen ein Symbol (0 oder 1) hinzufügt, und zwar 1 mit der Wahrscheinlichkeit p) die Geschwindigkeit der Berechnung beeinflußt (zeitliche Kompliziertheit). Er knüpft dabei an Beiträge von JABLONSKI, BARZDIN, TURAKAINEN, RABIN und anderen an.

Der Herausgeber war bemüht, den einzelnen Beiträgen eine einheitliche Bezeichnungsweise zugrunde zu legen. Diese schwierige Aufgabe ist nicht immer gelungen. Insbesondere in den Diagrammen bzw. Abbildungen kommen vom Text abweichende Buchstabentypen vor, was allerdings kaum zu Mißverständnissen führen dürfte.

Der Herausgeber

Classifying Topoi

by J. C. Cole, Brighton

In this paper, I wish to present, (with no claim to originality) in asnon-technical way as possible, the basic results in the theory of classifying topoi. I therefore restrict the discussion to topoi defined over "the" category of sets, $\mathsf{S} = \mathsf{Set}$. It should, however, be borne in mind, that all these constructions may be made, with valid results, over any base topos. For the special results that are required in the general case, the reader is referred to Wraith's excellent discursive article [11], or, more technically, to Kock and Wraith [5].

Let us recall Lawvere and Tierney's definition: an *elementary topos* is a category, E, which has

(i) finite limits, i.e. a terminal object, 1, and pullbacks;
(ii) exponentiation, i.e., for A, B, there is an object B^A, and for each C a bijection between maps $C \to B^A$ and maps $C \times A \to B$;
(iii) a subobject classifier, $t\colon 1 \to \Omega$, with the property that for any monic $A' \rightarrowtail A$, there is a unique characteristic map $A \xrightarrow{\chi} \Omega$ such that

$$\begin{array}{ccc} A' & \rightarrowtail & A \\ \downarrow & & \downarrow{\scriptstyle \chi} \\ 1 & \xrightarrow{t} & \Omega \end{array}$$

is a pullback diagram.

Clearly, these properties are "set-like": to have finite limits expresses the fact that we wish to talk of n-tuples of elements, and to have the theory of equality; exponentiation gives us the higher-order facilities that set-theoretic language has; and the final axiom says that Ω is an object of "truth-values", whence we may identify subsets and properties — as with the aussonderungsaxiom of set theory.

However, there are many different topoi — in each case best thought of as categories of variable sets. We obtain as examples, sets, finite sets, Boolean-valued sets, the sets in a Kripke-model of set-theory, but much more generally, any functor-category $\mathsf{S}^{\mathfrak{C}}$ where $\mathfrak{C}$ is a small category; the category of sheaves on a topological space is a topos, and the trivial category 1 is also a topos (sheaves on the empty space).

Just as logic may be developed within set-theory, we may do so inside a topos. The object of truth-values, Ω, gives propositional logic, characteristic maps are predicates, and exponentiation gives us higher-order logic: "the power-set" of A is Ω^A. Of course, since we are dealing with variable sets, the logic is not classical, but intuitionistic. Notice that Ω is the "power-set" of 1, and since subobjects of a fixed object have an ordering, we find Ω is a partially-ordered object in E. The properties of E are in fact reflected in Ω itself: there is a product, $\wedge: \Omega \times \Omega \to \Omega$ and an exponentiation, $\Rightarrow: \Omega \times \Omega \to \Omega$, and of course, the terminal object is the "element" t: $1 \to \Omega$. It can be shown that E also has finite colimits (since E^{op} is a category algebraic over E), and so we have a coproduct, $\vee: \Omega \times \Omega \to \Omega$, and an initial element, $f: 1 \to \Omega$ (f for false is the characteristic map of the unique $0 \rightarrowtail 1$, where 0 is E's initial object). Osius' two papers in [8], [9] give two different expositions of the fully developed logic inside an arbitrary topos.

If $\{R \rightarrowtail A^n, f: A^m \to A\}$ are relations and functions in any topos, we may interpret any formula of the appropriate language, and characterise sentences as true, false, or even "partially true" — the truth value would be a subject $U \rightarrowtail 1$ (or its characteristic map, $u: 1 \to \Omega$).

The notion of sheaf on a space is obtained from presheaf, a functor $O_X^{op} \to \mathsf{Set}$, where O_X is the category of open sets and inclusions in the space X, by saying that "things defined locally, on a covering, glue together to form things defined globally". If we have a notion of covering, we may speak of sheaves on a category — these will be those functors which can't tell the difference between coverings and isomorphisms. This idea was made precise by Grothendieck and we present here the elementary-topos version, in "logical" terms. Suppose we are given a certain set of truth-values, $J \rightarrowtail \Omega$ which we wish to identify with $t: 1 \to \Omega$. Then a subobject, $i: A' \rightarrowtail A$ will be "as good as the whole object" if its characteristic map factors not through $t: 1 \to \Omega$, but through $J \rightarrowtail \Omega$. We call these the J-covering monos. In fact, we may calculate the truth-value $[i$ is an isomorphism$]$ which is an "element" $1 \to \Omega$; i is a J-covering iff this element factors as $1 \to J \to \Omega$. Notice that J itself has a characteristic map, $j: \Omega \to \Omega$. For the above ideas to make sense, J (and hence j) must satisfy certain properties. We say j is a Grothendieck topology if the following diagrams commute:

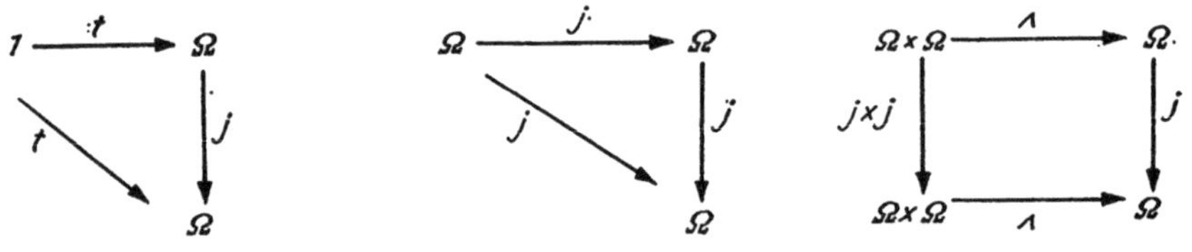

which say: an isomorphism is J-covering, a J-covering of a J-covering is a J-covering, and finally, J is a filter — as the intersection of two J-coverings is a J-covering. Note that the word "topology" is misleading — while j acts as a "closure operator" on truth-values, it is telling us the largest truth-value covered by a given truth-value, and is not giving the topological closure at all. The word is used because it allows us to think spatially about a structure which may well not be spatial in origin.

If j is a Grothendieck topology, we have defined a J-covering mono (also called J-dense, or j-dense) to be a subobject $A' \overset{i}{\rightarrowtail} A$ for which $\chi_i: A \to \Omega$ factors as $A \to J \rightarrowtail$ $\rightarrowtail \Omega$; equivalently, the square

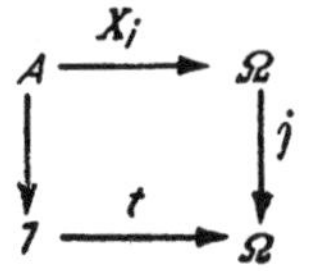

commutes. A j-sheaf (or J-sheaf) is an object, F, for which, for any j-covering mono, $i: A' \rightarrowtail A$, the induced $\mathsf{E}(A, F) \to \mathsf{E}(A', F)$ is bijective. That is, for any $A' \to F$ there is a unique extension $A \to F$ making

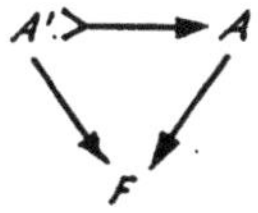

commute. Thus, as far as F is concerned, $A' \rightarrowtail A$ might as well be an isomorphism. The full subcategory of j-sheaves itself turns out to be a topos. In fact it is a reflective subtopos — to each object, X of E there is a "best approximating" sheaf, $R(X)$, with a map $X \to R(X)$ such that for any sheaf F, every map $X \to F$ factors uniquely as $X \to R(X) \to F$: as far as sheaves are concerned, X might as well be isomorphic to $R(X)$. $R(-)$ is in fact a functor, sheafification, and the universal mapping property states that it is left adjoint to the inclusion:

$$\mathrm{Sh}_j(\mathsf{E}) \underset{\hookrightarrow}{\overset{R}{\leftarrow}} \mathsf{E}\,.$$

There is another way of describing this situation. We may calculate the "truth-value" [f is an isomorphism] of any map, $f: A \to B$, to obtain a subobject of 1, [f iso] $=$ $= U \rightarrowtail 1$. If $U \rightarrowtail 1$ is a j-covering, then J "thinks" that f is indeed an isomorphism. It is in fact possible to adjoin a formal inverse, $f^{-1}: B \to A$ to each such map, f, to obtain a category, $\mathsf{E}(J^{-1})$, by analogy with the process of formally inverting certain elements of a ring. $\mathsf{E}(J^{-1})$ is actually equivalent to $\mathrm{Sh}_j(\mathsf{E})$, and the functor $\mathsf{E} \to \mathsf{E}(J^{-1})$ corresponds to sheafification under this equivalence. An important fact is that the sheafification functor is left exact, that is, it preserves all finite limits. Since it is a left adjoint functor it automatically preserves all colimits.

Now, suppose E is the topos $\mathbf{S}^{\mathcal{O}X^{op}}$ of presheaves (of sets) on a space, X. We take j-coverings to be actual coverings in the topological sense: if $\{U_i \subseteqq U\}$ is a collection of open sets in X, which cover U, then we obtain an inclusion of presheaves $\coprod_i \hat{U}_i \rightarrowtail \hat{U}$ where

$$\Big(\coprod_i \hat{U}_i\Big)(V) = \begin{cases} 1 \text{ if } \; V \subseteqq U_i & \text{for some } i, \\ \varphi \text{ otherwise}, \end{cases}$$

$$\hat{U}(V) = \begin{cases} 1 & \text{if } \; V \subseteqq U \\ \varphi & \text{otherwise}, \end{cases}$$

and we declare such inclusions of presheaves to be j-coverings. The resulting j-sheaves are the classical sheaves on the space X. It can be shown that for any topos, E, and for any topology, j in E, $\mathrm{Sh}_j(\mathsf{E})$ is actually a topos.

Now, suppose we are given two topologies, j, k; we form the topoi $\mathrm{Sh}_j(\mathsf{E})$ and $\mathrm{Sh}_k(\mathsf{E})$. Their intersection is again a reflective subcategory of E, with left-exact reflection functors. It may be shown that all such subcategories are sheaves for some topology, 1, with more coverings than j and k. Conversely, we may consider those coverings which are both j- and k-coverings. Again, these are given by a topology. We see that the topologies form a lattice. In the case of a topos which has all limits and colimits (which will be the case for most of our work), this is even a complete lattice — the intersection of a family of topologies is a topology. We may thus speak of the weakest topology for which a particular collection of maps are all coverings. In the topos of sheaves for this topology, these maps will have been "forced" to be isomorphisms.

If $f: X \to Y$ is a continuous map of topological spaces, we obtain functors, f_* the direct image, f^* the inverse image between the (classical) sheaf-categories,

$$\mathrm{Sh}(X) \underset{f_*}{\overset{f^*}{\leftrightarrows}} \mathrm{Sh}(Y)$$

with f^* left-exact and left adjoint to f_*. Recall that for $\mathrm{Sh}_j(\mathsf{E})$, R is left-exact and left adjoint to i. We are thus led to define a morphism of topoi $f: \mathsf{E} \to \mathsf{F}$ to be a pair of functors

$$\mathsf{E} \underset{f_*}{\overset{f^*}{\leftrightarrows}} \mathsf{F}$$

with f^* left-exact and left adjoint to f_*.

If we take $\mathsf{F} = \mathsf{S}$, it is easy to show that $f_*(X) \cong \mathsf{E}(1, X)$, and $f^*(S) \cong \coprod_S 1$, from which we may deduce not only that coproducts of 1 exist, but all set-indexed colimits, and limits as well. Conversely, if E has coproducts of 1 indexed by arbitrary sets, then it has a unique morphism to S, and hence has all limits and colimits. We shall from now on restrict our attention to those topoi which are defined over S in this sense, (although S may in fact be replaced by any topos), and speak of S-topoi.

There are not only morphisms between topoi, but also natural transformations between topos-morphisms. We define a 2-cell $\alpha: f \Rightarrow g$, where $f, g: \mathsf{E} \to \mathsf{F}$ to be a natural transformation between the inverse-image functors, $\alpha: f^* \Rightarrow g^*$. (Note that these are in bijective correspondence with natural transformations $\beta: g_* \Rightarrow f_*$.) Thus the topos-morphisms from E to F actually form a category, $\mathrm{Topoi}/\mathsf{S}(\mathsf{E}, \mathsf{F})$, whence $\mathrm{Topoi}/\mathsf{S}$ is itself a 2-category. The construction of the topos of sheaves on a space actually gives an embedding of the category of spaces and continuous maps into $\mathrm{Topoi}/\mathsf{S}$, whence we may think of a topos both as a generalised set-theory, and as a generalised space. It is the interplay between these aspects which gives a powerful tool for studying structures which vary (e.g. over a space), as in algebraic geometry, differentiable manifolds, etc. — in most cases, the variable structure is the ring of germs of (continuous, algebraic, differentiable, etc.) functions at each point. Since S is the topos of sheaves on the one-point space, we may identify the points of a topos, E, with the topos-morphisms $\mathsf{S} \to \mathsf{E}$. A warning should be given, however: a topos may have far more structure than can be deduced from knowing its points. In particular, many non-trivial topoi have no points at all! (see, e.g. WRAITH in [11] for an example).

We have seen the two basic ways of constructing topoi: there are presheaf topoi, of the form $\mathsf{S}^{\mathfrak{C}}$, or, more generally $\mathsf{E}^{\mathfrak{C}}$, with $\mathfrak{C}$ a small category, and there are sheaf subtopoi. We shall see that each has its universal property with regard to topos-morphisms, and use these properties, together with the internal logic, to construct topoi with a "universal", or "generic" model of a given first-order theory of a certain kind, that is, we construct the classifying topos of the theory. We recall first some technical results which underly almost all work with categories: there seem to be only three basic principles in "pure" category theory, and we use all three — the present author takes the view that category theory should best be understood as a language of abstract structure, exceedingly rich in applications, rather than a pure theory in its own right, poor in results.

The first is the celebrated YONEDA Lemma:

YL. *If* $F \in \mathbf{S}^{\mathfrak{A}^{op}}$ *and* $A \in \mathfrak{A}$, *then there is a natural ismorphism*

$$F(A) \cong \mathbf{S}^{\mathfrak{A}^{op}}(\mathfrak{A}(-, A), F(-)).$$

We use the hom-functor $\mathfrak{A}(-, A)$ to represent A itself as an object of $\mathbf{S}^{\mathfrak{A}^{op}}$. When F is also of this form, we obtain the immediate corollary that this representation, $Y: \mathfrak{A} \to \mathbf{S}^{\mathfrak{A}^{op}}$ is full and faithful — we may think of $\mathfrak{A}$ as a full subcategory of $\mathbf{S}^{\mathfrak{A}^{op}}$. More importantly, it allows us to think of the functor F as having "elements at time A", since an element $x: 1 \to F(A)$ corresponds bijectively (and naturally) to an "element" $Y(A) \to F$ in the topos $\mathbf{S}^{\mathfrak{A}^{op}}$. A map $t: B \to A$ in $\mathfrak{A}$ gives us a transition $F(t)$: $F(A) \to F(B)$ from "elements at time, A" to "elements at the later time, B", corresponding by YL to $Y(B) \to Y(A) \to F$. This idea leads to "KRIPKE-JOYAL" semantics (OSIUS [9]). But remember when we know the elements of $F(A)$ for each A, we know F, so we must be able to recover F from the hom-sets $\mathbf{S}^{\mathfrak{A}^{op}}(Y(A), F)$: in fact, F is the colimit over all values of A. We formalise this by forming a category F of "elements of F", which has two descriptions, via YL. The objects of F are pairs, (x, A) where x is an element of $F(A)$, i.e., the objects are maps $1 \xrightarrow{x} F(A)$. A morphism, $(x, A) \to (y, B)$ is to be a map $B \xrightarrow{f} A$ in $\mathfrak{A}$ for which $F(f)\,(x) = y$, that is, the morphisms are commutative triangles,

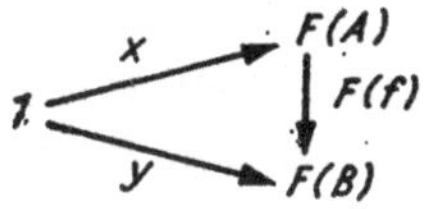

An equivalent description is to say that an object of F is a map $Y(A) \xrightarrow{x} F$ in $\mathbf{S}^{\mathfrak{A}^{op}}$, and a morphism $(x, A) \to (y, B)$ is a commuting triangle,

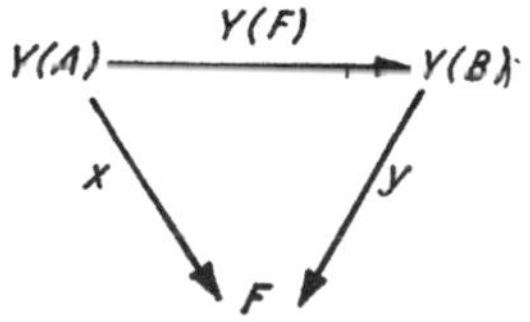

in $\mathbf{S}^{\mathfrak{A}^{op}}$. The first description speaks of the set-theoretic elements, and the second speaks of the "elements" in $\mathbf{S}^{\mathfrak{A}^{op}}$. There is an obvious "forgetful" functor, $\mathsf{F}^{op} \to \mathfrak{A}$, which, when followed by Y gives us a diagram $\mathsf{F}^{op} \to \mathfrak{A} \to \mathbf{S}^{\mathfrak{A}^{op}}$ whose colimit is F.

Lemma. $F \cong \varinjlim\,(\mathsf{F}^{op} \to \mathfrak{A} \xrightarrow{Y} \mathbf{S}^{\mathfrak{A}^{op}})$.

We have displayed F as a colimit of representable functors — which is the categorical way of stating that F is determined by its values.

The next result, that of KAN extensions, is a natural consequence of this idea. Suppose we have a functor $F: \mathfrak{A} \to \mathsf{E}$, where E is some category with all colimits. We may ask how to extend F to $\mathbf{S}^{\mathfrak{A}^{op}} \to \mathsf{E}$, thinking of $\mathfrak{A}$ as a subcategory of $\mathbf{S}^{\mathfrak{A}^{op}}$. We know how to define such an extension on representable functors (those of the form $Y(A)$), and every functor, we have seen, is a colimit of representables. Suppose $G \in \mathbf{S}^{\mathfrak{A}^{op}}$. Then we form the category, G, of "elements of G" as above, so that $G = \varinjlim\,(\mathsf{G}^{op} \to \mathfrak{A} \to \mathbf{S}^{\mathfrak{A}^{op}})$. We define $G \otimes_{\mathfrak{A}} F = \varinjlim\,(\mathsf{G}^{op} \to \mathfrak{A} \xrightarrow{F} \mathsf{E})$, which exists, since E has colimits, and this actually extends F. Furthermore, we have a functor $\mathsf{E} \to \mathbf{S}^{\mathfrak{A}^{op}}$, which takes $X \in \mathsf{E}$ to $\mathsf{E}(F(\ ), X): \mathfrak{A}^{op} \to \mathbf{S}$.

KE. *For any complete* E, *any small* $\mathfrak{A}$, *there is a correspondence (bijective up to isomorphism) between functors* $F\colon \mathfrak{A} \to$ E *and adjoint pairs* E $\underset{f_*}{\overset{f^*}{\leftrightarrows}}$ $S^{\mathfrak{A}^{op}}$ *(f^* left adjoint to f_*), given by sending F to* $() \otimes_{\mathfrak{A}} F$ *and its right adjoint,* $E(F(\,), -)$, *and sending* (f^*, f_*) *to the composite* $Yf^*\colon \mathfrak{A} \to S^{\mathfrak{A}^{op}} \to$ E. *Furthermore the triangle*

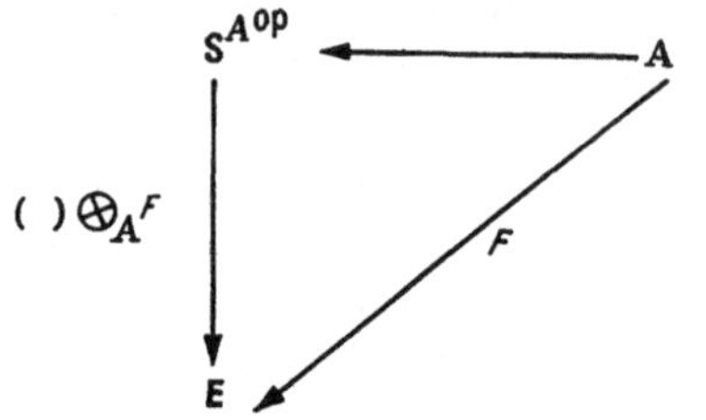

commutes up to isomorphism in E.

In fact this correspondence is actually an equivalence of categories: Cat $(\mathfrak{A}, E) \cong$ $\cong$ Adj $(E, S^{\mathfrak{A}^{op}})$.

[The "tensor-product" notation is no coincidence: replace S by Ab, let $\mathfrak{A}$ be a ring = = monoid-object in Ab = single-object category object in Ab, make all functors additive, and we are describing the tensor product of a right $\mathfrak{A}$-module, F, with a left-module, G. We thus see that Kan extensions are indeed generalised tensor products. Below, we give a Morita-like theorem.]

Needless to say, with a construction so natural, it too may be described in many different ways. Let us now suppose that E is in fact an S-topos, so that we may think of its objects as "variable sets". Clearly, the "constant sets" in E are those of the form $\coprod_S 1$. Recall that E comes equipped with a topos-map $p\colon E \to S$, and $p^*(S) = \coprod_S 1$, so p^* is the "constant set"-functor. It is also left-exact. Now, since $\mathfrak{A}$ is a small category, it has a set A_0, of objects, a set, A_1, of maps, and so we may describe $\mathfrak{A}$ by a diagram

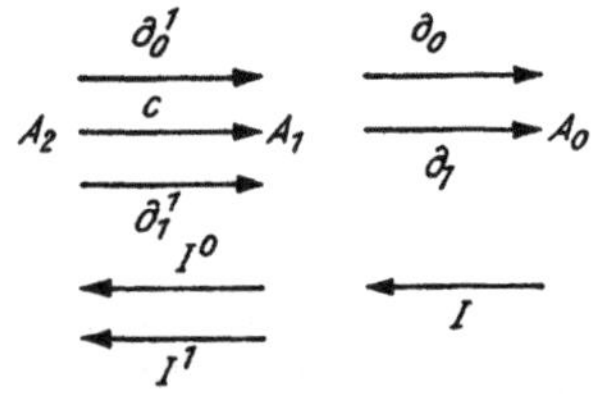

where ∂_0, ∂_1 are "domain" and "codomain" maps, I takes an object to its identity-map, A_2 is the set of composable pairs of maps (the pullback of ∂_0 with ∂_1), c is the composition map, and so on. Applying p^* to this diagram gives us a similar diagram in E, which we call the category-object, $p^*(\mathfrak{A})$ in E. Just as we did for set-valued functors, we may describe $F\colon \mathfrak{A} \to$ E by means of a category-object of "E-elements", $F^{op} \to p^*(\mathfrak{A})$, with its obvious projection functor. If $G \in S^{\mathfrak{A}^{op}}$, we may describe $G \otimes_A F \in$ $\in$ E as follows. Form G $\to \mathfrak{A}$, the category of "elements of G". Map it to E by p^*, to obtain p^*G $\to p^*(\mathfrak{A})$. Form the pullback (of functors!) in E:

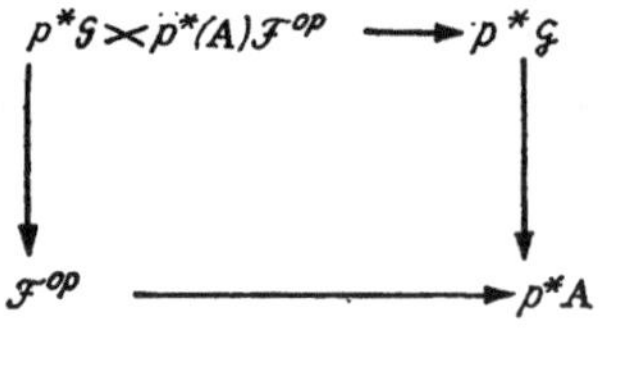

which may be interpreted by considering $p^*\mathsf{G} \underset{p^*(\mathfrak{A})}{\times} \mathsf{F}^{op} \to \mathsf{F}^{op}$ as the "category of E-elements" of a functor $\mathsf{J} \to \mathsf{E}$. We now take its colimit in E.

Lemma. $G \otimes_{\mathfrak{A}} F = \lim (\mathsf{J} \to \mathsf{E})$.

We shall not go into the proof here, since it essentially involves "internalising", i.e. working in E "as if" it were "the" category of sets. The point of describing this construction (in fact it is the "coend" construction given in MAC LANE [7]) is to be able to use the third fundamental result in category theory to find a relationship between F and $(\ ) \times_{\mathfrak{A}} F$. Recall that amongst partial orderings, the directed partial orderings have particularly nice properties. Correspondingly, filtered categories also have good behaviour. A category is *filtered* if

(i) For each pair of objects A_1, A_2 there are maps

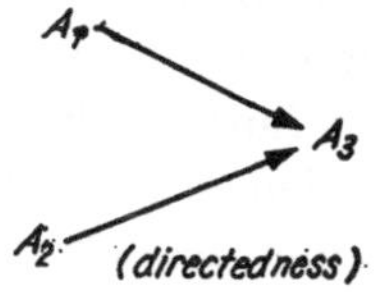

(ii) For each pair of maps, $f, g\colon A_1 \rightrightarrows A_2$, there exists a map $h\colon A_2 \to A_3$ such that $fh = gh\colon A_1 \rightrightarrows A_2 \to A_3$ commutes.

Now suppose $F\colon \mathfrak{C} \to \mathsf{S}$ is a functor and $\mathfrak{C}$ is filtered. Then the colimit of F is "described by its finite initial segments" — that is, in forming the colimit as a quotient of the disjoint union of the values of F, we only identify those elements which are already idenfitied somewhere in the diagram $F\colon \mathfrak{C} \to \mathsf{S}$. In particular, we may compute the colimit of a product of such functors $(F \times G)(C) = F(C) \times G(C)$ as the product of the colimits, and similarly for all finite limits.

Proposition. *If $\mathfrak{C}$ is filtered, and I is finite, then for any diagram $F\colon \mathfrak{C} \times I \to \mathsf{S}$,*

$$\varinjlim_{\mathfrak{C}} \varprojlim_I F = \varprojlim_I \varinjlim_{\mathfrak{C}} F\,.$$

I.e., if the values are written $F_i(c)$, we have for example, $\varinjlim_{\mathfrak{C}} \prod_i F_i(c) \cong \prod_i \varinjlim_{\mathfrak{C}} F_i(c)$ and similarly for equalizers, terminal object and pullbacks.

This theorem is also true of categories other than S, and is in particular true, when stated in the correct manner, in any topos, E. Now, to connect this with KAN extensions, suppose first that $\mathfrak{A}$ has finite limits, and $F\colon \mathfrak{A} \to \mathsf{S}$ is left-exact. It is easy to see then that the category of elements, F, is co-filtered (F^{op} is filtered), and conversely, if F is co-filtered, then F is left-exact. We may go further. If $F\colon \mathfrak{A} \to \mathsf{E}$ is left-exact, the corresponding $\mathsf{F}^{op} \to p^*\mathfrak{A}$ has F^{op} filtered, in a sense appropriate to categories in E. When we pull F^{op} back along $p^*\mathsf{G} \to p^*\mathfrak{A}$, it remains filtered. We have thus described $(\ ) \otimes_{\mathfrak{A}} F$ as the functor:

$$\mathsf{S}^{\mathfrak{A}^{op}} \xrightarrow{p^*} \mathsf{E}^{p^*\mathfrak{A}^{op}} \xrightarrow{q^*} \mathsf{E}^{\mathsf{F}} \overset{\lim}{\rightrightarrows} \mathsf{E}$$

where p^* is the "constant set" functor applied to the diagrams representing the categories of "elements of G", q^* is "pulling-back along $\mathsf{F}^{op} \to p^*\mathfrak{A}$", and the colimit is a filtered colimit precisely when F is left-exact. Since p^* is left-exact (it is the inverse-image functor of the topos-map $p\colon \mathsf{E} \to \mathsf{S}$), and "pulling-back" is always left-exact, we see that the KAN extension $(\ ) \otimes_{\mathfrak{A}} F\colon \mathsf{S}^{\mathfrak{A}^{op}} \to \mathsf{E}$ is left-exact, if and only if F is left-exact. Putting this together and thinking of a left-exact functor as a flat module we obtain the non-additive Morita theorem:

Classification Theorem (Grothendieck, Giraud, Diaconescu [4], [3], [1]).

For a small category, $\mathfrak{A}$, having finite limits, and an S*-topos,* E*, there is an equivalence of categories*

$$\mathrm{Lex}\,(\mathfrak{A}, \mathsf{E}) \simeq \mathrm{Topoi/S}\,(\mathsf{E}, \mathsf{S}^{\mathfrak{A}^{\mathrm{op}}})\,.$$

KE told us that functors $\mathfrak{A} \to \mathsf{E}$ corresponded to adjoint pairs $\mathsf{E} \leftrightarrows \mathsf{S}^{\mathfrak{A}^{\mathrm{op}}}$, and the above argument says that f^* is left-exact (and hence gives a topos-map $\mathsf{E} \to \mathsf{S}^{\mathfrak{A}^{\mathrm{op}}}$) precisely when the corresponding F is left-exact. All the constructions may be made functorial, which gives us the further information about morphisms. In fact, the Diaconescu result is rather stronger, and was obtained by "lifting" Giraud's strengthening to the elementary topos case. We may take S to be any elementary topos (not just those defined over sets), and $\mathfrak{A}$ may be any category-object in S, not just those having finite limits — but in this case a "left-exact functor, $\mathfrak{A} \to \mathsf{E}$" has a special definition (in fact we ask simply that its category of E-elements is cofiltered).

We are now ready to start constructing some classifying topoi, namely those for algebraic structures. Recall, for example, that we may describe a group as a set, G, together with functions $m\colon G \times G \to G$, $e\colon 1 \to G$, $i\colon G \to G$ making the diagrams:

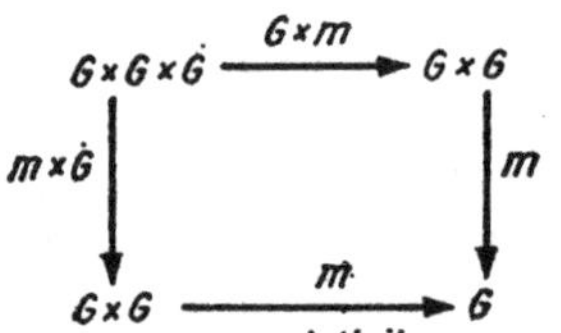

associativity

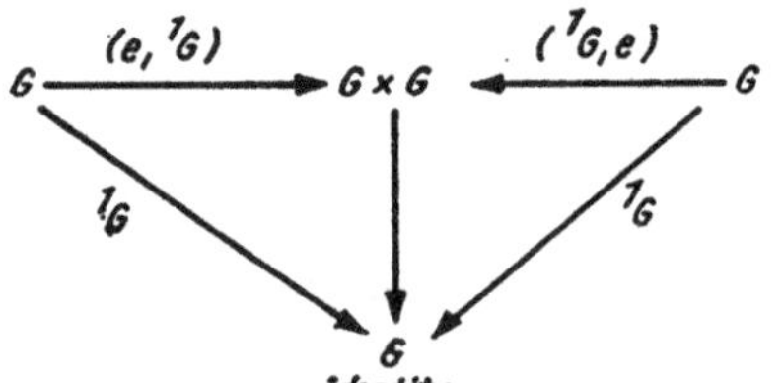

identity

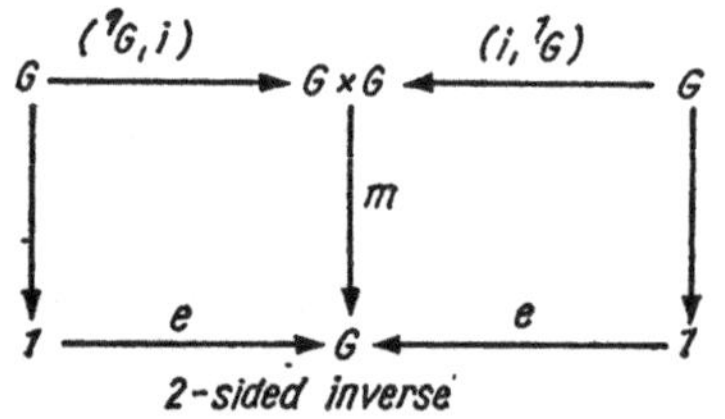

2-sided inverse

commute. Since the only categorical structure imposed on the diagrams is the notion "product", we are able to find a category, G, containing objects 1, G, G^2, — together with maps, m, e, i, and all maps which come from them by products, compositions and projections, so that a group is given precisely by a functor preserving finite products, $\mathsf{G} \to \mathsf{S}$. (Note: we are considering the terminal object 1 to be the empty product.) In fact, G has a concrete description as the dual of the finitely generated free groups. The maps $G^n \to G^m$ in G will be m-tuples of n-ary operations defined from m, e, i so we need only to know the n-ary operations $G^n \to G$, that is, the equivalence classes (under the group-theoretic axioms) of functions of n variables definable in the theory of groups — that is, equivalence classes of words in n variables. But such a (equivalence class of) word is just an element of $F(n)$, the free group on n generators. And an element $x \in F(n)$ is specified by a group-homomorphism $F(1) \to F(n)$ taking the generator of $F(1)$ to the element $x \in F(n)$. Thus $\mathsf{G}(G^n, G^m) = \mathsf{Gp}(F(m), F(n))$, so $\mathsf{G}^{op} \simeq$ the full subcategory of finitely generated free groups. The thesis of LAWVERE [6] showed how this process may be carried out for any finitary algebraic theory (it can also be generalised to the infinitary case — one needs infinite products, and hence infinitely generated free algebras to express the infinitary operations). Thus $\mathsf{Gp} = \mathrm{Prod}\,(\mathsf{G}, \mathsf{Set})$, and we may define a group in any category E with finite products to be a product-preserving functor $\mathsf{G} \to \mathsf{E}$.

Now, left-exact is a little more than just product-preserving — we do not yet speak of equalizers. We enlarge G a little, to obtain a category which gives groups as left-exact functors; this is achieved by "freely adjoining equalizers" to G — a procedure which may be carried out generally, but which in our case has again a concrete description. Adjoining equalizers to G is the same as adjoining coequalizers to G^{op}, so we consider what is the coequalizer of a pair of maps:

$$F(n) \underset{k}{\overset{h}{\rightrightarrows}} F(m)\ .$$

We must identify the elements $h(i)$, $k(i)$ $(i = 1, \dots, h)$, so we see that we are taking the quotient of $F(m)$ by the equations $h(i) = k(i)$. Thus, we obtain in this way all the finitely presented groups. It is not hard then to show that for any category, E, with finite limits, that $\mathsf{Gp}(\mathsf{E}) \simeq \mathrm{Lex}\,(\mathrm{fp}\ \mathsf{Gp}^{op}, \mathsf{E})$, by noting that the underlying object of the group is the value at $F(1)$ of the left-exact functor. Again, the whole procedure extends to all finitary algebraic theories. We can in fact extend it further — for example, the theory of categories is itself algebraic in the wider sense: it is many-sorted (objects and maps) and its operations are defined on subsets of products, given by equations, i.e. on finite limits of the sort-objects, so that a category "is" a left-exact functor $\mathsf{C}^{op} \to \mathsf{Set}$: the category C^{op} may be described as the category of simplicial types degenerate above dimension 2, or equivalently, as the dual of the category of finitely presented categories (those free on a finite graph, subject to certain diagram commutativity equations). Now, let A be a finitary algebraic theory, FP A be its category of finitely presented algebras, and let E be any S-topos.

Theorem. $\mathsf{A}\text{-alg}\,(\mathsf{E}) \cong \mathrm{Lex}\,(\mathrm{FP}\ \mathsf{A}^{op}, \mathsf{E})$ *(as seen above)*
$\simeq \mathrm{Topoi}/\mathsf{S}(\mathsf{E}, \mathsf{S}^{\mathrm{FP}\mathsf{A}})$.

Thus an A-algebra in E is given by a topos-map $\mathsf{E} \to \mathsf{S}^{\mathrm{FP}\mathsf{A}}$. We call $\mathsf{S}^{\mathrm{FP}\mathsf{A}}$ the *A-algebra classifying topos*, by analogy with classifying spaces. Thinking spatially, the *points* of $\mathsf{S}^{\mathrm{FP}\mathsf{A}}$ are the topos-maps $\mathsf{S} \to \mathsf{S}^{\mathrm{FP}\mathsf{A}}$, i.e. A-algebras in S. Thus $\mathsf{S}^{\mathrm{FP}\mathsf{A}}$ is the "space of all A-algebras", and, for example, to give a group, continuously varying over a space, X, i.e. a sheaf of groups, or a group in $\mathrm{Sh}(X)$, is the same as giving a continuous map from X to the space of all groups, i.e. a topos-map $\mathrm{Sh}(X) \to \mathsf{S}^{\mathrm{FP}\mathsf{Gp}}$.

The following are some particular examples that have been considered. S^G for G a group, which was the original example, classifying principal homogeneous G-spaces, or G-torsors, although this does not exactly fall into our present algebraic scheme (GROTHENDIECK). Taking A to be the trivial theory, with no operations, whose models are just sets and whose finitely presented models are finite sets, we see that $\mathsf{S}^{\mathsf{Fin\,Set}}$ is the object-classifier: $\mathrm{Topoi}/\mathsf{S}(\mathsf{E}, \mathsf{S}^{\mathsf{Fin\,Set}}) \simeq \mathrm{Lex}\,(\mathsf{Fin\,Set}^{op}, \mathsf{E}) \simeq \mathrm{Obj}\,(\mathsf{E}) \simeq \mathsf{E}$, noting that $\mathsf{Fin\,Set}$ is in fact a topos within a topos, we see that $\mathsf{S}^{\mathsf{Fin\,Set}}$ is even a topos-object in the category of topoi, whence $\mathrm{Topoi}/\mathsf{S}(-, \mathsf{S}^{\mathsf{Fin\,Set}}) : (\mathrm{Topoi}/\mathsf{S})^{op} \to \mathsf{Cat}$ actually takes topoi as its values (WRAITH). $\mathsf{S}^{\mathrm{FP\,Ring}}$ is the ring classifier, a particularly interesting topos, since a scheme may be thought of as a functor from finitely presented rings to sets (Hakim).

We may ask, where do the algebras classified come from? Consider the diagram:

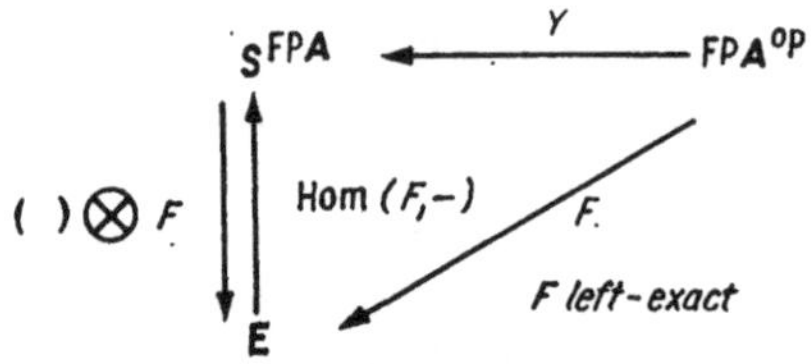

F determines an A-algebra in E, but, since Y is left-exact, it too determines an A-algebra in S^{FPA}. Since $Y(\,) \otimes F \cong F$, we see that F may be thought of as the inverse image of the A-algebra Y under the topos-map $E \to S^{FPA}$. Thus we consider Y to be the *universal*, or *generic* A-algebra. Its underlying object is in fact the forgetful functor FP A $\to$ S.

We turn next to considering theories richer than purely equational — we examine first-order theories, possibly many-sorted, since these give no more difficulties than one-sorted (apart from notational difficulties).

Suppose $\mathcal{L}$ is a first order predicate language (possibly with finitely many different sorts), so that we are given variables $x_1, x_2, \ldots,$ (one such set for each sort), a binary relation called equality, $=$, (one for each sort), relation symbols, $R(, \ldots ,)$ with fixed arity (and fixed sorts, $R((\,)^A, (\,)^B, \ldots, (\,)^N)$), function symbols $f(, \ldots ,)$ of fixed arity (and fixed sorts $f^A(()^B, ()^C, \ldots, ()^N)$), together with propositional connections, $\wedge$, $\vee$, $\Rightarrow$, and quantifiers $\forall$, $\exists$ (one for each sort). We consider constant symbols to be 0-ary function symbols, and we ask that there be 0-ary relation symbols, $\top$ (true) and $\bot$ (false), i.e. propositional constants. Terms are constructred inductively as usual — a variable is a term, and if $t_1 \ldots t_n$ are terms, f an n-ary function symbol, then $f(t_1 \ldots t_n)$ is a term. (If we have many sorts, each term has a sort: the variable x^A is of sort A, and if $t_1 \ldots t_n$ are of sorts $A, B, \ldots, N$, and f takes sorts $A, B, \ldots, N$ for arguments and is of sort K, then $f(t_1 \ldots t_n)$ is a term of sort K). Notice that a constant symbol is a term. Formulae are likewise defined inductively: if R is an n-ary relation symbol and $t_1 \ldots t_n$ are terms, then $R(t_1 \ldots t_n)$ is a formula; if φ, ψ are formulae, so are $\varphi \wedge \psi$, $\varphi \vee \psi$, $\varphi \Rightarrow \psi$; if φ is a formula and x a variable, then $\forall x \varphi$, $\exists x \varphi$ are formulae. (In the many-sorted case, the sorts of the terms must match up with the sorts of R, and variables must be quantified by quantifiers of the appropriate sort.) Notice that equality-formulae occur (only for terms of the same sort), and that $\top$, $\bot$ are formulae. A *theory* is determined by a set of *axioms*, that is, some collection of formulae. We make deductions according to the usual formal rules of logic — intuitionistic logic in the general case, since we may add as axioms the formulae $\neg\neg\varphi \Rightarrow \varphi$ for every φ, to obtain classical logic ($\neg\varphi$ is an abbreviation for $\varphi \Rightarrow \bot$). We write $\mathcal{T} \vdash \varphi$ if φ can be deduced from the axioms, $\mathcal{T}$. So far, our procedure is as usual amongst logicians, apart from the technical details of including $\bot$, $\top$ as formula, the possibility having many sorts, and the intuitionistic logic. But if we are to do categorical logic, we should be guided by the success of categorical algebra, and form a category from our theory, so that the models of the theory should be a certain class of functors, taking values in Set, in the first instance. So what in our category should correspond to a formula? In each model, M, of $\mathcal{T}$, each formula of one free variable defines a subset $\{a \in M : M \models$

$\varphi(a)\}$, and a formula $\varphi(x_1 \dots x_n)$ gives a subset $\{\langle a_1 \dots a_n\rangle \in M^n : M \models \varphi(a_1 \dots a_n)\} \subseteq M^n$. Thinking of formulae as extensions, we see that $\varphi(x) \wedge \psi(x)$ "is" the intersection of the "subsets" $\varphi(x)$, $\psi(x)$, disjunction "is" union, and so on. Formulae with no free variables (called sentences) determine subsets of $M^0 = 1$, i.e., they are true, or false in a particular model. The extension of $\top$ will always be the one-point set, and of $\bot$, the empty set. If f is an n-ary function symbol, we have $f\colon M^n \to M$ in each model, M, and similarly for each term. But there may be functions definable in the theory which are not named by terms. On the other hand, a term τ gives a formula $y = \tau(x)$ which is clearly functional. We define the category $\mathfrak{T}$ to have as objects all formulae of $\mathscr{L}$, and a map of $\mathfrak{T}$, $\varphi(x_1 \dots x_n) \to \psi(x_1 \dots x_m)$ will be an equivalence class of formulae $\chi(x_1 \dots x_{n+m})$ which are functional with domain and codomain φ and ψ, in the sense that

$$\mathscr{T} \vdash \forall x_1 \dots x_n \varphi(x_1 \dots x_n) \Rightarrow \exists x_{n+1} \dots x_{n+m} \chi(x_1 \dots x_{n+m}) \wedge \psi(x_{n+1} \dots x_{n+m}) ,$$

$$\mathscr{T} \vdash \chi(x_1 \dots x_{n+m}) \wedge \chi(x_1 \dots x_n, x_{n+m+1} \dots x_{n+2m}) \Rightarrow x_{n+1} = x_{n+m+1} \wedge \dots \wedge x_{n+m} = x_{n+2m} ,$$

$$\mathscr{T} \vdash \chi(x_1 \dots x_{n+m}) \Rightarrow \varphi(x_1 \dots x_n) \wedge \psi(x_{n+1} \dots x_{n+m}) .$$

The equivalence relation is simply that of logical equivalence — χ and χ' define the same map if

$$\mathscr{T} \vdash \chi(x_1 \dots x_{n+m}) \Rightarrow \chi'(x_1 \dots x_{n+m}) \quad \text{and} \quad \mathscr{T} \vdash \chi'(x_1 \dots x_{n+m}) \Rightarrow \chi(x_1 \dots x_{n+m}) .$$

Note — we insist that φ, ψ and χ have no more free variables than those shown.

It is intuitively clear that $\mathfrak{T}$ is indeed a category, and that set-theoretic arguments with elements may be translated into categorical arguments valid for $\mathfrak{T}$. Thus, for example, we prove that $\mathfrak{T}$ has finite limits, by taking the product of $\varphi(x_1 \dots x_n)$ and $\psi(x_1 \dots x_m)$ to be $\varphi(x_1 \dots x_n) \wedge \psi(x_{n+1} \dots x_{n+m})$ (notice the relabelling — any disjoint sets of free variables would do), $\top$ is the terminal object, and equalizers are formed in the obvious way: the equalizer of χ, χ' is the formula (or rather, equivalence class determined by) $\exists x_{n+1} \dots x_{n+m} \chi(x_1 \dots x_{n+m}) \wedge \chi'(x_1 \dots x_{n+m})$, so that we are picking out those $x_1 \dots x_n$ for which the "functions" χ, χ' agree. Clearly, $\bot$ is an initial object, a subobject $\varphi \rightarrowtail \psi$ is (isomorphic to) a formula with the same free variables as φ for which $\mathscr{T} \vdash \varphi(x_1 \dots x_n) \Rightarrow \psi(x_1 \dots x_n)$. The union of subobjects is just their disjunction, and the image of a map $\chi\colon \varphi \to \psi$ is the formula $\exists x_1 \dots x_n \chi(x_1 \dots x_n, x_{n+1} \dots x_{n+m})$, and is the coequalizer of the equivalence relation given by χ, in the usual "set-theoretic" way. The initial object, finite unions of subobjects, and images may similarly be shown to be stable under pulling-back along any map, again by arguments analogous to those used for sets. We show thus that $\mathfrak{T}$ is a regular category with stable $\bot$, $\vee$, and images. Such categories are often known as geometric categories. In fact, $\mathfrak{T}$ has even more structure: images are constructed from $\exists$, and form a functor left adjoint to pulling back subobjects; similarly, $\forall$ gives a right adjoint to pulling back; $\varphi \Rightarrow (-)$ may be characterised as a right adjoint to $\varphi \wedge (-)$ in the lattice of subobjects of an object, and in fact may be considered as a special case of a $\forall$-functor. The construction and properties of $\mathfrak{T}$ may be found in great detail in DIONNE's thesis [2].

Now, we may form the topos $\mathbf{S}^{\mathfrak{T}}$, which will certainly give us models of the algebraic part of the theory $\mathscr{T}$, since the YONEDA functor Y, is left-exact. Unfortunately, Y fails to preserve the extra structure corresponding to $\bot$, $\vee$, $\exists$, let alone $\Rightarrow$, $\forall$. However, for any topos-morphism $f: \mathsf{E} \to \mathsf{F}$, the inverse-image functor f^*, does preserve 0, images, and unions, since it is a left adjoint, hence preserves colimits. So we hope to "force Y to preserve $\perp$, $\vee$, $\exists$" in order to obtain a universal model for at least this part of the logic. Since f^* does not in general preserve $\Rightarrow$, $\forall$, we cannot expect classifying topoi for theories involving these connectives in their axioms. We consider now theories whose axioms are all of the form $\forall x_1 \ldots x_n \varphi \Rightarrow \psi$, where φ, ψ use only the logical symbols $\wedge$, $\vee$, $\exists$, $\top$, $\perp$ $=$, and call these geometric theories.

Proposition. *If $\mathcal{T}$ is a geometric theory, then models of $\mathcal{T}$ are in bijective correspondence with geometric functors M: $\mathfrak{T} \to \mathsf{S}$, i.e. functors which are left-exact, and preserve* 0, *images and finite unions of subobjects.*

The proof is by induction on the construction of the axioms — the formula $x = x$ is taken to the underlying set of the model, and, inductively, we show that each formula is in fact taken to (a set isomorphic to) its extension. Although $\forall$, $\Rightarrow$ are not necessarily preserved, it is clear that if $\forall x_1 \ldots x_n \varphi \Rightarrow \psi$ is an axiom of $\mathcal{T}$, then φ is a subobject of ψ in $\mathfrak{T}$, whence $M \models \forall x_1 \ldots x_n \varphi \Rightarrow \psi$ actually holds: those implications which are true in $\mathcal{T}$ are automatically preserved, and similarly the universal closure of any true formula will be preserved since $\top$ is preserved.

We use this proposition to define what we mean by a model of $\mathcal{T}$ in any other geometric category, and check that this semantics does not conflict with the other possible semantics — e.g. Kripke-Joyal semantics, or the "internal" semantics of Mitchell-Osius. We also note that a *map* of models, defined as a natural transformation between the geometric functors, does not conflict with our intuitive notion, namely, a function between the underlying sets which preserves the terms and relations of the model and hence preserves the truth of all geometric formulae (those not involving $\forall$, $\Rightarrow$). So, a model of $\mathcal{T}$ in a topos E is a geometric functor $\mathfrak{T} \to \mathsf{E}$, and a map of models is a natural transformation between such functors.

Now, how should we "force" the Yoneda embedding $\mathfrak{T} \to \mathsf{S}^{\mathfrak{T}^{op}}$ to become a model? We want the "comparison" morphisms, e.g. $Y(\varphi) \wedge Y(\psi) \to Y(\varphi \wedge \psi)$, $\mathrm{Im}_x(Y\varphi) \to$ $\to Y(\exists x\varphi)$, to become isomorphisms — so we define a topology for which these maps are coverings, so that they become isomorphisms when we sheafify. It may be shown that topologies $j: \Omega \to \Omega$ in $\mathsf{S}^{\mathfrak{T}^{op}}$ may be defined by covering families in $\mathfrak{T}$, so we define the *finite cover* topology by saying that a family $\{A_i \xrightarrow{f_i} A\}$ covers A if some finite subfamily $\{A_{i_k} \xrightarrow{f_{i_k}} A\}_{k=1}^{n}$ has the property that $\bigvee_{k=1}^{n} \mathrm{Im}\,(f_{i_k}) \cong A$. Thus the injections $\varphi \to \varphi \vee \psi$, $\psi \to \varphi \vee \psi$ form a covering, a single image map, $\varphi \to \exists x\varphi$ is a covering, and the empty set of maps covers the initial object. We say that a functor F: $\mathfrak{T} \to \mathsf{E}$ preserves coverings if for each covering family $\{A_i \xrightarrow{f_i} A\}$ in $\mathfrak{T}$, the family $\{F(A_i) \xrightarrow{F(f_i)} F(A)\}$ is a collectively epimorphic family in E (that is $F(f_i) \cdot h = F(f_i) \cdot k$ for each i implies $h = k$).

It should be noted that we may use this definition because in a topos, E, epimorphisms behave well; in other categories, a more "surjective" kind of epimorphism is needed.

Lemma. *A functor $\mathfrak{T} \to \mathsf{E}$ is geometric (i.e. is left-exact and preserves* 0, *images and finite unions) if and only if it is left-exact and cover-preserving.*

The proof is obvious.

Now, it may be shown that for any topology given by covering families in $\mathfrak{C}$, we obtain a topology $j\colon \Omega \to \Omega$ in $\mathsf{S}^{\mathfrak{C}^{op}}$ whose sheaves are those $F\colon \mathfrak{C}^{op} \to \mathsf{S}$ which satisfy the following "glueing condition":

F is a sheaf iff for every covering $\{C_i \to C\}$ in $\mathfrak{C}$, the diagram

$F(C) \to \prod_i F(C_i) \rightrightarrows \prod_{i,j} F(C_i \times_C C_j)$ is an equaliser diagram

This says that if $s_i \in F(C_i)$, $i \in I$, is a family of "local" elements, which agree on the overlaps: $s_i/C_i \times_C C_j = s_j/C_i \times_C C_j$, then the s_i's glue together to give a unique $s \in F(C)$ with $s/C_i = s_i$.

Consider the following diagram:

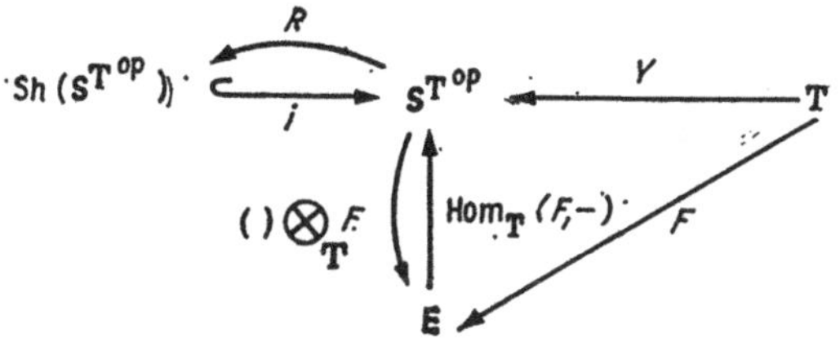

in which F is left-exact, and hence gives rise to the topos-morphism $\mathsf{E} \to \mathsf{S}^{\mathfrak{T}^{op}}$, by the Classification Theorem. We wish to show that this topos-morphism factors through the sheaf inclusion $\mathrm{Sh}(\mathsf{S}^{\mathfrak{T}^{op}}) \hookrightarrow \mathsf{S}^{\mathfrak{T}^{op}}$ exactly when F is geometric. We use a general theorem due to Grothendieck.

Theorem [4]. Topoi/S$(\mathrm{Sh}_J(\mathsf{S}^{\mathfrak{C}^{op}}), \mathsf{E}) \cong$ Lex Cov pres $(\mathfrak{C}, \mathsf{E})$, *where J is any topology on $\mathfrak{C}$, and the right-hand category consists of those functors which are left-exact and preserve coverings.*

Proof. Given a left-exact $F\colon \mathfrak{C} \to \mathsf{E}$, we form the topos-map

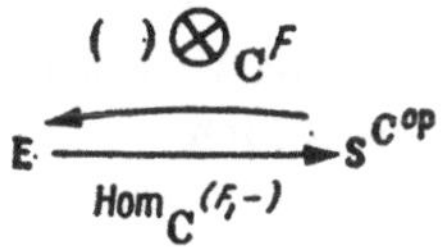

from the Classification Theorem. We ask that $\mathrm{Hom}_{\mathfrak{C}}(F, X)$ is a J-sheaf for each X. That is, for each J-covering family $\{C_i \to C\}$ in $\mathfrak{C}$, we have an equalizer diagram,

$$\mathsf{E}(F(C), X) \to \prod_i \mathsf{E}(F(C_i), X) \rightrightarrows \prod_{i,j} \mathsf{E}(F(C_i \times_C C_j), X)\,.$$

But this must hold for all X, so we have an equalizer diagram of functors:

$$\mathsf{E}(F(C), -) \to \prod_i \mathsf{E}(F(C_i), -) \rightrightarrows \prod_{i,j} \mathsf{E}(F(C_i \times_C C_j), -)\,.$$

Recall from the definitions of coequalizers that this is precisely equivalent to saying that

$$\coprod_{i,j} F(C_i \times_C C_j) \rightrightarrows \coprod_i F(C_i) \to F(C)$$

is a coequalizer in E. But F is left-exact, so $F(C_i \times_C C_j) \cong F(C_i) \times_{F(C)} F(C_j)$. Now, in a topos, the "pulling-back" functor along any morphism has a right adjoint, whence pulling-back preserves colimits.

Hence $\coprod_{i,j} F(C_i \times_C C_j) \cong \coprod_{i,j} F(C_i) \times_{F(C)} F(C_j) \cong \coprod_i F(C_i) \times_{F(C)} \coprod_i F(C_i)$, so our sheaf-condition is equivalent to stating that $\coprod_i F(C_i) \times_{F(C)} \coprod_i F(C_i) \rightrightarrows \coprod_i F(C_i) \to F(C)$ is a coequalizer diagram. But the two left-hand maps are just the kernel pair of the right-hand map, so we ask that $\coprod_i F(C_i) \to F(C)$ be the coequalizer of its kernel pair, i.e., that it be an epimorphism. This is clearly equivalent to stating that $\{F(C_i) \to F(C)\}$ forms a collectively epimorphic family, and so we have shown that $\mathrm{Hom}_{\mathfrak{C}}(F, -)$ factors through the inclusion functor, $\mathrm{Sh}(\mathsf{S}^{\mathfrak{C}^{op}}) \hookrightarrow \mathsf{S}^{\mathfrak{C}^{op}}$ iff F preverses coverings. It is a trivial matter to check that the inverse image is given by $\mathrm{Sh}(\mathsf{S}^{\mathfrak{C}^{op}}) \hookrightarrow$ $\hookrightarrow \mathsf{S}^{\mathfrak{C}^{op}} \xrightarrow{()\otimes_{\mathfrak{C}} F} \mathsf{E}$, whence we have a topos-map $\mathsf{E} \to \mathrm{Sh}(\mathsf{S}^{\mathfrak{C}^{op}})$. Notice in particular that we have also shown that the functor $\mathfrak{C} \xrightarrow{Y} \mathsf{S}^{\mathfrak{C}^{op}} \xrightarrow{R} \mathrm{Sh}_J(\mathsf{S}^{\mathfrak{C}^{op}})$ is itself left-exact and cover-preserving, so that composition with the inverse image functor of any topos-map $\mathsf{E} \to \mathrm{Sh}_J(\mathsf{S}^{\mathfrak{C}^{op}})$ will give us the inverse passage from topos-maps to left-exact cover-preserving functors.

We now have all the pieces that we require.

$$\begin{aligned}(\text{REYES [10]})\ \mathcal{T}\text{-models}\,(\mathsf{E}) &\simeq \mathrm{Geom}\,(\mathfrak{T}, \mathsf{E}) \\ &\simeq \mathrm{Lex\ Cov\ pres}\,(\mathfrak{T}, \mathsf{E}) \\ &\simeq \mathrm{Topoi}/\mathsf{S}\,(\mathsf{E}, \mathrm{Sh}(\mathsf{S}^{\mathfrak{T}^{op}}))\,,\end{aligned}$$

so that $\mathrm{Sh}(\mathsf{S}^{\mathfrak{T}^{op}})$ is in fact the $\mathcal{T}$-model classifier. The universal, or generic $\mathcal{T}$-model is again the (sheafified) YONEDA functor, $\mathfrak{T} \to \mathsf{S}^{\mathfrak{T}^{op}} \to \mathrm{Sh}(\mathsf{S}^{\mathfrak{T}^{op}})$, and every $\mathcal{T}$-model in every S-topos is the inverse image of this one, by some topos-map. In fact, REYES has shown that the sheafified YONEDA functor also preserves $\forall, \Rightarrow$, so that the generic model will be a model even for non-geometric first order theories. It will, however, lose the property of being generic, since the inverse image functors of topos-morphisms fail, in general, to preserve $\forall, \Rightarrow$. We can, however, build classifying topos for a wide range of theories-orderings, existential-algebraic, and so on. Of particular interest is the ZARISKI topos which classifies local rings — those rings for which

$$(1 = 0) \Rightarrow \perp \quad \text{and} \quad \exists y(xy = 1) \vee ((1 - x)\, y = 1)\,,$$

since, for spaces at least, a ring of functions whose stalks are local rings is local in the sense of topos-semantics. Thus the coordinate rings of schemes, algebraic spaces, analytic spaces, manifolds, and so on, are all classified by a topos-map to the ZARISKI topos. Very many constructions may be best understood through the use of classifying topoi. For example, let $\mathcal{T}$ be the theory of a localisation of a ring, in the following sense (due to KOCK): it is a 2-sorted theory, with sorts A, B, and a function symbol $f\colon A \to B$, with axioms saying A is a ring, B is a local ring, f is a ring homomorphism, and

$$\forall b_1 \in B\ \exists a_1, a_2 \in A : b_1 \cdot f(a_1) = f(a_2) \wedge \exists b_2 \in B : f(a_1) \cdot b_2 = 1$$

and

$$\forall a_1 \in A\ f(a_1) = 0 \Rightarrow \exists a_2 \in A : a_1 \cdot a_2 = 0 \wedge \exists b\, f(a_2) \cdot b = 1\,.$$

Now, if we take the classifying topos Sp, for this theory, it contains a generic localisation, $f\colon A \to B$, with A a ring and B a local ring. But A must be classified by some topos-map to the ring-classifier, $\mathfrak{R}$. Similarly, B is classified by a map to the local ring classifier, Zar:

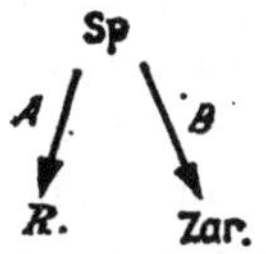

Suppose R is any ring (in S). Then it is classified by a point of $\mathfrak{R}$: $\mathsf{S} \xrightarrow{R} \mathfrak{R}$. We may pull back the map A along R, to obtain a topos:

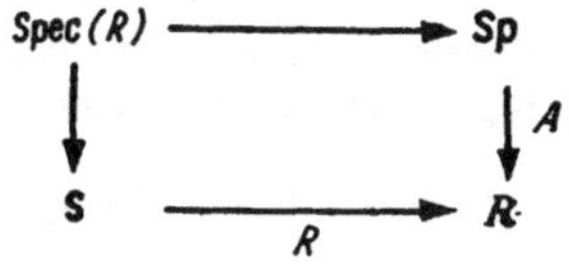

which is in fact the topos of sheaves on the (usual) spectrum of R. The generic localisation in Sp gives a localisation $\widehat{R} \to \widetilde{R}$ in Spec(R), which is exactly what we expect: $\widehat{R}$ is the constant sheaf, and $\widetilde{R}$ is the sheaf of local rings with which the spectrum is endowed. Studying the properties of R, Zar and Sp, thus gives us a great deal of information about affine schemes, and by "glueing", about all schemes. Thus we are able to use the logic of geometric theories to further the development of algebraic geometry.

I shall end by indicating some of the directions in which we can improve still further on the existence of classifying topoi. Since inverse image functors actually preserve all colimits, we can introduce infinitary disjunctions into our theories. We are also able to use infinitely many sorts: suppose I is an index set for our sorts, A_i. We categorise this as $A \to I$ where $A = \coprod_i A_i$. If forming the $\mathcal{T}$-classifier is thought of as "adjoining a generic $\mathcal{T}$-model to S", then it is possible to adjoin, say, a generic I-indexed collection of objects, $A \to I$. In fact, if we have a finite diagram, with certain objects specified, we may adjoin to S such a diagram, to obtain a classifying topos for such diagrams. We may then use a topology, and sheafify to make certain parts of the diagram into limits, colimits and so on. And this process may be carried out over an arbitrary topos. (One must take care to state things in a way which is intuitionistically correct, and to make the infinitary statements meaningful.) For example, by adjoining to E a generic left-exact cover-preserving functor $\mathfrak{C} \to \mathsf{E}^{\mathfrak{A}}$, for a given notion of covering, we may form a topos which classifies these — but by the theorem above, this topos actually classifies topos-morphisms $\mathsf{E}^{\mathfrak{A}} \to \mathrm{Sh}_{\mathsf{E}}(\mathfrak{C}, J)$, which amounts to saying that it is the exponential object $\mathrm{Sh}(\mathfrak{C}, J)^{(\mathsf{E}^{\mathfrak{A}})}$ in the category of topoi. In a similar way, Spec(R) is obtained by adjoining to the topos in which R lies, a generic localisation of R.

Finally, it should be noted that generic structures often have special properties. It may be shown that the generic ring is a field in the sense that

$$\neg (x = 0) \Rightarrow \exists y\, (xy = 1)\,.$$

Forcing is used in set-theory to construct generic sets, whose special properties lead to independence proofs. Thus, for example, to show the independence of the continuum hypothesis, one adjoins to a given model of set theory, a generic I-indexed collection of subsets of the natural numbers. This collection has the special property that each of the subsets is different: the associated $I \to 2^N$ is monic. By choosing I large, (bigger than $\aleph_1$) we negate the continuum hypothesis in the new topos, which actually "is" the associated BOOLEan-valued model of set theory. A "collapsing" by an ultrafilter gives an ordinary 2-valued model in which CH fails. In the opposite direction, a number of algebro-geometric results have been proved by model-theoretic methods: ultra-products, model-theoretic forcing, and so on. It seems that the connections between logic and geometry are ready for a deep and wide-ranging exploitation, using topoi as a major tool.

References

[1] R. DIACONESCU: Change of basis for some toposes, Ph.D. Thesis, Dalhousie University, 1973; published under the title "Change of base for toposes with generators" in J. Pure and Appl. Algebra 6 (1975), 191–218.

[2] A. DIONNE: Des théories élémentaires aux categories conceptuelles, M. Sc. Thesis, Université de Montreal, 1973.

[3] J. GIRAUD: Classifying topos, in "Toposes, Algebraic Geometry and Logic", Lecture Notes in Math. 274 (1972), 43–56.

[4] M. ARTIN; A. GROTHENDIECK; J. L. VERDIER: Théorie des topos et cohomologie étale des schémas, Lecture Notes in Math. 269 (1972).

[5] A. KOCK; G. C. WRAITH: Elementary Toposes, Aarhus Lecture Notes Series, No. 30. September 1971.

[6] F. W. LAWVERE: Functorial Semantics of Algebraic Theories, Thesis, Columbia University, N. Y., 1963.

[7] S. MAC LANE: Categories for the Working Mathematician, Berlin—Heidelberg—New York, 1971.

[8] G. OSIUS: Logic and set theoretical tools in elementary topoi, in: Model Theory and Topoi, Lecture Notes in Math. 445 (1975), 297–346.

[9] G. OSIUS: A note on Kripke-Joyal semantics for the internal language of topoi, in: Model Theory and Topoi, Lecture Notes in Math. 445 (1975), 349–354.

[10] G. E. REYES: From Sheaves to Logic, Université de Montréal, 1972, published in Studies in algebraic logic, pp. 143–204, Studies in Math., Vol. 9, Math. Assoc. Amer., Washington, D.C., 1974.

[11] G. C. WRAITH: Lectures on elementary topoi, in: Model Theory and Topoi, Lecture Notes in Math. 445 (1975), 114–206.

Finite limit-colimit completions of small categories

by H. Kaphengst and H. Reichel, Dresden

0. Introduction

Let C be a small category. We call a functor $J\colon \mathsf{C} \to L(\mathsf{C})$ a free limit extension of C, if

(i) $L(\mathsf{C})$ is closed with respect to small limits,

(ii) every functor $F\colon \mathsf{C} \to \mathsf{B}$ in a complete category B can up to natural equivalence uniquely be extended to a limit preserving functor $L_J(F)\colon L(\mathsf{C}) \to \mathsf{B}$ with $J \cdot L_J(F) = F$.

By the properties (i) and (ii) the category $L(\mathsf{C})$ is characterized up to equivalence.

Let $\boldsymbol{D}_i$ denote directed graphs, and let $\Sigma = \{H_i\colon \boldsymbol{D}_i \to \mathsf{C};\, i \in I\}$ be a set of diagrams in C such that for every diagram $H_i\colon \boldsymbol{D}_i \to \mathsf{C}$ belonging to Σ the limit $\varprojlim H_i$ in C exists. We call a functor $F\colon \mathsf{C} \to \mathsf{B}$ a Σ-limit-preserving functor, if F preserves all limits with diagram from Σ. Respectively we call a functor $J\colon \mathsf{C} \to L_\Sigma(\mathsf{C})$ a Σ-free limit extension of C, if

(i) $L_\Sigma(\mathsf{C})$ is complete and J is Σ-limit-preserving,

(ii) every Σ-limit-preserving functor $F\colon \mathsf{C} \to \mathsf{B}$ in a complete category can up to natural equivalence uniquely be extended to a limit preserving functor

$$L_{J,\Sigma}(F)\colon L_\Sigma(\mathsf{C}) \to \mathsf{B}$$

with $\quad J \cdot L_{J,\Sigma}(F) = F\,.$

It is well known (see Ulmer; Gabriel 1971, Kaphengst; Reichel 1972) that for every small category C it holds:

A) The Yoneda-embedding $Y_\mathsf{C}\colon \mathsf{C} \to \hom[\mathsf{C}^{\mathrm{op}}, \mathsf{Set}]$ is a free colimit extension of C. For every functor $F\colon \mathsf{C} \to \mathsf{B}$ where B is cocomplete the corresponding colimit preserving extension is given by the Kan-functor extension

$$E_{Y_\mathsf{C}}(F)\colon \hom[\mathsf{C}^{\mathrm{op}}, \mathsf{Set}] \to \mathsf{B}$$

of F along the Yoneda-embedding.

B) Let $\hom_\Sigma[\mathsf{C}^{\mathrm{op}}, \mathsf{Set}]$ be the full subcategory of Σ^{op}-limit preserving functors, $Y_{\mathsf{C},\Sigma}\colon \mathsf{C} \to \hom_\Sigma[\mathsf{C}^{\mathrm{op}}, \mathsf{Set}]$ denotes the factorization of the Yoneda-embedding by the inclusion of $\hom_\Sigma[\mathsf{C}^{\mathrm{op}}, \mathsf{Set}]$. Then $Y_{\mathsf{C},\Sigma}\colon \mathsf{C} \to \hom_\Sigma[\mathsf{C}^{\mathrm{op}}, \mathsf{Set}]$ is a Σ-free colimit extension of C. Let $F\colon \mathsf{C} \to \mathsf{B}$ be a Σ-colimit-preserving functor of C into a cocomplete category B, then the corresponding colimit preserving extension of F is given by the Kan-functor extension

$$E_{Y_{\mathsf{C},\Sigma}}(F)\colon \hom_\Sigma[\mathsf{C}^{\mathrm{op}}, \mathsf{Set}] \to \mathsf{B}$$

which is a right adjoint functor to the restricted YONEDA-functor $Y_{F,\Sigma}$: $\mathsf{B} \to$ $\to \hom_\Sigma[\mathsf{C}^{op}, \mathsf{Set}]$ of $F: \mathsf{C} \to \mathsf{B}$ with $Y_{F,\Sigma}(B) = \hom_{\mathsf{B}}[-F, B]$ for every $B \in$ obj B. It arises the question whether for every small category C there exists a free limit-colimit extension. From [4] it follows that for small limits and colimits such a free limit-colimit extension does not exist. But for finite limits and colimits the problem seems to be open.

In the following we will show that for finite limits and colimits the free $\aleph_0$-limit-colimit extension exists.

This extension can be constructed by the help of the theory of heterogeneous, equational-partial algebraic systems. A categorical approach to this theory is described in [3] and in a more general context also in [2]. A more classical description of this theory is given in [1] and a summarized form can be found in [5], pp. 300–315.

Intuitively we interpret a category C which is closed with respect to finite limits and colimits as an algebraic structure with two carriers and several full and equational-partial operations.

First we remember that a small category C is closed with respect to finite limits and colimits, iff C has an initial object, a terminal object, pullbacks and pushouts.

The enriched categories which we will introduce in this paper may be called pullback-pushout-categories or briefly pp-categories. The two carriers of the pp-categories will an be called 'obj' and 'mor' just like in the case of general categories. The existence of initial object and a terminal object is represented by 0-ary full operations $\nu_0: \to$ obj, $\tau_0: \to$ obj. The construction of the left object and the right object for morphisms is given by full operations δ_0, δ_1: mor $\to$ obj. The full operation ε: obj $\to$ mor represents the building of the identity morphism for every object and φ denotes the binary, partial multiplication of categories which is defined for all ordered pairs (x, y) of morphisms with $\delta_1(x) = \delta_0(y)$. The canonical morphisms from the initial object and to the terminal object are represented by the full operations ν: obj $\to$ mor, τ: obj $\to$ $\to$ mor. The pullback construction will be represented by two partial, binary operations π_2, π_3 such that

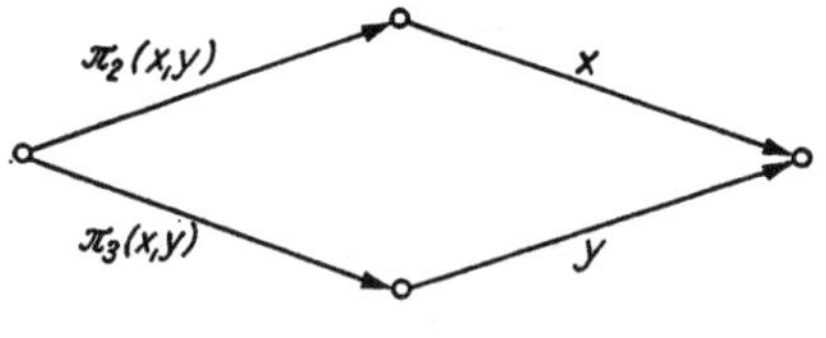

forms a pullback diagram for every pair (x, y) of morphisms with $\delta_1(x) = \delta_1(y)$, and by a 4-ary, partial operation μ such that

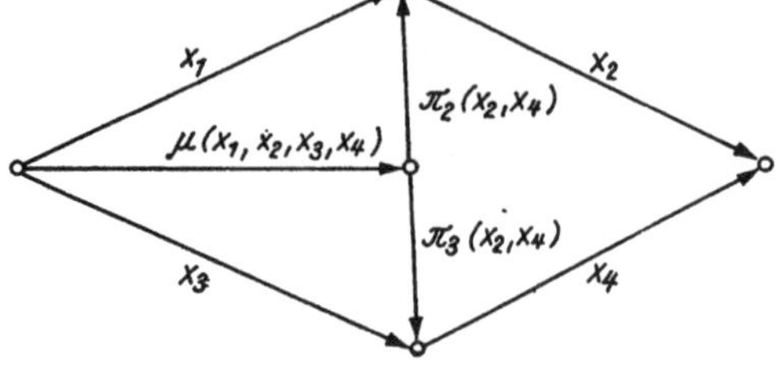

is a commutative diagram for every quadruple (x_1, x_2, x_3, x_4) of morphisms with $\delta_0(x_1) = \delta_0(x_3)$, $\delta_1(x_1) = \delta_0(x_2)$, $\delta_1(x_3) = \delta_0(x_4)$, $\delta_1(x_2) = \delta_1(x_4)$, and $\varphi(x_1, x_2) = \varphi(x_3, x_4)$. In an analogous way we represent the construction of pushouts by two binary, partial operations σ_2, σ_3 such that

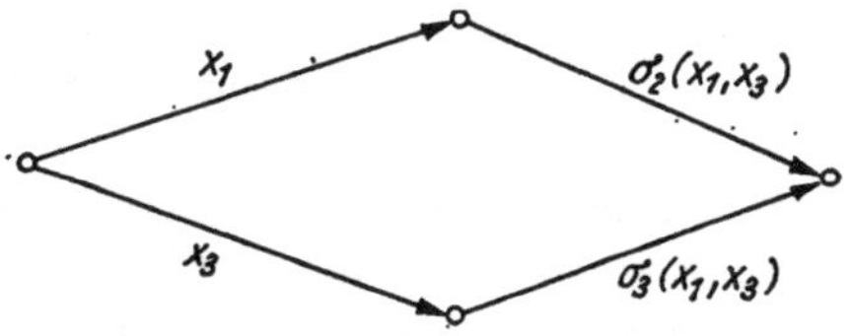

is a pushout diagram for every pair (x_1, x_3) of morphisms with $\delta_0(x_1) = \delta_0(x_3)$ and by a 4-ary, partial operation ϱ such that for every commutative square (x_1, x_2, x_3, x_4) the following diagram

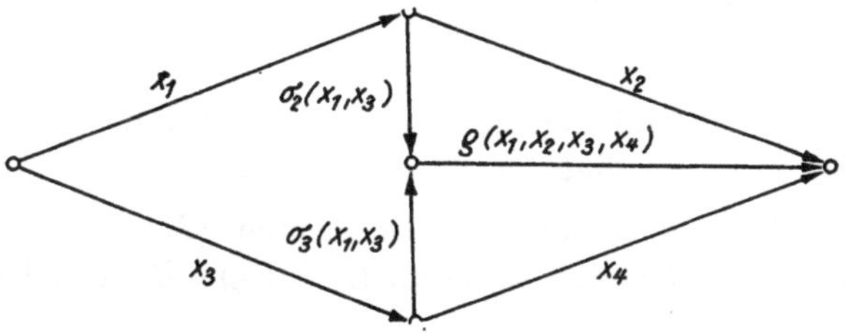

is commutative.

In the terminology of [3] we have to define a small category T_{pp} together with a system Σ_{pp} of finite diagrams in T_{pp} such that every diagram of Σ_{pp} has a colimit in T_{pp}. The operative theory $\langle \mathsf{T}_{pp}, \Sigma_{pp} \rangle$ must be defined in such a way that a pp-category P may be understood as a Σ_{pp}^{op}-limit preserving functor $P\colon \mathsf{T}_{pp}^{op} \to \mathsf{Set}$.

In our case the category T_{pp} is an infinite category with a finite system Σ_{pp} of diagrams. But in some sense the category T_{pp} is finitely representable. Before defining the category T_{pp} we shall explain in an exact way what we understand by a finite representation of an operative theory.

1. Representations of operative theories

First we recall the definition of an operative theory from [3].

Definition 1. Let $\mathfrak{r}$ be any regular cardinality. An *operative theory* of rank $\mathfrak{r}$ is an ordered pair $\langle \mathsf{T}, \Sigma \rangle$ consisting of a small category T and a set Σ of small diagrams $H\colon \boldsymbol{D} \to \mathsf{T}$ such that

(1) for every $H\colon \boldsymbol{D} \to \mathsf{T}$ from Σ there exists in T a colimit of H,
(2) for every $H\colon \boldsymbol{D} \to \mathsf{T}$ from Σ it holds $\mathrm{card}(\boldsymbol{D}) < \mathfrak{r}$.

For the definition of a representation of an operative theory we need some preparing notations.

Definition 2. A directed graph G is called a *rooty graph*, if there exists a node R_G of G, which is called the root of G, such that

(1) for every node $X \neq R_G$ of G there exists exactly one edge from R_G to X, which will be called the X-th projection of the rooty graph G,
(2) there is no edge in G ending in the root R_G of G.

It is evident that any rooty graph has only one root. Let G be a rooty graph then by $G \setminus R_G$ we denote the directed graph which arises from G by removing the root R_G and all projections of G.

A graph homomorphism $F\colon G \to M$ (or a diagram $F\colon G \to \mathsf{C}$) where G is a rooty graph will also be called a rooty diagram in M (or in C). The restriction of F to $G \setminus R_G$ will be called the underlying diagram of F and will be denoted by $F \setminus R_G$.

Definition 3. A *representation* of an operative theory is an ordered triple $\langle G, \mathfrak{D}, \mathfrak{A}\rangle$ consisting of a directed graph G, of a set $\mathfrak{D}$ of rooty diagrams in G, and of a set $\mathfrak{A}$ of pairs (w_1, w_2) of pathes over G such that w_1 and w_2 have the same initial and terminal node.

A representation $\langle G, \mathfrak{D}, \mathfrak{A}\rangle$ is called *finite*, if G is a finite, directed graph and $\mathfrak{D}$ and $\mathfrak{A}$ are both finite sets and every rooty diagram from $\mathfrak{D}$ is a finite diagram in G. If only any rooty diagram from $\mathfrak{D}$ is a finite diagram in G then we call the *representation finitary.*

Definition 4. Let C be a category and $\langle G, \mathfrak{D}, \mathfrak{A}\rangle$ a representation of an operative theory. A *model* of $\langle G, \mathfrak{D}, \mathfrak{A}\rangle$ in C is a diagram $M\colon G \to \mathsf{C}$ such that

(1) for every rooty diagram $H\colon D \to G$ out of $\mathfrak{D}$ the system $\big(M(H(p_X))\big)$, $X \neq R_D$ node of D, is a limit of the underlying diagram $(H \setminus R_D) \cdot M\colon D \setminus R_D \to \mathsf{C}$ of $H \cdot M\colon D \to \mathsf{C}$,

(2) let $\mathrm{Pa}_F\colon \mathsf{Pa}_G \to \mathsf{C}$ be the extension of the diagram $M\colon G \to \mathsf{C}$ to the functor of the free path category over G. Then for every pair $(f, g) \in \mathfrak{A}$ it must hold $\mathrm{Pa}_F(f) = \mathrm{Pa}_F(g)$.

Definition 5. We denote by $\mathsf{Mod}\langle G, \mathfrak{D}, \mathfrak{A}\rangle$ the category of lla $\langle G, \mathfrak{D}, \mathfrak{A}\rangle$-models in Set. Then we can say that the representation $\langle G, \mathfrak{D}, \mathfrak{A}\rangle$ is a *representation of the operative theory* $\langle \mathsf{T}, \Sigma\rangle$, if the corresponding model categories $\mathsf{Mod}\langle G, \mathfrak{D}, \mathfrak{A}\rangle$ and $\hom_\Sigma[\mathsf{T}^{op}, \mathsf{Set}]$ are equivalent categories.

It is easy to see that for any operative theory $\langle \mathsf{T}, \Sigma\rangle$ there exists at least one representation of the given operative theory, because $\langle \mathsf{T}, \Sigma\rangle$ may be interpreted as a representation $\langle T, \Sigma^*, \mathfrak{A}_T\rangle$ where

- T is the underling graph of the category T^{op},
- Σ^* is the system of rooty diagrams which arises from Σ by extending each diagram $H^{op}\colon D^{op} \to \mathsf{T}^{op}$ to a rooty diagram $H_R^{op}\colon D_R^{op} \to \mathsf{T}^{op}$ such that $\big(H_R(p_X)\big)$, X node of D, is a limit of $H^{op}\colon D^{op} \to \mathsf{T}^{op}$ and
- where $\mathfrak{A}_T \subseteqq \mathsf{Pa}_T \times \mathsf{Pa}_T$ is the equivalence relation which is induced by the canonical functor $Q_T\colon \mathsf{Pa}_T \to \mathsf{T}^{op}$ from the free path category over the underlying graph T of T^{op} to T^{op}.

The reason for the introduction of representations of operative theories is the fact, that for operative theories there exist representations which can be more easier described than the given operative theories.

By the following theorem we will show that for any representation $\mathfrak{r} = \langle G, \mathfrak{D}, \mathfrak{A}\rangle$ there exists an operative theory $\langle \mathsf{T}_\mathfrak{r}, \Sigma_\mathfrak{r}\rangle$ such that $\mathfrak{r}$ is a representation of $\langle \mathsf{T}_\mathfrak{r}, \Sigma_\mathfrak{r}\rangle$. The category $\mathsf{T}_\mathfrak{r}$ may be understood as the weakest category C such there exists a model of $\mathfrak{r}$ in C.

Before formulating and proving the theorem we introduce two useful notations. Let $H: \boldsymbol{D} \to \mathsf{C}$ be any rooty diagram in a category C. Then we call $H: \boldsymbol{D} \to \mathsf{C}$ a *commuting diagram* in C, if for every edge $k: D_1 \to D_2$ with $D_1 \neq R_D$ and $D_2 \neq R_D$ the equation $H(p_{D_1}) \cdot H(k) = H(p_{D_2})$ is true. The rooty diagram $H: \boldsymbol{D} \to \mathsf{C}$ will be called a *limit diagram* in C, if the system $(H(p_X))$, $X \neq R_D$ node of $\boldsymbol{D}$, forms a limit for the underlying diagram $H \setminus R_D: \boldsymbol{D} \setminus R_D \to \mathsf{C}$.

The corresponding dual notions are corooty diagram, commuting codiagram, and colimit diagram.

Theorem 1. *For every finitary representation* $\mathfrak{r} = \langle G, \mathfrak{D}, \mathfrak{A} \rangle$ *there exists a category* $\mathsf{T}_\mathfrak{r}$ *and a model* $J: G \to \mathsf{T}_\mathfrak{r}$ *of* $\mathfrak{r}$ *in* $\mathsf{T}_\mathfrak{r}$ *with the following property: For every model* $F: G \to \mathsf{C}$ *of* $\mathfrak{r}$ *there exists exactly one functor* $F_\mathfrak{r}: \mathsf{T}_\mathfrak{r} \to \mathsf{C}$ *with* $F = J \cdot F_\mathfrak{r}$.

Proof. Let $\sigma(\mathfrak{D}, \mathfrak{A})$ be the smallest congruence relation in the free path category Pa_G over G such that $\mathfrak{A} \subseteqq \sigma(\mathfrak{D}, \mathfrak{A})$ and for every rooty diagram $H: \boldsymbol{D} \to G$ of $\mathfrak{D}$ and any edge $k: D_1 \to D_2$ of $\boldsymbol{D} \setminus R_D$ the pair $(H(p_{D_1}) \cdot H(k); H(p_{D_2}))$ is contained in $\sigma(\mathfrak{D}, \mathfrak{A})$. We denote by T_0 the factor category $\mathsf{Pa}_G/\sigma(\mathfrak{D}, \mathfrak{A})$ and by $Q_{\mathfrak{D},\mathfrak{A}}: \mathsf{Pa}_G \to \mathsf{Pa}_G/\sigma(\mathfrak{D}, \mathfrak{A})$ the corresponding natural functor. If we denote by $I_G: G \to \mathsf{Pa}_G$ the canonical inclusion of G into the free path category over G then we can define $Q_0: G \to \mathsf{T}_0$ as the resulting diagram $Q_0 = I_G \cdot Q_{\mathfrak{D},\mathfrak{A}}$.

Now let be $H: \boldsymbol{D} \to G$ any rooty diagram belonging to $\mathfrak{D}$, $H': \boldsymbol{D} \to \mathsf{T}_0$ any commuting diagram in T_0 with $H' \neq H \cdot Q_0$ and $H' \setminus R_D = (H \cdot Q_0) \setminus R_D$. For any such pair of diagrams we adjoin in the weakest way a morphism $\varphi_{H,H'}: H'(R_D) \to (H \cdot Q_0) \setminus (R_D)$ to T_0 such that the condition $\varphi_{H,H'}(H \cdot Q_0)(p_X) = H'(p_X)$ for every node $X \neq R_D$ from $\boldsymbol{D}$ is satisfied.

The category T_1^* arises from the category T_0 by adjoining in the above way all such morphisms $\varphi_{H,H'}$ to T_0. By L_1^* we denote the resulting inclusion functor of T_0 into T_1^*.

In T_1^* we define a relation $\varrho_1 \subseteqq \mathsf{T}_1^* \times \mathsf{T}_1^*$ by $(f, g) \in \varrho_1$, if there exists a rooty diagram $H: \boldsymbol{D} \to G$ of $\mathfrak{D}$ with $f \cdot Q_1^*(H(p_X)) = g \cdot Q_1^*(H(p_X))$ for all nodes $X \neq R_D$ of $\boldsymbol{D}$ and $Q_1^* = Q_0 \cdot L_1^*$. If we denote by ϱ_1^* the congruence relation in T_1^* generated by ϱ_1 then we define T_1 as the factor category $\mathsf{T}_1 = \mathsf{T}_1^*/\varrho_1^*$, $L_1^{**}: \mathsf{T}_1^* \to \mathsf{T}_1$ as the corresponding natural functor and finally we set $L_1 = L_1^* \cdot L_1^{**}$ and $Q_1 = Q_0 \cdot L_1$.

In the next step we will complete the chain $\mathsf{T}_0 \xrightarrow{L_1} \mathsf{T}_1$ to a countable, infinite chain

$$\mathsf{T}_0 \xrightarrow{L_1} \mathsf{T}_1 \xrightarrow{L_2} \cdots \xrightarrow{L_n} \mathsf{T}_n \xrightarrow{L_{n+1}} \mathsf{T}_{n+1} \to \cdots .$$

Firstly we give an intuitive interpretation of the constructed categories T_0 and T_1: By construction of the category T_0 and the diagram $Q_0: G \to \mathsf{T}_0$ we have got the weakest category in which all equations of $\mathfrak{A}$ are satisfied and which makes any rooty diagram $H: \boldsymbol{D} \to G$ of $\mathfrak{D}$ to a commuting diagram $H \cdot Q_0: \boldsymbol{D} \to G \to \mathsf{T}_0$. But the commuting diagrams $H \cdot Q_0: \boldsymbol{D} \to \mathsf{T}_0$ do not fulfil the universal property of a limit diagram. By construction of the category T_1, of the functor $L_1: \mathsf{T}_0 \to \mathsf{T}_1$ and of the diagram $Q_1: G \to \mathsf{T}_1$ we have secured that for any rooty diagram $H: \boldsymbol{D} \to G$ of $\mathfrak{D}$ the commuting diagram $H \cdot Q_1: \boldsymbol{D} \to \mathsf{T}_1$ has the universal property at least with respect to those commuting diagrams $H': \boldsymbol{D} \to \mathsf{T}_1$ which can be factorized through $L_1: \mathsf{T}_0 \to \mathsf{T}_1$.

Now we start with the inductive definition of the infinite chain. For any T_n there is given a diagram $Q_n: G \to \mathsf{T}_n$ by $Q_n = Q_{n-1} \cdot L_n$. The category T_n arises from T_{n-1} by adjoining in the weakest way a morphism $\varphi_{H,H'}: H'(R_D) \to H \cdot Q_{n-1}(R_D)$ to T_{n-1} for any rooty diagram $H: \boldsymbol{D} \to \boldsymbol{G}$ of $\mathfrak{D}$ and any commuting diagram $H': \boldsymbol{D} \to \mathsf{T}_{n-1}$ with $H' \neq H \cdot Q_{n-1}$ and $H' \setminus R_{\boldsymbol{D}} = (H \setminus R_{\boldsymbol{D}}) \cdot Q_{n-1}$, which cannot be factorized by $L_{n-1}: \mathsf{T}_{n-2} \to \mathsf{T}_{n-1}$, in such a way that the equation $\varphi_{H,H'} \cdot (H \cdot Q_{n-1}(p_X)) = H'(p_X)$ is true for any node $X \neq R_{\boldsymbol{D}}$ of $\boldsymbol{D}$. We denote by L_n^* the resulting inclusion functor of T_{n-1} into T_n^* and by Q_n^* the product $Q_{n-1} \cdot L_n^*$.

In T_n^* we define a binary relation ϱ_n by $(f, g) \in \varrho_n$ if there exists a rooty diagram H: $\boldsymbol{D} \to \boldsymbol{G}$ of $\mathfrak{D}$ with $f \cdot Q_n^*(H(p_X)) = g \cdot Q_n^*(H(p_X))$ for all nodes $X \neq R_{\boldsymbol{D}}$ of $\boldsymbol{D}$. Let ϱ_n^* be the congruence relation in T_n^* generated by ϱ_n, then T_n is given by the factor category $\mathsf{T}_n^*/\varrho_n^*$. We denote by $L_n^{**}: \mathsf{T}_n^* \to \mathsf{T}_n^*/\varrho_n^*$ the corresponding natural functor onto the factor category and get the functor $L_n: \mathsf{T}_{n-1} \to \mathsf{T}_n$ as the product $L_n = L_n^* \cdot L_n^{**}$. Then the wanted category $\mathsf{T}_{\mathfrak{t}}$ is the directed colimit of the following infinite diagram in Cat:

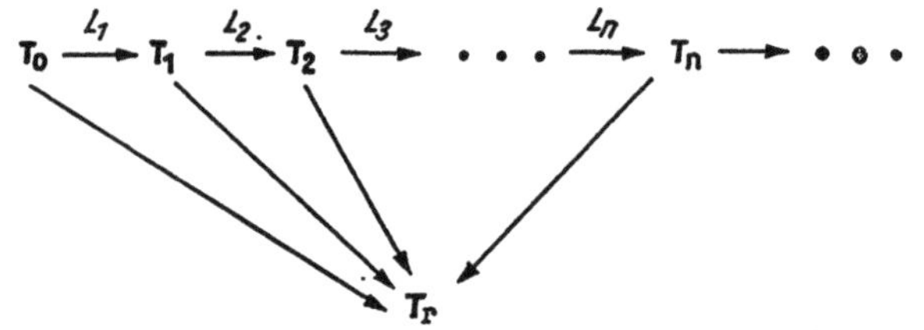

By $J_i: \mathsf{T}_i \to \mathsf{T}_{\mathfrak{t}}$, $i \in \{0, 1, 2, \ldots\}$, we denote the injections of the colimit and get the wanted model by

$$J = Q_0 \cdot J_0 = Q_1 \cdot J_1 = \ldots = Q_n \cdot J_n = \ldots .$$

In the following it is useful to remark that any functor $L_n: \mathsf{T}_{n-1} \to \mathsf{T}_n$ is one-to-one and on the object set bijective. Thus it may be assumed that the set of nodes of $\boldsymbol{G}$ and the set of objects of $\mathsf{T}_{\mathfrak{t}}$ are the same and that every injection $J_n: \mathsf{T}_n \to \mathsf{T}_{\mathfrak{t}}$ is one-to-one.

We have now to prove that $J = Q_0 \cdot J_0: \boldsymbol{G} \to \mathsf{T}_{\mathfrak{t}}$ is a model of $\mathfrak{t}$ in $\mathsf{T}_{\mathfrak{t}}$. Since J_0: $\mathsf{T}_0 \to \mathsf{T}_{\mathfrak{t}}$ is a functor, the unique extension of $Q_0 \cdot J: \boldsymbol{G} \to \mathsf{T}_{\mathfrak{t}}$ is given by $Q_{\mathfrak{D},\mathfrak{A}} \cdot J_0$: $\mathsf{Pa}_{\boldsymbol{G}} \to \mathsf{T}_{\mathfrak{t}}$. From the construction of $Q_{\mathfrak{D},\mathfrak{A}}: \mathsf{Pa}_{\boldsymbol{G}} \to \mathsf{Pa}_{\boldsymbol{G}}/\sigma(\mathfrak{D}, \mathfrak{A})$ it follows that for any pair $(f, g) \in \mathfrak{A}$ the equation $Q_{\mathfrak{D},\mathfrak{A}} \cdot J_0(f) = Q_{\mathfrak{D},\mathfrak{A}} \cdot J_0(g)$ holds and that for any diagram $H: \boldsymbol{D} \to \boldsymbol{G}$ of $\mathfrak{D}$ the diagram $H \cdot J = H \cdot Q_0 \cdot J_0: \boldsymbol{G} \to \mathsf{T}_{\mathfrak{t}}$ is a commuting diagram in $\mathsf{T}_{\mathfrak{t}}$. Let $H: \boldsymbol{D} \to \boldsymbol{G}$ be any diagram of $\mathfrak{D}$ and let $H': \boldsymbol{D} \to \mathsf{T}_{\mathfrak{t}}$ be any commuting diagram in $\mathsf{T}_{\mathfrak{t}}$ with the same underlying graph as $H \cdot J: \boldsymbol{D} \to \mathsf{T}_{\mathfrak{t}}$. Since $\boldsymbol{D}$ is a finite directed graph and $\mathsf{T}_{\mathfrak{t}}$ is a colimit of a directed graph in Cat there is a natural number n_0 such that the commuting diagram $H': \boldsymbol{D} \to \mathsf{T}_{\mathfrak{t}}$ may be factorized by $J_{n_0}: \mathsf{T}_{n_0} \to \mathsf{T}_{\mathfrak{t}}$. Let be $H' = H'' \cdot J_{n_0}$. In T_{n_0+1} there exists a morphism $\varphi_{H,H''}: L_{n_0+1}(H''(R_{\boldsymbol{D}})) \to L_{n_0+1}(Q_{n_0}(H(R_{\boldsymbol{D}})))$ with $\varphi_{H,H''} \cdot Q_{n_0+1}(H(p_X)) = L_{n_0+1}(H''(p_X))$ for all nodes $X \neq R_{\boldsymbol{D}}$ of $\boldsymbol{D}$. Thus we get $J_{n_0+1}(\varphi_{H,H''}) \cdot J(H(p_X)) = H'(p_X)$ for every node $X \neq R_{\boldsymbol{D}}$ of $\boldsymbol{D}$.

Let φ be any other morphism of $\mathsf{T}_{\mathfrak{t}}$ with $\varphi \cdot J(H(p_X)) = H'(p_X)$ for every node $X \neq R_{\boldsymbol{D}}$ of $\boldsymbol{D}$. In T_{n_1} there exist a natural number $n_1 > n_0 + 1$ and a morphism φ' with $\varphi = J_{n_1}(\varphi')$ and $\varphi' \cdot Q_{n_1}(H(p_X)) = L_{n_0+1,n_1}(\varphi_{H,H''}) \cdot Q_{n_1}(H(p_X))$ for every node $X \neq R_{\boldsymbol{D}}$ of $\boldsymbol{D}$, where $L_{n_0+1,n_1}: \mathsf{T}_{n_0+1} \to \mathsf{T}_{n_1}$ is the product $L_{n_0+1,n_1} = L_{n_0+1} \cdot L_{n_0+2} \ldots L_{n_1}$. By construction of T_{n_1+1} it holds $L_{n_1+1}(\varphi') = L_{n_1+1}(L_{n_0+1,n_1}(\varphi_{H,H''})) = L_{n_0+1,n_1+1}(\varphi_{H,H''})$ and $\varphi = J_{n_1}(\varphi') = J_{n_1+1}(L_{n_1+1}(\varphi')) = J_{n_1+1}(L_{n_0+1,n_1+1}(\varphi_{H,H''})) = J_{n_0+1}(\varphi_{H,H''})$.

We have proved that $H \cdot J\colon \boldsymbol{D} \to \mathsf{T}_{\mathfrak{r}}$ is a limit diagram for every diagram $H\colon \boldsymbol{D} \to \boldsymbol{G}$ from $\mathfrak{D}$. At last we show that for any model $F\colon \boldsymbol{G} \to \mathsf{C}$ of $\mathfrak{r}$ there exists exactly one functor $F_{\mathfrak{r}}\colon \mathsf{T}_{\mathfrak{r}} \to \mathsf{C}$ with $F = J \cdot F_{\mathfrak{r}}$.

First we prove that for any model $F\colon \boldsymbol{G} \to \mathsf{C}$ there exists one and only one family (F_i), $i \in \{0, 1, 2, \ldots\}$, of functors $F_i\colon \mathsf{T}_i \to \mathsf{C}$ with $Q_i \cdot F_i = F$ for $i \in \{0, 1, 2, \ldots\}$ and $F_i = L_{i+1} \cdot F_{i+1}$ for every natural number i.

Since $F\colon \boldsymbol{G} \to \mathsf{C}$ is a model of $\mathfrak{r} = \langle \boldsymbol{G}, \mathfrak{D}, \mathfrak{A} \rangle$ in C it follows that the congruence relation $\sigma(\mathfrak{D}, \mathfrak{A})$ in $\mathsf{Pa}_{\boldsymbol{G}}$ is smaller than the congruence relation $\sigma(\mathrm{Pa}_F)$ generated by the extension Pa_F of F to the free path category $\mathsf{Pa}_{\boldsymbol{G}}$ over $\boldsymbol{G}$. Thus we get exactly one functor $F_0\colon \mathsf{T}_0 \to \mathsf{C}$ with $\mathrm{Pa}_F = Q_{(\mathfrak{D}, \mathfrak{A})} \cdot F_0$ and $Q_0 \cdot F_0 = I_{\boldsymbol{G}} \cdot Q_{(\mathfrak{D}, \mathfrak{A})} \cdot F_0 = I_{\boldsymbol{G}} \cdot \mathrm{Pa}_F = F$.

Now we assume that for every $i \leqq n$ we have a functor $F_i\colon \mathsf{T}_i \to \mathsf{C}$ with $Q_i \cdot F_i = F$. The category T^*_{n+1} arises from T_n by adjoining a set of morphisms $\varphi_{H,H'}$ in the above described way. Let $\varphi_{H,H'}$ be such an adjoined morphism of T^*_{n+1}, $H\colon \boldsymbol{D} \to \boldsymbol{G}$ the corresponding rooty diagram of $\mathfrak{D}$, and $H'\colon \boldsymbol{D} \to \mathsf{T}_n$ the corresponding commuting diagram. By presumption $H \cdot F = H \cdot Q_n \cdot F_n\colon \boldsymbol{D} \to \mathsf{C}$ is a limit diagram in C such that there is exactly one morphism $\psi_{H,H'}\colon F_n(H'(R_{\boldsymbol{D}})) \to F(H(R_{\boldsymbol{D}}))$ in C with $\psi_{H,H'} F(H(p_X)) = F_n(H'(p_X))$ for every node $X \neq R_{\boldsymbol{D}}$ of $\boldsymbol{D}$. This means that $F_n\colon \mathsf{T}_n \to \mathsf{C}$ can be extended in a unique way to a functor $F^*_{n+1}\colon \mathsf{T}^*_{n+1} \to \mathsf{C}$ with $F_n = L^*_{n+1} \cdot F^*_{n+1}$ by setting $F^*_{n+1}(\varphi_{H,H'}) = \psi_{H,H'}$. If $(f, g) \in \mathsf{T}^*_{n+1} \times \mathsf{T}^*_{n+1}$ is a pair of morphisms for which there is a rooty diagram $H\colon \boldsymbol{D} \to \boldsymbol{G}$ from $\mathfrak{D}$ with $f \cdot Q^*_{n+1}(H(p_X)) = g \cdot Q^*_{n+1}(H(p_X))$ for all nodes $X \neq R_{\boldsymbol{D}}$ of $\boldsymbol{D}$, then we have $F^*_{n+1}(f \cdot Q^*_{n+1}(H(p_X))) = F^*_{n+1}(g \cdot Q^*_{n+1}(H(p_X)))$ and $F^*_{n+1}(f) \cdot F(H(p_X)) = F^*_{n+1}(g) \cdot F(H(p_X))$ for all nodes $X \neq R_{\boldsymbol{D}}$ of $\boldsymbol{D}$. Since $H \cdot F\colon \boldsymbol{D} \to \mathsf{C}$ is a limit diagram in C for every $H\colon \boldsymbol{D} \to \boldsymbol{G}$ of $\mathfrak{D}$ we get $F^*_{n+1}(f) = F^*_{n+1}(g)$ for every pair $(f, g) \in \varrho_{n+1} \subseteqq \mathsf{T}^*_{n+1} \times \mathsf{T}^*_{n+1}$. In this way we get the unique existence of a functor $F_{n+1}\colon \mathsf{T}_{n+1} \to \mathsf{C}$ with $L^{**}_{n+1} \cdot F_{n+1} = F^*_{n+1}$ and $L_{n+1} \cdot F_{n+1} = L^*_{n+1} \cdot L^{**}_{n+1} \cdot F_{n+1} = L^*_{n+1} \cdot F^*_{n+1} = F$.

Since $\mathsf{T}_{\mathfrak{r}}$ was constructed as a colimit in Cat of the chain

$$\mathsf{T}_0 \xrightarrow{L_1} \mathsf{T}_1 \to \ldots \xrightarrow{L_n} \mathsf{T}_n \to \ldots$$

we get exactly one functor $F_{\mathfrak{r}}\colon \mathsf{T}_{\mathfrak{r}} \to \mathsf{C}$ with $J_i \cdot F_{\mathfrak{r}} = F_i$ for every natural number i. This means $J \cdot F_{\mathfrak{r}} = Q_0 \cdot J_0 \cdot F_{\mathfrak{r}} = Q_0 \cdot F_0 = F$.

It is obvious that for any finitary representation $\mathfrak{r} = \langle \boldsymbol{G}, \mathfrak{D}, \mathfrak{A} \rangle$ up to an isomorphism there is only one category $\mathsf{T}_{\mathfrak{r}}$ with a model $J\colon \boldsymbol{G} \to \mathsf{T}_{\mathfrak{r}}$ of $\mathfrak{r}$ in $\mathsf{T}_{\mathfrak{r}}$ which has the universal property of Theorem 1. This model $J\colon \boldsymbol{G} \to \mathsf{T}_{\mathfrak{r}}$ of $\mathfrak{r}$ will be called the completion of the representation $\mathfrak{r} = \langle \boldsymbol{G}, \mathfrak{D}, \mathfrak{A} \rangle$.

Corollary 1.1. *Let $\mathfrak{r} = \langle \boldsymbol{G}, \mathfrak{D}, \mathfrak{A} \rangle$ be any finitary representation of an operative theory. Then $\mathfrak{r}$ is a representation of the operative theory $\langle \mathsf{T}^{\mathrm{op}}_{\mathfrak{r}}, \Sigma_\infty \rangle$ where*

$$\Sigma_\infty = \{(H \setminus R_{\boldsymbol{D}}) \cdot J^{\mathrm{op}}\colon (\mathsf{G} \setminus R_{\boldsymbol{D}})^{\mathrm{op}} \to \mathsf{T}^{\mathrm{op}}_{\mathfrak{r}} \mid H \in \mathfrak{D}\}$$

and $J\colon \boldsymbol{G} \to \mathsf{T}_{\mathfrak{r}}$ is the completion of $\mathfrak{r}$.

Proof. If we take $\mathsf{C} = \mathsf{Set}$ in Theorem 1 we get an isomorphism between $\hom_{\Sigma_\infty}[\mathsf{T}_{\mathfrak{r}}, \mathsf{Set}]$ and $\mathsf{Mod}_{\mathfrak{r}}$.

2. Pullback-pushout-categories (pp-categories)

In this section we give a finite representation $\mathfrak{r}_{\mathrm{pp}} = \langle \boldsymbol{G}_{\mathrm{pp}}, \mathfrak{D}_{\mathrm{pp}}, \mathfrak{A}_{\mathrm{pp}} \rangle$ for the wanted operative theory $\langle \mathsf{T}_{\mathrm{pp}}, \Sigma_{\mathrm{pp}} \rangle$.

The finite directed graph G_{pp} is given in the following way: Besides of the nodes and edges of the graphical representation in G_{pp} there are the following edges:

λ_0, λ_1: mor → pr, λ_2, λ_3: ass → pr, λ_4, λ_5: fp → pr,
λ_6, λ_7: fs → pf, λ_8: csq → fp, λ_9: csq → fs,
$\lambda_{10}, \lambda_{11}, \lambda_{12}, \lambda_{13}$: csq → pr, λ_{14}: fs → obj, λ_{15}: fp → obj;

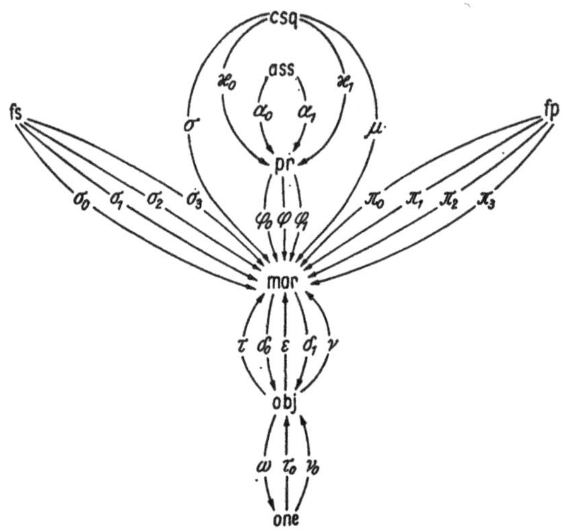

The set $\mathfrak{D}_{pp}$ of rooty diagrams in G_{pp} consists of ten diagrams $H_0, H_1, \ldots, H_9$. The diagram H_0: $D_0 \to G_{pp}$ is given by the graph $D_0 = 1$ which has only one node 1 and no edge, and by $H_0(1) =$ one.

The diagrams H_i for $i = 1, 2, \ldots, 9$ all have the same domain D. D has three nodes D_1, D_2, D_3 which are different from the additional root R_D and two edges $k_1: D_1 \to D_3$, $k_2: D_2 \to D_3$ which are different from the three additional projections $\pi_{D_1}, \pi_{D_2}, \pi_{D_3}$.

In the following we give the diagrams $H_1, H_2, \ldots, H_9$ by representing the image of D in G_{pp} by H_i, $i = 1, 2, \ldots, 9$. We remark that all diagrams represent pullback diagrams in G_{pp} (note that the images of the projection π_{D_3}: $R_D \to D_3$ are not contained in the representation of the diagrams $H_1, H_2, \ldots, H_9$ and in the representation of the directed graph G_{pp}).

The diagram H_0 determines 'one' as the terminal object. The diagrams H_1, H_3, H_4, H_5 serve to define domains for equational-partial operations over the object 'mor' that is one of the two base objects 'obj' and 'mor'. H_1 defines the domain 'pr' of the binary operation 'φ' where 'φ_0' and 'φ_1' are the restrictions of the first and second projection. H_2 defines the object 'ass' which we need to express the associativity of 'φ'. H_3 defines the domain for the operations 'π_2' and 'π_3', this means, for building the pullback construction. H_4 defines the domain of 'σ_2' and 'σ_3'. H_5 defines the domain 'csq' of 'μ' and 'σ' which represent the universal property of pullbacks and pushouts, respectively. Intuitively 'csq' represents the set of commutative squares of a category. By the help of the diagrams H_6 and H_7 we express the condition that for every object of a category with an initial object and a terminal object there exist exactly one initial morphism and exactly one terminal morphism. Diagram H_8 and H_9 also serve to express the universal property of pullbacks and pushouts.

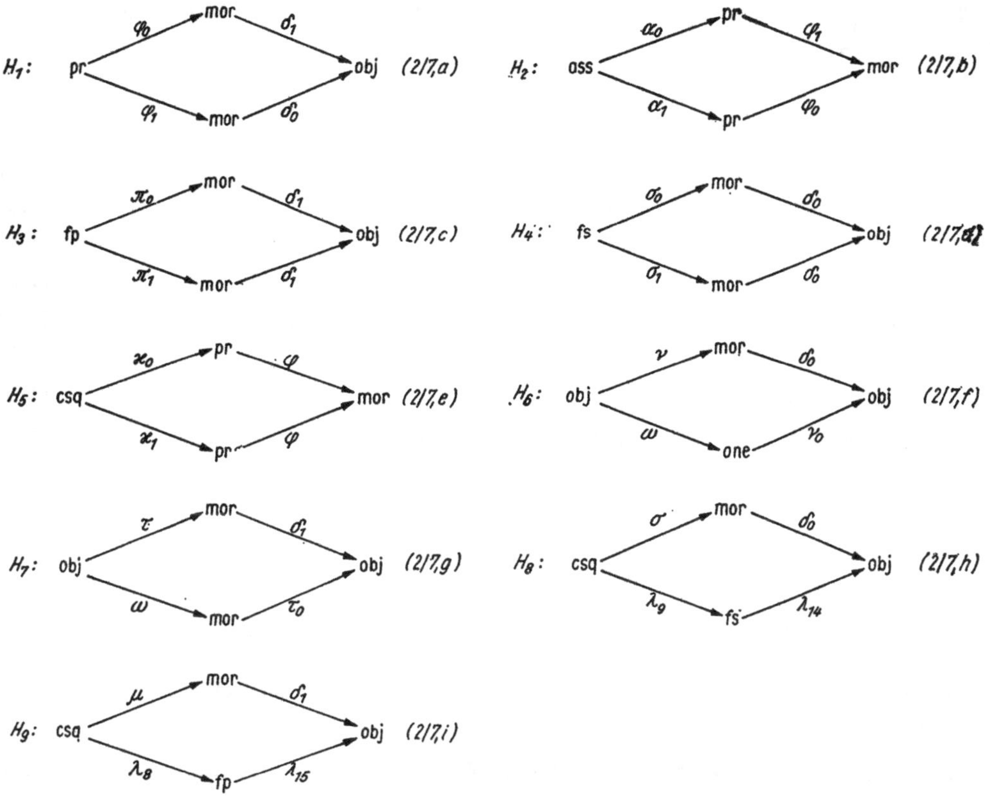

Before giving the set $\mathfrak{A}_{pp}$ of equations, we introduce a subset $\mathfrak{A}_{pp}^0 \subseteq \mathfrak{A}_{pp}$ which permits us to give a clear idea of the meaning of the edges $\lambda_0, \lambda_1, \ldots, \lambda_{15}$. $\mathfrak{A}_{pp}^0$ contains the following equations:

$$\lambda_0 \cdot \varphi_0 = 1_{\text{mor}}, \qquad \lambda_0 \cdot \varphi_1 = \delta_1 \cdot \varepsilon, \tag{0.0}$$
$$\lambda_1 \cdot \varphi_0 = \delta_0 \cdot \varepsilon, \qquad \lambda_1 \cdot \varphi_1 = 1_{\text{mor}}, \tag{0.1}$$
$$\lambda_2 \cdot \varphi_0 = \alpha_0 \cdot \varphi, \qquad \lambda_2 \cdot \varphi_1 = \alpha_1 \cdot \varphi_1, \tag{0.2}$$
$$\lambda_3 \cdot \varphi_0 = \alpha_0 \cdot \varphi_0, \qquad \lambda_3 \cdot \varphi_1 = \alpha_1 \cdot \varphi, \tag{0.3}$$
$$\lambda_4 \cdot \varphi_0 = \pi_2, \qquad \lambda_4 \cdot \varphi_1 = \pi_0, \tag{0.4}$$
$$\lambda_5 \cdot \varphi_0 = \pi_3, \qquad \lambda_5 \cdot \varphi_1 = \pi_1, \tag{0.5}$$
$$\lambda_6 \cdot \varphi_0 = \sigma_0, \qquad \lambda_6 \cdot \varphi_1 = \sigma_2, \tag{0.6}$$
$$\lambda_7 \cdot \varphi_0 = \sigma_1, \qquad \lambda_7 \cdot \varphi_1 = \sigma_3, \tag{0.7}$$
$$\lambda_8 \cdot \pi_0 = \varkappa_0 \cdot \varphi_1, \qquad \lambda_8 \cdot \pi_1 = \varkappa_1 \cdot \varphi_1, \tag{0.8}$$
$$\lambda_9 \cdot \sigma_0 = \varkappa_0 \cdot \varphi_0, \qquad \lambda_9 \cdot \sigma_1 = \varkappa_1 \cdot \varphi_0, \tag{0.9}$$
$$\lambda_{10} \cdot \varphi_0 = \mu, \qquad \lambda_{10} \cdot \varphi_1 = \lambda_8 \cdot \pi_2, \tag{0.10}$$
$$\lambda_{11} \cdot \varphi_0 = \mu, \qquad \lambda_{11} \cdot \varphi_1 = \lambda_8 \cdot \pi_3, \tag{0.11}$$
$$\lambda_{12} \cdot \varphi_0 = \lambda_9 \cdot \sigma_2, \qquad \lambda_{12} \cdot \varphi_1 = \sigma, \tag{0.12}$$
$$\lambda_{13} \cdot \varphi_0 = \lambda_9 \cdot \sigma_3, \qquad \lambda_{13} \cdot \varphi_1 = \sigma, \tag{0.13}$$
$$\lambda_{14} = \sigma_2 \cdot \delta_1, \tag{0.14}$$
$$\lambda_{15} = \pi_2 \cdot \delta_0. \tag{0.15}$$

The set $\mathfrak{A}_{pp}$ consists of the equations (0.0), ..., (0.15) and of the following equations (1.0), ..., (1.14):

$$\varepsilon \cdot \delta_0 = \varepsilon \cdot \delta_1 = 1_{obj}, \tag{1.0}$$

$$\tau \cdot \delta_0 = \nu \cdot \delta_1 = 1_{obj}, \tag{1.1}$$

$$\lambda_0 \cdot \varphi = \lambda_1 \cdot \varphi = 1_{mor}, \tag{1.2}$$

$$\varphi \cdot \delta_1 = \varphi_1 \cdot \delta_1, \qquad \varphi \cdot \delta_0 = \varphi_0 \cdot \delta_0, \tag{1.3}$$

$$\lambda_2 \cdot \varphi = \lambda_3 \cdot \varphi, \tag{1.4}$$

$$\pi_2 \cdot \delta_1 = \pi_0 \cdot \delta_0, \qquad \pi_2 \cdot \delta_0 = \pi_3 \cdot \delta_0, \qquad \pi_3 \cdot \delta_1 = \pi_1 \cdot \delta_0, \tag{1.5}$$

$$\sigma_2 \cdot \delta_1 = \sigma_3 \cdot \delta_1, \qquad \sigma_2 \cdot \delta_0 = \sigma_2 \cdot \delta_1, \qquad \sigma_3 \cdot \delta_0 = \sigma_1 \cdot \delta_1, \tag{1.6}$$

$$\lambda_4 \cdot \varphi = \lambda_5 \cdot \varphi, \tag{1.7}$$

$$\lambda_6 \cdot \varphi = \lambda_7 \cdot \varphi, \tag{1.8}$$

$$\mu \cdot \delta_0 = \varkappa_0 \cdot \varphi \cdot \delta_0, \qquad \mu \cdot \delta_1 = \lambda_9 \cdot \pi_2 \cdot \delta_0, \tag{1.9}$$

$$\lambda_{10} \cdot \varphi = \varkappa_0 \cdot \varphi_0, \tag{1.10}$$

$$\lambda_{11} \cdot \varphi = \varkappa_1 \cdot \varphi_0, \tag{1.11}$$

$$\sigma \cdot \delta_1 = \varkappa_0 \cdot \varphi_1 \cdot \delta_1, \qquad \sigma \cdot \delta_0 = \lambda_9 \cdot \sigma_2 \cdot \delta_1, \tag{1.12}$$

$$\lambda_{12} \cdot \varphi = \varkappa_0 \cdot \varphi_1, \tag{1.13}$$

$$\lambda_{13} \cdot \varphi = \varkappa_1 \cdot \varphi_1 . \tag{1.14}$$

Theorem 2. *Any model M: $G_{pp} \to$ Set of $\langle G_{pp}, \mathfrak{D}_{pp}, \mathfrak{A}_{pp}\rangle$ in Set is a category with finite limits and finite colimits.*

Proof. Let M: $G_{pp} \to$ Set be a model of $\langle G_{pp}, \mathfrak{D}_{pp}, \mathfrak{A}_{pp}\rangle$. The two sets $M(\text{obj})$ and $M(\text{mor})$ together with the mappings $M(\delta_0)$, $M(\delta_1)$: $M(\text{mor}) \to M(\text{obj})$, $M(\varepsilon)$: $M(\text{obj}) \to M(\text{mor})$, and $M(\varphi)$: $M(\text{pr}) \to M(\text{mor})$ form a category M, the so-called underlying category of M: $G_{pp} \to$ Set. This is easy to see, because H_1M and H_2M are limit diagrams in Set (pullbacks) and M fulfils the equations (1.0), (1.2), (1.3), and (1.4). Since H_1M and H_2M are pullbacks in Set, we can assume that $M(\text{pr})$ is the set of all ordered pairs $(x, y) \in M(\text{mor}) \times M(\text{mor})$ with $M(\delta_1)(x) = M(\delta_0)(y)$, $M(\varphi_0)(x, y) = x$, and $M(\varphi_1)(x, y) = y$, and that $M(\text{ass}) \subseteq M(\text{mor}) \times M(\text{mor}) \times M(\text{mor})$ is the set of all triples (x, y, z) with $M(\delta_0)(y) = M(\delta_1)(x)$, $M(\delta_0)(z) = M(\delta_1)(y)$, and $M(\alpha_0)(x, y, z) = (x, y)$, $M(\alpha_1)(x, y, z) = (y, z)$. By the equations (0.2) and (0.3) we get $M(\lambda_2)(x, y, z) = (M(\varphi)(x, y), z)$ and $M(\lambda_3)(x, y, z) = (x, M(\varphi)(y, z))$ such that the equation (1.4), i.e. $\lambda_2 \cdot \varphi = \lambda_3 \cdot \varphi$, means the associativity of the equational-partial operation φ.

With the help of the 0-ary operations $M(\tau_0)$, $M(\nu_0)$: $M(\text{one}) \to M(\text{obj})$, the 1-ary operations $M(\tau)$, $M(\nu)$: $M(\text{obj}) \to M(\text{mor})$, the pullback diagrams H_7M, H_6M and the equations (1.1) we can show that M is a category with a terminal object $M(\tau_0)(\emptyset) \in M(\text{obj})$ and an initial object $M(\nu_0)(\varphi) \in M(\text{obj})$. Since H_3M, H_4M and H_5M are pullback diagrams in Set we can assume:

(i) $M(\text{fp}) = \{(x, y) \in M(\text{mor})^2 \mid M(\delta_1)(x) = M(\delta_1)(y)\}$,

$M(\pi_0)(x, y) = x$, $\quad M(\pi_1)(x, y) = y$;

(ii) $M(\text{fs}) = \{(x, y) \in M(\text{mor})^2 \mid M(\delta_0)(x) = M(\delta_0)(y)\}$,

$M(\sigma_0)(x, y) = x$, $\quad M(\sigma_1)(x, y) = y$;

(iii) $$M(\mathrm{csq}) = \{(x_1, x_2, x_3, x_4) \in M(\mathrm{mor})^4 \mid M(\delta_0)(x_1) = M(\delta_0)(x_3),$$
$$M(\delta_1)(x_2) = M(\delta_1)(x_4),\ M(\delta_1)(x_1) = M(\delta_0)(x_2),$$
$$M(\delta_1)(x_3) = M(\delta_0)(x_4),\ M(\varphi)(x_1, x_2) = M(\varphi)(x_3, x_4)\},$$
$$M(\varkappa_0)(x_1, x_2, x_3, x_4) = (x_1, x_2), \qquad M(\varkappa_1)(x_1, x_2, x_3, x_4) = (x_3, x_4).$$

The equations (1.5), (1.7), (1.9), (1.10), (1.11), (0.4), (0.5), (0.8), (0.10), (0.11) together with the pullback diagram H_9M guarantee that for any $(x, y) \in M(\mathrm{fp})$ the diagram

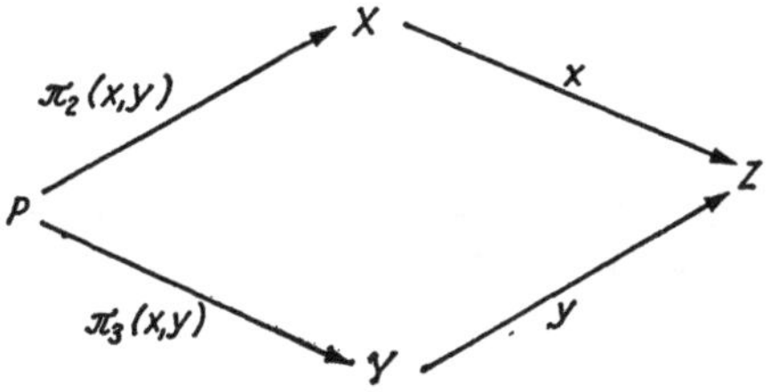

is a pullback diagram in the category M.

The equations (1.5), (1.7), (0.4) and (0.5) mean that $(\pi_2(x, y), x, \pi_3(x, y), y)$ is a commutative square, where (0.4) and (0.5) only express that $M(\lambda_4)$, $M(\lambda_5)$ are the mappings with $M(\lambda_4)(x, y) = (\pi_2(x, y), x)$ and $M(\lambda_5)(x, y) = (\pi_3(x, y), y)$.

Let (a, x, b, y) be any other commutative square in M. Then there is a morphism $\mu(a, x, b, y)\colon A \to P$, where $a\colon A \to X$, $b\colon A \to Y$, such that $a = M(\varphi)(\mu(a, x, b, y), \pi_2(x, y))$, $b = M(\varphi)(\mu(a, x, b, y), \pi_3(x, y))$. Since H_9M is a pullback diagram in Set there is only one morphism $z\colon A \to P$ in M with $a = M(\varphi)(z, \pi_2(x, y))$ and $b = M(\varphi)(z, \pi_3(x, y))$.

In the same way one can see that for any $(x, y) \in M(\mathrm{fs})$ the commutative square $(x, \sigma_2(x, y), y, \sigma_3(x, y))$ is a pushout diagram in M.

Thus we have shown that M is a category with an initial object and a terminal object which has pullbacks and pushouts.

Let $\mathfrak{t}_c = \langle G_c, \mathfrak{D}_c, \mathfrak{A}_c\rangle$ be a representation of the operative theory of small categories. Without restrictions of generality we may assume that G_c is a subgraph of G_{pp}, and $\mathfrak{D}_c \subseteq \mathfrak{D}_{\mathrm{pp}}$, $\mathfrak{A}_c \subseteq \mathfrak{A}_{\mathrm{pp}}$. By restriction of $\mathfrak{t}_{\mathrm{pp}}$-models $M\colon G_{\mathrm{pp}} \to \mathrm{Set}$ to $\mathfrak{t}_c$ we get a forgetful functor

$$V\colon \mathrm{Mod}\langle G_{\mathrm{pp}}, \mathfrak{D}_{\mathrm{pp}}, \mathfrak{A}_{\mathrm{pp}}\rangle \to \mathrm{Mod}\langle G_c, \mathfrak{D}_c, \mathfrak{A}_c\rangle .$$

Corollary. *A small category* C *is closed with respect to finite limits and finite colimits if and only if there is an $\mathfrak{t}_{\mathrm{pp}}$-model $M\colon G_{\mathrm{pp}} \to \mathrm{Set}$ with $\mathrm{C} = V(M)$.*

Proof. By Theorem 2 we obtain that the underlying category $\mathrm{M} = V(M)$ for any $\mathfrak{t}_{\mathrm{pp}}$-model $M\colon G_{\mathrm{pp}} \to \mathrm{Set}$ is closed with respect to finite limits and finite colimits. Let C be any category with finite limits and finite colimits. We can define the 0-ary operations $\tau_0, \nu_0\colon \{\varnothing\} \to \mathrm{obj}(\mathrm{C})$ by choosing a terminal respectively an initial object of C. If we have chosen these objects, then the operations $\tau, \nu\colon \mathrm{obj}(\mathrm{C}) \to \mathrm{mor}(\mathrm{C})$ are given in a unique manner. By the axiom of choice for every pair $X \xrightarrow{x} Z \xleftarrow{y} Y$ we can choose a pair $(\pi_2(x, y), \pi_3(x, y))$ of morphisms such that $(\pi_2(x, y), x, \pi_3(x, y), y)$

forms a pullback diagram

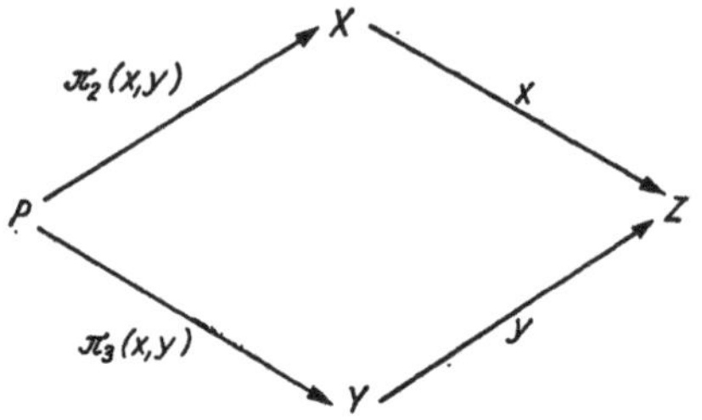

This means that with the help of the axiom of choice we can define operations π_2 and π_3 and analogous operationsly σ_2 and σ_3. By the universal property of pullbacks and pushouts the operations μ and σ are uniquely determined, if we have chosen the operations π_2, π_3, σ_2, and σ_3.

It is also obvious that for every $\mathfrak{r}_{pp}$-morphism $f\colon M_1 \to M_2$ of $\mathsf{Mod}\,\mathfrak{r}_{pp}$ the underlying functor $V(f)\colon \mathsf{M}_1 \to \mathsf{M}_2$ is a finite limit and finite colimit preserving functor in Cat.

Now we are able to prove the existence of the free $\aleph_0$-limit-colimit extension for every small category C.

Theorem 3. *For any small category* C *there exists a small category* $\mathsf{F(C)}$ *and a functor* $J_{\mathsf{C}}\colon \mathsf{C} \to \mathsf{F(C)}$ *with*

(i) $\mathsf{F(C)}$ *has finite limits and finite colimits,*

(ii) *every functor* $F\colon \mathsf{C} \to \mathsf{K}$ *in a category* K *with finite limits and finite colimits can up to natural equivalence be extended in a unique way to a finite limit and finite colimit preserving functor*

$$\mathsf{F}_{J_{\mathsf{C}}}(F)\colon \mathsf{F(C)} \to \mathsf{K} \text{ with } J_{\mathsf{C}} \cdot \mathsf{F}_{J_{\mathsf{C}}}(F) = F\,.$$

Proof. Let $I_c\colon G_c \to \mathsf{T}_c$ and $I_{pp}\colon G_{pp} \to \mathsf{T}_{pp}$ be the completions of the representations $\mathfrak{r}_c$ and $\mathfrak{r}_{pp}$. By Theorem 1 there is exactly one functor $I\colon \mathsf{T}_c \to \mathsf{T}_{pp}$ with $I_c \cdot I = I_{pp}$. The functor I^{op} is an embedding of the operative theory $\langle \mathsf{T}_c^{op}, \Sigma_c^{op}\rangle$ of small categories into the operative theory $\langle \mathsf{T}_{pp}^{op}, \Sigma_{pp}^{op}\rangle$ of small pp-categories.

By $Y_{\mathsf{T}_{pp},\Sigma_{pp}}\colon \mathsf{T}_{pp}^{op} \to \hom_{\Sigma_{pp}}[\mathsf{T}_{pp}, \mathsf{Set}]$ (equivalent to $\mathsf{Mod}\,\mathfrak{r}_{pp}$) and $Y_{\mathsf{T}_c,\Sigma_c}\colon \mathsf{T}_c^{op} \to \hom_{\Sigma_c}[\mathsf{T}_c, \mathsf{Set}]$ (equivalent to $\mathsf{Mod}\,\mathfrak{r}_c$) we denote the factorizations of the YONEDA-embeddings through the corresponding categories of operative systems (see [3]). From [3] it follows that $Y_{\mathsf{T}_{pp},\Sigma_{pp}}$ and $Y_{\mathsf{T}_c,\Sigma_c}$ are dense, full, and faithful functors. In [3] it was shown that the functor $I^{op} \cdot Y_{\mathsf{T}_{pp},\Sigma_{pp}}\colon \mathsf{T}_c^{op} \to \hom_{\Sigma_{pp}}[\mathsf{T}_{pp}, \mathsf{Set}]$ is left adjoint to the forgetful functor $V\colon \hom_{\Sigma_{pp}}[\mathsf{T}_{pp}, \mathsf{Set}] \to \hom_{\Sigma_c}[\mathsf{T}_c, \mathsf{Set}]$ relativ to $Y_{\mathsf{T}_c,\Sigma_c}$ and that $\hom_{\Sigma_{pp}}[\mathsf{T}_{pp}, \mathsf{Set}]$ is closed with respect to arbitrary colimits. From a theorem of ULMER [2] (see also Theorem 4 in the appendix of [3]) it follows that the KAN-functor extension of $I^{op} \cdot Y_{\mathsf{T}_{pp},\Sigma_{pp}}$ along $Y_{\mathsf{T}_c,\Sigma_c}$, which will be denoted by $E\colon \hom_{\Sigma_c}[\mathsf{T}_c, \mathsf{Set}] \to \hom_{\Sigma_{pp}}[\mathsf{T}_{pp}, \mathsf{Set}]$, is a left adjoint functor the forgetful functor V.

If $J_{\mathsf{C}}\colon \mathsf{C} \to V(E(\mathsf{C}))$ denotes the natural functor which results from the adjointness of E and V, then $J_{\mathsf{C}}\colon \mathsf{C} \to \mathsf{F(C)}$ with $\mathsf{F(C)} = V(E(\mathsf{C}))$ is a free $\aleph_0$-limit-colimit extension of C. For the proof of this assertion let $F\colon \mathsf{C} \to \mathsf{K}$ be a functor in a category K with finite limits and finite colimits. By the corollary there exists a pp-category K with $V(\mathsf{K}) = \mathsf{K}$. From the adjointness of E and V it follows the existence of a unique pp-morphism $F^*\colon E(\mathsf{C}) \to \mathsf{K}$ with $J_{\mathsf{C}} \cdot V(F^*) = F$. It is obvious that $V(F^*)\colon \mathsf{F}(\mathsf{C}) \to \mathsf{K}$ preserves finite limits and finite colimits.

Let $F^{**}\colon \mathsf{F}(\mathsf{C}) \to \mathsf{K}$ be any other functor which preserves finite limits and finite colimits, and for which $J_{\mathsf{C}} \cdot F^{**} = F$ holds. Then there exists a pp-category K^* with $V(\mathsf{K}^*) = \mathsf{K}$ such that $F^{**}\colon E(\mathsf{C}) \to \mathsf{K}^*$ is a pp-morphism. It is easy to show that the largest subcategory $\mathsf{S} \subseteqq \mathsf{F}(\mathsf{C})$, for which the restrictions $V(F^*)/\mathsf{S}$ and F^{**}/S are natural equivalent, contains the image of C by $J_{\mathsf{C}}\colon \mathsf{C} \to \mathsf{F}(\mathsf{C})$, and that S is closed with respect to all additional operations of the pp-category $E(\mathsf{C})$. Since the pp-category $E(\mathsf{C})$ is generated by the image $J_{\mathsf{C}}(\mathsf{C})$, it follows that $\mathsf{S} = \mathsf{F}(\mathsf{C})$.

The above construction shows that for any regular cardinal $\mathfrak{r}$ there exists the free $\mathfrak{r}$-limit-colimit extension for any small category.

References

[1] H. Kaphengst; H. Reichel: Algebraische Algorithmentheorie, VEB Robotron, Zentrum für Forschung und Technik, WIB (1971).
[2] F. Ulmer; P. Gabriel: Lokal präsentierbare Kategorien, Lecture Notes in Math. 221 (1971).
[3] H. Kaphengst; H. Reichel: Operative Theorien und Kategorien von operativen Systemen, in: Studien zur Algebra und ihre Anwendungen 1, Akademie-Verlag, Berlin 1972, S. 41–56.
[4] M. Dehen: Classification de catégories, Sém. Choquet: 1968/69, Initiation à l'Analyse, Exp. 3, 96 pp., Secrétariat mathématique, Paris 1969.

Klassifikation von Kategorien von Korrespondenzen

von M. Š. CALENKO, Moskau

Die Theorie der binären Relationen im allgemeinen und die Algebra binärer Relationen im besonderen gestaltet sich in zunehmendem Maße zu einem „angewandten" Teilgebiet der modernen Mathematik. Ihr Anwendungsbereich erstreckt sich von der Funktionalanalysis und der homologischen Algebra bis hin zur mathematischen Linguistik und der Modellierung automatisierter Informationssysteme. Basierend auf einer breiten Anwendung kategorientheoretischer Ideen und Methoden wurde durch die Arbeiten von S. MAC LANE [11] und D. PUPPE [13] eine neue Etappe zum Studium der Algebra binärer Relationen eingeleitet. In dem Zeitraum von 1962 bis 1973 wurden von verschiedenen Autoren (D. PUPPE, M. CALENKO, M. BURGIN, D. RAJKOV, P. A. GRILLET, Y. KAWAHARA u. a.) sogenannte Kategorien von Korrespondenzen über abelschen, exakten, regulären und anderen Klassen von Kategorien konstruiert. Kategorien von Korrespondenzen stellen eine wichtige Unterklasse der Klasse von Kategorien mit Involution, oder I-Kategorien, dar. Da sich zu Beginn der siebziger Jahre eine Tendenz zum Aufbau einer allgemeinen Theorie der I-Kategorien abzeichnete, entsteht ganz natürlich die Frage zur axiomatischen Charakterisierung der Kategorien von Korrespondenzen innerhalb der Klasse der I-Kategorien. In der vorliegenden Arbeit wird auf diese Frage eine Antwort gegeben und eine Klassifikation der Kategorien von Korrespondenzen in Abhängigkeit von den Eigenschaften ihrer Unterkategorien von funktionalen Morphismen vorgeschlagen.

Der Verfasser dankt Herrn R. PETERS, Köthen, für die Übersetzung ins Deutsche.

1. Eine beliebige binäre Relation zwischen den Mengen A und B, d. h. eine Untermenge R des kartesischen Produktes $A \times B$, wird eindeutig durch ein Paar von Abbildungen $f\colon R \to A$ und $g\colon R \to B$, welche Einschränkungen der Projektionen des Produktes auf R sind, dargestellt. Mit anderen Worten, anstelle der Untermenge $R \subseteqq A \times B$ kann man auch das Paar von Abbildungen (f, g) betrachten.

Wenn zum anderen A und B Gruppen (oder Moduln) sind und wenn R Untergruppe des direkten Produktes $A \times B$ ist, so existiert ein anderes Verfahren zur Vorgabe von R. Die Untergruppenpaare

$$D_R = \{a \mid \exists b(a, b) \in R\}\,, \qquad K_R = \{a \mid (a, 1) \in R\}$$

von A und

$$B_R = \{b \mid \exists a(a, b) \in R\}\,, \qquad I_R = \{b \mid (1, b) \in R\}$$

von B definieren nämlich zusammen mit dem von GOURSAT [3] angegebenen Isomorphismus $D_R/K_R \cong B_R/I_R$ ebenfalls die Untergruppe R eindeutig.

Beide Darstellungsverfahren für R dienten als Grundlage für die Konstruktion verschiedener Klassen von Kategorien mit Involution, die jedoch die gemeinsame Bezeichnung Kategorien von Korrespondenzen erhielten ([2], [4], [5], [8], [14]—[16]).

In der vorliegenden Arbeit wird gezeigt, daß die Darstellung von Relationen mit Hilfe von Paaren von Morphismen und die mit diesen Darstellungen verbundene Konstruktion Folgerungen dreier zusätzlicher Axiome bezüglich der Kategorien mit Involution sind (siehe unten Axiome I0—I2). Bezogen auf ihre Unterkategorien funktionaler Morphismen spielen Kategorien, in denen diese Axiome erfüllt sind, eine ähnliche Rolle wie linke Quotientenringe im Sinne von Ore [7]. Jeder funktionale Morphismus ist voll regulär im Sinne von von Neumann (volle Regularität ist das Analogon zur Umkehrbarkeit), und jeder Morphismus hat die Form $f^{\#}g$, wobei f und g funktional sind und $\#$ die Involution bedeutet (Involution ist das Analogon zur Bildung des inversen Elementes).

Die zweite Darstellungsart von Relationen führt, allgemein gesprochen, zu einer anderen Klasse von Kategorien mit Involution. Über Zusammenhänge zwischen beiden Klassen wird weiter unten etwas gesagt.

2. Sei E eine Klasse von Morphismen einer gewissen Kategorie K. Das kommutative Quadrat

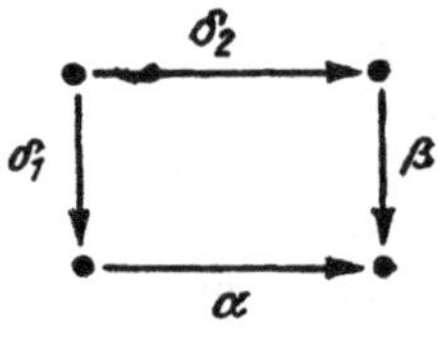

(1)

heißt E-*universell bezüglich* α *und* β, wenn es für jedes kommutative Quadrat

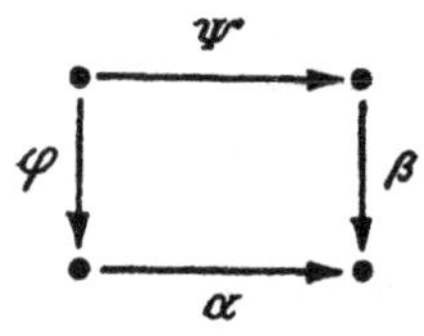

(2)

solche Morphismen $\varepsilon \in \mathsf{E}$ und γ gibt, daß

$$\gamma\delta_1 = \varepsilon\varphi\,, \qquad \gamma\delta_2 = \varepsilon\psi \tag{3}$$

gilt. Ein E-universelles Quadrat wird durch $[\delta_1, \alpha, \beta, \delta_2]_{\mathsf{E}}$ bezeichnet.

Ein Morphismus $\mu\colon K \to A$ heißt E-*Kern des Paares von Morphismen* $\alpha, \beta\colon A \to B$, wenn $\mu\alpha = \mu\beta$ gilt und wenn es für jeden Morphismus φ, für den $\varphi\alpha = \varphi\beta$ gilt, solche Morphismen $\varepsilon \in \mathsf{E}$ und γ gibt, daß $\varepsilon\varphi = \gamma\mu$ gilt.

Eine Unterkategorie E einer Kategorie K heißt *retraktiv* (man vgl. mit [8]), wenn sie folgenden Axiomen genügt:

Axiom E 1. Iso $\mathsf{K} \subseteq \mathsf{E}$.

Axiom E 2. Ist $\alpha\beta \in \mathsf{E}$, so gilt $\beta \in \mathsf{E}$.

Axiom E 3. Für ein beliebiges Paar von Morphismen mit gemeinsamem Ziel existiert ein E-universelles Quadrat.

Axiom E 4. Wenn das Quadrat (1) E-universell ist und $\alpha \in \mathsf{E}$, so ist $\delta_2 \in \mathsf{E}$.

Im weiteren wird noch das folgende Axiom benötigt.

Axiom E 5. Für jedes Paar von Morphismen mit gemeinsamer Quelle und gemeinsamem Ziel existiert ein E-Kern.

Wir erinnern daran, daß eine Kategorie R Kategorie mit Involution oder I-Kategorie heißt, wenn jede Morphismenmenge $H_{\mathsf{R}}(A, B)$ durch eine Relation $\subset$ partiell geordnet ist und wenn für beliebige $A, B \in \text{Ob } \mathsf{R}$ eine Abbildung (Involution)

$$H_{\mathsf{R}}(A, B) \ni \alpha \mapsto \alpha^{\#} \in H_{\mathsf{R}}(B, A)$$

gegeben ist, wobei folgende Bedingungen erfüllt sind:

a) $\alpha^{\#\#} = \alpha$;

b) $(\alpha\beta)^{\#} = \beta^{\#}\alpha^{\#}$;

c) aus $\alpha \subset \beta$ folgt $\alpha^{\#} \subset \beta^{\#}$;

d) aus $\alpha \subset \beta$ folgt $\alpha\gamma \subset \beta\gamma$ für beliebiges γ.

Ein Morphismus $\alpha\colon A \to B$ einer I-Kategorie R heißt D-*regulär* (B-, K-, I-regulär), wenn gilt $\alpha\alpha^{\#} \supset 1_A$ ($\alpha^{\#}\alpha \supset 1_B$, $\alpha\alpha^{\#} \subset 1_A$, $\alpha^{\#}\alpha \subset 1_B$). DI-reguläre Morphismen heißen *funktional* (eigentlich); sie bilden eine Unterkategorie F(R) der Kategorie R. DIB-reguläre Morphismen heißen *Projektionen*; sie bilden eine Unterkategorie Pr R; DIK-reguläre Morphismen heißen *Injektionen*; sie bilden eine Unterkategorie In R.

Ein Morphismus $\theta\colon A \to A$ heißt *Äquivalenz* (Koäquivalenz) des Objektes A, wenn gilt $\theta = \theta^{\#} = \theta^2 \supset 1_A$ ($\theta = \theta^{\#} = \theta^2 \subset 1_A$). Morphismen der Art $ff^{\#}$ ($f^{\#}f$) mit $f \in \mathsf{F}(\mathsf{R})$ heißen *Kongruenzen* (Kokongruenzen). Eine Kongruenz (Kokongruenz) ist exakt realisierbar, wenn man voraussetzen kann, daß $f \in \text{Pr R}$ ($f \in \text{In R}$).

Eine I-Kategorie R heißt *Kategorie von Korrespondenzen*, wenn sie den folgenden Axiomen genügt:

Axiom I 0. Jeder Morphismus φ aus R ist in der Form

$$\varphi = f^{\#}g \tag{4}$$

für gewisse funktionale Morphismen f und g darstellbar.

Axiom I 1. Für beliebige Objekte A, B aus R ist die Menge $H_{\mathsf{R}}(A, B)$ bezüglich Durchschnittsbildung ein Halbverband.

Axiom I 2. Für beliebige Morphismen $\alpha \in H_{\mathsf{R}}(A, B)$, $\varphi \in H_{\mathsf{R}}(B, C)$, $\beta \in H_{\mathsf{R}}(A, C)$ gilt

$$(\alpha \cap \beta\varphi^{\#})\varphi \supset \alpha\varphi \cap \beta\,. \tag{5}$$

Satz 1. *Es sei* R *eine Kategorie von Korrespondenzen, und es seien* $\varphi, \psi \in H_{\mathsf{R}}(A, B)$.

Die Inklusion $\varphi \subset \psi$ *gilt dann und nur dann, wenn sich für beliebige Darstellungen von* φ *und* ψ *in der Art* $\varphi = f^{\#}g$, $\psi = x^{\#}y$ *mit* $f, g, x, y \in \mathsf{F}(\mathsf{R})$ *eine Projektion* p *und ein funktionaler Morphismus* h *derart angeben lassen, daß* $pf = hx$ *und* $pg = hy$ *gelten.*

Theorem 1. *Es sei* R *eine Kategorie von Korrespondenzen.*

Dann ist die Unterkategorie der Projektionen Pr R *eine retraktive Unterkategorie der Kategorie der funktionalen Morphismen* F(R) *und enthält nur reguläre Epimorphismen aus* F(R).

3. Sei K eine gewisse Kategorie, in der eine Unterkategorie E, die den Axiomen E 1 bis E 3 genügt, ausgezeichnet ist. Für ein beliebiges Paar von Objekten $A, B \in \mathrm{Ob}\,\mathsf{K}$ werden wir mit $P(A, B)$ die Klasse aller Paare von Morphismen (α, β) mit $\alpha: X \to A$ und $\beta: X \to B$ bezeichnen. Ein Paar der Art $(\gamma\alpha, \gamma\beta)$ soll auch durch die Schreibweise $\gamma(\alpha, \beta)$ dargestellt werden. Für $(\alpha, \beta), (\alpha_1, \beta_1) \in P(A, B)$ setzen wir $(\alpha, \beta) <_{\mathsf{E}} (\alpha_1, \beta_1)$, wenn $\varepsilon(\alpha, \beta) = \gamma(\alpha_1, \beta_1)$ für gewisse Morphismen $\varepsilon \in \mathsf{E}$ und γ ist. Durch $\sim_{\mathsf{E}}$ soll die Äquivalenzrelation, die durch die transitive Hülle der Relation $<_{\mathsf{E}}$ induziert wird, bezeichnet werden. Die Äquivalenzklasse mit dem Repräsentanten (α, β) wird mit $[\alpha, \beta]_{\mathsf{E}}$ bezeichnet und E-Korrespondenz zwischen A und B genannt. Die Klasse aller E-Korrespondenzen zwischen A und B bezeichnen wir mit $\mathsf{Rel}_{\mathsf{E}}(A, B)$.

Ist $[\alpha, \beta]_{\mathsf{E}}, [\alpha_1, \beta_1]_{\mathsf{E}} \in \mathsf{Rel}_{\mathsf{E}}(A, B)$, $[\gamma, \delta]_{\mathsf{E}} \in \mathsf{Rel}_{\mathsf{E}}(B, C)$, so setzen wir

$$[\alpha, \beta]_{\mathsf{E}} \subset [\alpha_1, \beta_1]_{\mathsf{E}},$$

wenn gilt

$$(\alpha, \beta) <_{\mathsf{E}} (\alpha_1, \beta_1); \tag{6}$$

$$[\alpha, \beta]_{\mathsf{E}}^{\#} = [\beta, \alpha]_{\mathsf{E}}; \tag{7}$$

$$[\alpha, \beta]_{\mathsf{E}} \circ [\gamma, \delta]_{\mathsf{E}} = [\delta_1\alpha, \delta_2\delta]_{\mathsf{E}}, \tag{8}$$

wobei $[\delta_1, \beta, \gamma, \delta_2]_{\mathsf{E}}$ ein E-universelles Quadrat ist.

Satz 2. *Die Formel* (8) *ist korrekt genau dann, wenn die Unterkategorie* E *retraktiv ist. Für die retraktive Unterkategorie* E *ist die Relation* $<_{\mathsf{E}}$ *transitiv und die durch* (8) *definierte Multiplikation assoziativ.*

Theorem 2. *In der Kategorie* K *sei eine retraktive Unterkategorie* E *ausgezeichnet; für beliebige Objekte* $A, B \in \mathrm{Ob}\,\mathsf{K}$ *möge die Klasse* $\mathsf{Rel}_{\mathsf{E}}(A, B)$ *eine Menge sein.*

Dann werden durch die Formeln (6)—(8) *die Klasse von Objekten* $\mathrm{Ob}\,\mathsf{K} = \mathrm{Ob}\,\mathsf{Rel}(\mathsf{K};\mathsf{E})$ *und die Mengen* $\mathsf{Rel}_{\mathsf{E}}(A, B)$, $A, B \in \mathrm{Ob}\,\mathsf{K}$, *zu einer Kategorie mit Involution* $\mathsf{Rel}(\mathsf{K}; \mathsf{E})$, *in der das Axiom* I 0 *erfüllt ist.*

Theorem 3. *Unter den Bedingungen von Theorem* 2 *erfüllt die Kategorie* $\mathsf{Rel}(\mathsf{K}; \mathsf{E})$ *das Axiom* I 1 *dann und nur dann, wenn in* K *das Axiom* E 5 *erfüllt ist. Ist aber in* K *das Axiom* E 5 *erfüllt, so wird auch in* $\mathsf{Rel}(\mathsf{K}; \mathsf{E})$ *das Axiom* I 2 *erfüllt.*

Also ist die Kategorie $\mathsf{Rel}(\mathsf{K}; \mathsf{E})$ genau dann eine Kategorie von Korrespondenzen, wenn die retraktive Unterkategorie E das Axiom E 5 erfüllt.

4. Möge die Kategorie K mit der ausgezeichneten retraktiven Unterkategorie E den Bedingungen von Theorem 2 genügen. Indem wir jedem Morphismus $\alpha: A \to B$ aus K seine Grafik $\Gamma_{\mathsf{E}}(\alpha) = [1_A, \alpha]_{\mathsf{E}}$ zuordnen, erhalten wir einen Funktor $\Gamma_{\mathsf{E}}: \mathsf{K} \to \mathsf{Rel}(\mathsf{K}; \mathsf{E})$.

Das kommutative Quadrat (2) in einer I-Kategorie heißt voll kommutativ, wenn gilt $\varphi^{\#}\psi = \alpha\beta^{\#}$. Ein Funktor $F: \mathsf{R} \to \mathsf{R}'$ zwischen I-Kategorien heißt I-Funktor, wenn $F(\alpha^{\#}) = F(\alpha)^{\#}$ ist und wenn aus $\alpha \subset \beta$ folgt $F(\alpha) \subset F(\beta)$.

Theorem 4. *Unter den Bedingungen von Theorem* 2 *besitzt der Funktor* $\Gamma_{\mathsf{E}}: \mathsf{K} \to \mathsf{Rel}(\mathsf{K}; \mathsf{E})$ *die folgenden Eigenschaften:*

a) $\Gamma_{\mathsf{E}}(\mathsf{K}) \subseteqq \mathsf{F}(\mathsf{Rel}(\mathsf{K}; \mathsf{E}))$;

b) Γ_{E} *überführt jedes* E-*universelle Quadrat in ein voll kommutatives Quadrat;*

c) *Für jeden kovarianten Funktor F aus* K *in eine* I-*Kategorie* R, *für den* $F(\mathsf{K}) \subseteqq \mathsf{F}(\mathsf{R})$ *gilt und der* E-*universelle Quadrate in voll kommutative Quadrate überführt, existiert ein solcher eindeutiger* I-*Funktor* $\overline{F}$: $\mathsf{Rel}(\mathsf{K};\,\mathsf{E}) \to \mathsf{R}$, *daß* $F = \Gamma_{\mathsf{E}}\overline{F}$ *ist.*

d) Γ_{E} *ist eine Einbettung dann und nur dann, wenn gilt* $\mathsf{E} \subseteqq \mathrm{Epi}\ \mathsf{K}$.

e) Γ_{E} *ist ein Isomorphismus zwischen* K *und* $\mathsf{F}(\mathsf{Rel}(\mathsf{K};\,\mathsf{E}))$ *dann und nur dann, wenn jeder Morphismus aus* E *ein regulärer Epimorphismus ist.*

Das Theorem 4 verschärft entsprechende Resultate aus den Arbeiten [8] und [16]. Aus den Theoremen 1 und 4 ergibt sich das folgende Theorem.

Theorem 5. *Jede Kategorie von Korrespondenzen* R *ist* I-*isomorph zur Kategorie* $\mathsf{Rel}(\mathsf{F}(\mathsf{R});\,\mathsf{Pr}\ \mathsf{R}))$.

Also existiert eine (bis auf Isomorphie) eineindeutige Zuordnung zwischen den Kategorien von Korrespondenzen und den Paaren (K; E), wobei E eine retraktive Unterkategorie der Kategorie K ist, die nur aus regulären Epimorphismen besteht und dem Axiom E 5 genügt.

5. In einer beliebigen Kategorie K können viele retraktive Unterkategorien, welche aus regulären Epimorphismen bestehen, existieren. Darum kann die Kategorie K als Unterkategorie aller funktionalen Morphismen verschiedener Kategorien von Korrespondenzen dienen. In jeder Kategorie K mit universellen Quadraten existiert eine kleinste retraktive Unterkategorie, die Unterkategorie der Retraktionen Ret K, d. h. der links umkehrbaren Epimorphismen. Von Interesse ist die folgende Frage: Unter welchen Bedingungen sind die Klassen $\mathsf{Rel}_{\mathsf{Ret}\,\mathsf{K}}(A,\,B)$ für beliebige Objekte A und B Mengen? Wenn zum Beispiel das Skelett der Kategorie K eine Menge ist, so sind die Klassen $\mathsf{Rel}_{\mathsf{Ret}\,\mathsf{K}}(A,\,B)$ Mengen. Darum existiert für die Kategorie G_0 der endlichen Gruppen eine Kategorie $\mathsf{Rel}(\mathsf{G}_0;\,\mathsf{Ret}\ \mathsf{G}_0)$, die von der gewöhnlichen Kategorie der binären Relationen verschieden ist.

Eine zweite Frage über retraktive Unterkategorien ist die folgende: Wie ist die Gesamtheit aller aus regulären Epimorphismen bestehenden retraktiven Unterkategorien einer gegebenen Kategorie strukturiert? So gibt es zum Beispiel in der Kategorie der Mengen oder in der Kategorie der Vektorräume nur eine retraktive Unterkategorie. Andererseits existieren in Varietäten universeller Algebren, in denen nicht jeder Epimorphismus links umkehrbar ist, wenigstens zwei retraktive Unterkategorien.

6. Wenn auch in einer Kategorie K viele retraktive Unterkategorien vorhanden sein können, so kann es in K nur eine bikategorielle Struktur (K, E, M) geben, in der E aus regulären Epimorphismen besteht. In diesem Fall ist $\mathsf{E} = \mathrm{Epi}_r\ \mathsf{K}$ die Klasse aller regulären Epimorphismen und $\mathsf{M} = \mathrm{Mon}\ \mathsf{K}$ die Klasse aller Monomorphismen. Ist $\mathrm{Epi}_r\ \mathsf{K}$ außerdem eine retraktive Unterkategorie, so erfüllt die Kategorie $\mathsf{Rel}(\mathsf{K};\,\mathrm{Epi}_r\mathsf{K})$ das folgende Axiom.

Axiom I 3. Jede Koäquivalenz ist eine exakt realisierbare Kokongruenz.

Satz 3. *Die Kategorie mit Involution* R *möge das Axiom* I 0 *erfüllen.*

Dann sind folgende Bedingungen äquivalent:

a) *In* R *ist Axiom I 3 erfüllt;*

b) *in* R *ist jede Kongruenz exakt realisierbar;*

c) *in* R *ist jede Kokongruenz exakt realisierbar.*

Die Unterkategorie $\mathsf{F}(\mathsf{R})$ der Kategorie R besitzt wegen Theorem 4 aus der Arbeit [16] bei Erfüllung von I 0 und I 3 die bikategorielle Struktur $(\mathsf{F}(\mathsf{R}), \mathsf{Pr}\,\mathsf{R}, \mathsf{In}\,\mathsf{R})$. Darüber hinaus erweist sich $\mathsf{F}(\mathsf{R})$ als Kategorie mit regulären Kobildern (in der Terminologie des Buches [17]).

Es ist natürlich, eine Kategorie R mit regulären Kobildern, deren Unterkategorie $\mathrm{Epi}_r\,\mathsf{K}$ retraktiv ist, regulär zu nennen. Diese Definition unterscheidet sich etwas von der von Barr [1], erscheint aber angesichts der Zusammenhänge mit Kategorien von Korrespondenzen gerechtfertigter.

Theorem 6. *Es existiert eine (bis auf Isomorphie) eindeutige Zuordnung zwischen regulären Kategorien* K, *in denen Axiom* E 5 *erfüllt ist und in denen für jedes Paar von Objekten* A, B *die Klasse* $\mathsf{Rel}_{\mathrm{Epi}_r\mathsf{K}}(A, B)$ *eine Menge ist, und I-Kategorien, welche die Axiome* I 0—I 3 *erfüllen.*

Es ist natürlich, I-Kategorien, die die Axiome I 0—I 3 erfüllen, Kategorien von Korrespondenzen über regulären Kategorien zu nennen.

7. Wir nennen die Kongruenzen Θ_1 und Θ_2 des Objektes A koordiniert, wenn das Produkt $\Theta_1\Theta_2$ eine Kongruenz ist.

Theorem 7. *Es sei* R *eine Kategorie von Korrespondenzen über einer regulären Kategorie.*

Die Unterkategorie $\mathsf{F}(\mathsf{R})$ *ist eine Kategorie mit universellen Quadraten genau dann, wenn der Durchschnitt koordinierter Kongruenzen eine Kongruenz ist.*

Theorem 7 rechtfertigt die Einführung des folgenden Axioms.

Axiom I 4. Der Durchschnitt koordinierter Kongruenzen ist eine Kongruenz.

Also definiert die Gruppe von Axiomen I 0—I 4 Kategorien von Korrespondenzen über regulären Kategorien mit universellen Quadraten.

8. Das in der Arbeit [16] eingeführte Axiom I 5 ist eine Verschärfung von Axiom I 4.

Axiom I 5. Der Durchschnitt von Kongruenzen ist eine Kongruenz.

Das Axiom I 5 zusammen mit den Axiomen I 0—I 3 erlaubt, für jeden Morhpismus φ eine bis auf Isomorphie definierte Zerlegung

$$\varphi = m^{\#}p^{\#}qn \tag{9}$$

mit $m, n \in \mathsf{In}\,\mathsf{R}$, $p, q \in \mathsf{Pr}\,\mathsf{R}$, zu konstruieren. Insbesondere kann gezeigt werden, daß $ff^{\#} \cap gg^{\#} = 1$ ist, mit $f = pm$, $g = qn$, und daß (f, g) ein faktorisierendes Paar[1]) ist. Man kann dann sagen, daß jeder Morphismus gemäß Formel (1) durch ein faktorisierendes Paar darstellbar ist.

Satz 4. *In einer Kategorie von Korrespondenzen über einer regulären Kategorie ist jeder Morphismus genau dann durch ein faktorisierendes Paar darstellbar, wenn Axiom I 5 erfüllt ist.*

[1]) Das Paar (f, g) heißt faktorisierend, wenn aus den Gleichungen $xf = yf$, $xg = yg$ für beliebige x, y die Gleichung $x = y$ folgt.

So reproduziert bei Erfüllung der Axiome I 0—I 3 und I 5 eine Kategorie von Korrespondenzen besonders genau die in § 1 beschriebene „klassische" Situation. Die Eigenschaften der Unterkategorie F(R) können ohne Zuhilfenahme E-universeller Quadrate und E-Kerne beschrieben werden. F(R) besitzt nämlich die folgenden Eigenschaften:

a) (F(R), $\mathrm{Epi_r}$ F(R), Mon F(R)) ist eine Bikategorie;

b) F(R) ist eine Kategorie mit universellen Quadraten;

c) F(R) ist eine Kategorie mit Kernen von Paaren von Morphismen;

d) wenn in dem universellen Quadrat (1) gilt $\alpha \in \mathrm{Epi_r}$ F(R), so gilt $\delta_2 \in \mathrm{Epi_r}$ F(R);

e) für ein beliebiges Objekt A ist die Menge $Q_p(A)$ der zulässigen Faktorobjekte bezüglich Vereinigungen ein Halbverband;

f) für ein beliebiges Paar von Objekten $A, B \in$ F(R) ist die Klasse $\mathrm{Rel}_{\mathrm{Epi_r} F(R)}(A, B)$ gleichmächtig zur Menge $H_R(A, B)$.

Die Gesamtheit der Eigenschaften a)—f) für F(R) ist gleichbedeutend mit dem Erfülltsein der Axiome I 0—I 3 und I 5 in R.

Kategorien, welche die Eigenschaften a)—e) erfüllen, nennen wir *streng regulär.* Damit wird es natürlich, I-Kategorien, welche den Axiomen I 0—I 3 und I 5 genügen, *Kategorien von Korrespondenzen über streng regulären Kategorien* zu nennen.

9. Wir nennen eine Kategorie K *endlich vollständig,* wenn für ein beliebiges Paar von Objekten $A, B \in$ K ein Produkt und für ein beliebiges Paar von Morphismen $\alpha, \beta\colon A \to B$ ein Kern existiert. Mit anderen Worten, bei der gewohnten Definition der endlichen Vollständigkeit wird die Forderung nach der Existenz einer rechten Null weggelassen.

Das folgende Axiom verschärft Axiom I 5 im Falle der Kategorien von Korrespondenzen.

Axiom I 6. In jeder Menge $H_R(A, B)$ existiert ein größtes Element $\Omega_{A,B}$, welches in der Form $\Omega_{A,B} = p_A^{\#} p_B$ darstellbar ist, wobei (p_A, p_B) ein faktorisierendes Paar von Projektionen sei.

Theorem 8. *Die Unterkategorie* F(R) *einer Kategorie von Korrespondenzen* R *ist dann und nur dann eine streng reguläre und endlich vollständige Kategorie, wenn in* R *die Axiome I 3 und I 6 erfüllt sind.*

So charakterisiert die Gruppe von Axiomen I 0—I 3 und I 6 Kategorien von Korrespondenzen, die von Grillet [5] konstruiert wurden. Diese Gruppe von Axiomen ist dem Axiomensystem von Kawahara [9] äquivalent.

10. Die in den Arbeiten von M. Calenko [15], M. Burgin [2] und M. Grandis [4] konstruierten Kategorien mit Involution besitzen die Eigenschaft, daß in ihnen jeder Morphismus eindeutig bis auf Isomorphie in ein Produkt der Art

$$\varphi = m^{\#} p q^{\#} n \tag{10}$$

zerlegt werden kann, wobei m, n Injektionen und p, q Projektionen sind. In diesen Kategorien ist jeder Morphismus difunktional, d. h., er genügt der Identität

$$x = x x^{\#} x\,. \tag{11}$$

I-Kategorien, in denen die Identität (11) gilt, heißen difunktional.

Die Zerlegung (10) entspricht genau der in § 1 beschriebenen, auf GOURSAT zurückgehenden Darstellung binärer Relationen in der Kategorie der Gruppen.

Satz 5. *Eine* I-*Kategorie, in der Axiom I 0 erfüllt ist, ist difunktional genau dann, wenn die Kongruenzen eines beliebigen Objektes vertauschbar sind.*

Theorem 9. *In einer* I-*Kategorie* R *sei das Axiom I 0 erfüllt, und jede Äquivalenz sei eine exakt realisierbare Kongruenz.*

Ein Morphismus φ *ist difunktional dann und nur dann, wenn er eine Zerlegung* (10) *besitzt.*

Das Theorem 9 enthält in sich das bekannte Theorem von RIGUET (siehe z. B. [12], S. 30) und erweitert es insbesondere auf alle Varietäten universeller Algebren.

Folgerung 1. *Für eine Kategorie von Korrespondenzen sind folgende Bedingungen äquivalent:*

a) *Jeder Morphismus ist gemäß Formel* (10) *darstellbar;*

b) *die Kategorie ist difunktional, und jede Äquivalenz ist eine exakt realisierbare Kongruenz.*

In difunktionalen I-Kategorien ist jeder reflexive Morphismus φ, d. h. jeder solche Morphismus, für den $\varphi \supset 1$ ist, eine Äquivalenz. Darum werden bei Erfüllung der Bedingung a) in Kategorien von Korrespondenzen durch die exakt realisierbaren Kongruenzen alle reflexiven Morphismen ausgeschöpft. Insbesondere enthält diese Bemerkung ein Resultat von FINDLEY für Ringe (siehe z. B. [10]).

Eine Kategorie mit Involution, in der jeder Morphismus eine bis auf Isomorphie eindeutige Zerlegung (10) besitzt, läßt eine konstruktive Beschreibung zu. Eine entsprechende Konstruktion wurde von M. CALENKO [15] vorgeschlagen und von M. BURGIN [2] und M. GRANDIS [4] verwendet. Eine der endgültigen Formulierung nahekommende Formulierung des Resultates ist von GRANDIS gegeben worden. Wir haben hier nicht die Möglichkeit, das Endresultat anzuführen.

Zum Abschluß möchte der Autor auf einen in der letzten Zeit von R. GUITARD [6] und V. B. GISIN aufgedeckten Zusammenhang zwischen Kategorien von Korrespondenzen und KLEISLI-Kategorien hinweisen, die entstehen, wenn man den Funktor, auf Mengen von Unterobjekten bezogene, als Monade in einem Topos betrachtet.

Literatur

[1] M. BARR: Exact categories. Lecture Notes in Math. **236** (1971), 1—120.

[2] М. С. Бургин: Категории соответствий и соответствия в γ-категориях. Тр. Моск. мат. об-ва. **22** (1970), 161—228.

[3] E. GOURSAT: Sur les Substitutions Orthogonales. Ann. Sci. Ecole Norm. Sup., (3), **6** (1889), 9—102.

[4] M. GRANDIS: Symétrisations de catégories. II. Publ. Ist. Mat. Univ. Genova **125** (1974).

[5] P.-A. GRILLET: Regular categories. Lecture Notes in Math. **236** (1971), 121—222.

[6] R. GUITARD: Monades involutives complémentées. Cahiers topologie géométrie différentielle XVI, **1** (1975), 17—102.

[7] I. N. HERSTEIN: Noncommutative rings. The Carus Mathematical Monographs, No. 15. Amer. Math. Ass., New York, 1968.

[8] Y. Kawahara: Relations in categories with pullbacks. Mem. Fac. Sci. Kyushu Univ. A 27, 1 (1973), 149—173.
[9] Y. Kawahara: Matrix calculus in I-category and an axiomatic characterisation of relations in a regular category. Mem. Fac. Sci. Kyushu Univ. A27, 2 (1973), 249—273.
[10] J. Lambek: Lectures on rings and modules. Blaisdell, 1966.
[11] S. Mac Lane: An algebra of additive relations. Proc. Nat. Acad. Sci. USA, 47, 6 (1961), 1043—1051.
[12] А. И. Мальцев: Алгебраические системы. Москва 1970.
[13] D. Puppe: Korrespondenzen in abelschen Kategorien. Math. Ann. 148, 1 (1962), 1—30.
[14] Д. А. Райков: Об одном классе категорий соответствий. Докл. АН СССР 205, 6 (1972), 1300—1303.
[15] М. Ш. Цаленко: Соответствия над квазиточной категорией. Докл. АН СССР 155, 2 (1964), 292—294.
[16] М. Ш. Цаленко: Категории соответствий над регулярными категориями. Докл. АН СССР 211, 2 (1973), 297—299.
[16a] М. Ш. Цаленко: Категории соответствий над регулярными категориями. Изв. Акад. Наук Арльянской ССР, 10, 2 (1975), 113—132.
[17] М. Ш. Цаленко, Е. Г. Шульгейфер: Основы теории категорий. Москва 1974.

Über Kongruenzen in regulären Kategorien

von Emil Schwab, Timişoara

1.

Ein Morphismus $f \in [A, B]$ einer Kategorie heißt regulär (im Sinne von J. von Neumann), wenn es einen Morphismus $g \in [B, A]$ gibt, so daß $fgf = f$ gilt. In der vorliegenden Note werden wir unter einer regulären Kategorie eine Kategorie verstehen, in der jeder Morphismus regulär ist.

In der Literatur wird die Bezeichnung „reguläre Kategorie" verschieden gebraucht. So versteht Y. Sai in [8] unter einer regulären Kategorie eine additive Kategorie mit endlichen Coprodukten, in der für jedes Objekt A der Ring $[A, A]$ im Sinne von J. von Neumann regulär ist. In [8] wird gezeigt, daß unter diesen Voraussetzungen jeder Morphismus der Kategorie regulär ist. Der Begriff der Regularität einer Kategorie aus den Arbeiten [3], [4] u. a. unterscheidet sich von der Regularität im Sinne von J. von Neumann. In der vorliegenden Note werden diese Kategorien im Unterschied zu den regulären Kategorien G-reguläre Kategorien genannt. Eine G-reguläre Kategorie — so wie sie in [3] definiert wird — ist eine endlich vollständige Kategorie, in der das Axiom der Faserprodukte in bezug auf die Klasse der G-regulären Epimorphismen gilt (ein Epimorphismus wird hier G-regulärer Epimorphismus genannt, wenn er der coequalizer eines Morphismenpaares ist) und in der jeder Morphismus f eine Faktorisierung der Form $f = uh$ zuläßt mit u als Monomorphismus und h als G-regulärem Epimorphismus, die im folgenden G-reguläre Faktorisierung genannt wird.

Wenn R eine reguläre Kategorie ist, in der $i = uh$ für jeden idempotenten Morphismus i gilt, wobei u ein Monomorphismus und h ein Epimorphismus sind, dann läßt in R jeder Morphismus f eine G-reguläre Faktorisierung zu (Satz 2.2.). Wenn wir voraussetzen, daß R auch mit Faserprodukten erscheint, dann ist unter anderem eine Reihe von Ergebnissen aus dieser Note (4. Abschnitt) analog zu den Ergebnissen, die sich auf Kongruenzen in G-regulären Kategorien aus [3] beziehen. Dies ist dem Umstande zuzuschreiben, daß die Bedingung, mit Faserprodukten versehen zu sein, sich als sehr „stark" erweist im Sinne der „Annäherung" von R an G-reguläre Kategorien (Satz 3.1, Bemerkung 3.2).

In dieser Note sind die Relationen nach der von Y. Kawahara gegebenen Konstruktion in Kategorien mit Faserprodukten berücksichtigt ([4]). Die Bezeichnungen bei Relationen sind diejenigen aus [4], im übrigen — und dies ist auch für die Reihenfolge der Komposition der Relationen gültig — wird die Terminologie in [5] gebraucht. In den Beweisen werden oft die Eigenschaften der rechtsregulären (linksregulären) Einheiten der regulären Morphismen verwendet ([9]).

2.

Ein Morphismus i wird rechtsreguläre Einheit des Morphismus f genannt, wenn $fi = f$ gilt und ein Morphismus g existiert mit $i = gf$. Ein Morphismus läßt eine rechtsreguläre Einheit dann und nur dann zu, wenn er ein regulärer Morphismus ist. Eine rechtsreguläre Einheit eines Morphismus ist ein idempotenter Morphismus.

Für zwei idempotente Morphismen $i, j \in [A, A]$ schreiben wir $j \leqq_r i$ ($j \leqq_l i$) dann und nur dann, wenn $ji = j$ ($ij = j$) gilt. Die Relationen $\leqq_r$ und $\leqq_l$ sind Quasiordnungsrelationen in der Menge der idempotenten Morphismen, die $[A, A]$ angehören.

Satz 2.1. *In einer regulären Kategorie ist jeder Epimorphismus ein G-regulärer Epimorphismus.*

Beweis. Sei h ein beliebiger Epimorphismus einer regulären Kategorie und i^h eine rechtsreguläre Einheit von h. Es gilt $i^h i^h = i^h i$. Wenn j ein idempotenter Morphismus mit $ji^h = ji$ ist, dann ist offenbar $j \leqq_r i^h$, und nach Satz 1.5. (2) aus [9] folgt $h = \text{coequ}\,(i^h, 1)$.

Sei R eine reguläre Kategorie, in der $i = uh$ für jeden idempotenten Morphismus i gilt, wobei u ein Monomorphismus und h ein Epimorphismus sind. Indem man das oben Gezeigte und Theorem 1.6. (1) aus [9] berücksichtigt, folgt

Satz 2.2. *In der Kategorie* R *läßt jeder Morphismus f eine G-reguläre Faktorisierung zu.*

3.

Im vorigen Abschnitt wurde eine reguläre Kategorie R betrachtet, in der $i = uh$ für jeden idempotenten Morphismus i gilt, wobei u ein Monomorphismus und h ein Epimorphismus sind. Im folgenden werden wir voraussetzen, daß R auch mit Faserprodukten erscheint. Es gilt

Satz 3.1. *Wenn* R *ein Nullobjekt hat, dann ist* R *eine G-reguläre Kategorie.*

Beweis. Da R eine Kategorie mit Nullobjekt und mit Faserprodukten ist, folgt, daß R eine endlich vollständige Kategorie darstellt. Offensichtlich stimmen in R die Klasse aller Epimorphismen (notiert als Epi(R)), die Klasse aller Retraktionen und die Klasse aller G-regulären Epimorphismen überein.

Wir setzen voraus, daß in R das Diagramm

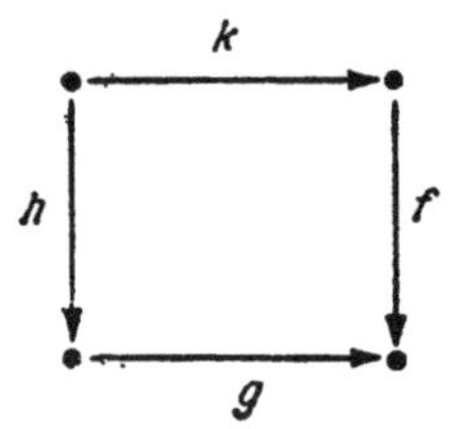

ein Faserprodukt-Diagramm mit f als Epimorphismus ist. Da auch

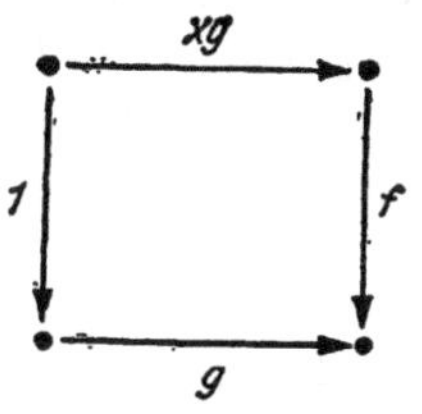

ein kommutatives Diagramm darstellt, in dem x ein Morphismus mit $fx = i$ ist, folgt, daß es einen eindeutig bestimmten Morphismus t mit $i = ht$ gibt und somit $h \in \mathrm{Epi}(\mathsf{R})$ gilt. In R gilt also das Axiom der Faserprodukte in bezug auf die Klasse der G-regulären Epimorphismen. Mit Satz 2.2 folgt, daß R eine G-reguläre Kategorie ist.

Bemerkung 3.2. Im oben formulierten Satz ist die Bedingung, daß R ein Nullobjekt besitzt, notwendig.

Sei S z. B. die Kategorie mit einem einzigen Objekt A, in der die Komposition der Morphismen durch die nachstehende Tabelle gegeben ist.

	1	a
1	1	a
a	a	1

Dann ist S eine Kategorie ohne Nullobjekt. Sie ist eine reguläre Kategorie mit G-regulärer Faktorisierung für die idempotenten Morphismen; sie ist eine Kategorie mit Faserprodukten, aber keine G-reguläre Kategorie.

4.

Wir werden uns jetzt auf Relationen und Kongruenzen in der Kategorie R beziehen, die im vorherigen Abschnitt betrachtet wurden.

In § 3 aus [4] werden für den Fall der Kategorien mit Faserprodukten die İ-Kategorie der Relationen in bezug auf eine retraktive Unterkategorie, in welcher die Involution und die Ordnungsrelation mit $*$ bzw. mit $\subset$ bezeichnet sind, und ein Funktor Γ, graphische Einbettung genannt, konstruiert.

Berücksichtigt man, daß die Klasse der Retraktionen einer Kategorie in jeder retraktiven Unterkategorie dieser Kategorie enthalten ist ([4], Satz 1.1.), dann folgt

Satz 4.1. $\mathrm{Epi}(\mathsf{R})$ *ist die einzige retraktive Unterkategorie von* R.

Damit kann die Präzisierung der retraktiven Unterkategorie im Falle der Kategorie R weggelassen werden; wir werden in diesem Falle die İ-Kategorie der Relationen einfach mit $\mathrm{Rel}(\mathsf{R})$ bezeichnen.

Ist $[f, g] \in [A, B]_{\mathrm{Rel}(\mathsf{R})}$ und $[h, k] \in [B, C]_{\mathrm{Rel}(\mathsf{R})}$, dann werden wir in $\mathrm{Rel}(\mathsf{R})$ für die Komposition der zwei Relationen, die eine Relation von A nach C ist, $[h, k]\,[f, g]$ schreiben und nicht $[f, g]\,[h, k]$ wie in [4], weil wir für die Komposition der Morphismen aus

R die in [5] gebrauchte Schreibweise benutzen wollen. Bezeichnungen, die sich auf Relationen beziehen, benutzen wir wie in [4]. Gilt $[f, g] \in [A, B]_{\mathsf{Rel(R)}}$ und $\Delta_A = [1_A, 1_A]$, $\Delta_B = [1_B, 1_B]$, dann folgt aus $[f, g]\,[f, g]^* \subset \Delta_B$ und $\Delta_A \subset [f, g]^*\,[f, g]$ mit Theorem 5.2 aus [4] die Existenz eines eindeutig bestimmten Morphismus $t \in \mathsf{R}$ mit $[f, g] = \Gamma(t)$. Berücksichtigt man auch die Formel 4.6. (e) aus [4], so gilt in R:

Satz 4.2. *Für eine Relation $[f, g] \in [A, B]_{\mathsf{Rel(R)}}$ existiert ein eindeutig bestimmter Morphismus $t \in \mathsf{R}$ mit $\Gamma(t) = [f, g]$ dann und nur dann, wenn $[f, g]\,[f, g]^* \subset \Delta_B$* und *$\Delta_A \subset [f, g]^*\,[f, g]$ gelten.*

Für einen Morphismus $f \in \mathsf{R}$ führen wir die Bezeichnung $\varrho_f = \Gamma(f)^*\,\Gamma(f)$ ein. Die Relation ϱ_f wird die vom Morphismus f erzeugte Kongruenz genannt (z. B. [3], I 5).

Gilt $f = uh$, wobei u ein Monomorphismus und h ein Epimorphismus sind, dann ist $\varrho_f = \varrho_h$. Da für einen Morphismus u nämlich $\Gamma(u)^*\,\Gamma(u) = \Delta$ gilt ([4], Formel 4.6.(c)), folgt

$$\varrho_f = \Gamma(f)^*\,\Gamma(f) = \Gamma(h)^*\,\Gamma(u)^*\,\Gamma(u)\,\Gamma(h) = \Gamma(h)^*\,\Gamma(h) = \varrho_h\,.$$

Für rechtsreguläre Einheiten der regulären Morphismen wurde in [9] folgendes Lemma bewiesen:

Lemma 4.3. *Seien f und g zwei reguläre Morphismen von derselben Quelle A sowie i^f und i^g rechtsreguläre Einheiten von f bzw. g. Dann sind folgende Aussagen äquivalent:*

(1) *Es gibt einen Morphismus t mit $f = tg$,*

(2) $fi^g = f$,

(3) $i^f \leqq_r i^g$.

Im weiteren beweisen wir folgenden Satz:

Satz 4.4. *Wenn f und g zwei Morphismen in R von derselben Quelle A sind, dann gilt $\varrho_g \subset \varrho_f$ dann und nur dann, wenn $i^f \leqq_r i^g$ ist, wobei i^f eine rechtsreguläre Einheit von f und i^g eine rechtsreguläre Einheit von g sind.*

Beweis. Sei $g = uh$ eine Faktorisierung von g mit einem Monomorphismus u und einem Epimorphismus h. Da $\varrho_h \subset \varrho_f$ ist und h ein G-regulärer Epimorphismus folgt nach Theorem 5.1 aus [4] die Existenz eines eindeutig bestimmten Morphismus $t \in \mathsf{R}$ mit $f = th$. Eine rechtsreguläre Einheit i^g von g ist aber eine Einheit von h, und mit Lemma 4.3 folgt $i^f \leqq_r i^g$.

Ist umgekehrt $i^f \leqq_r i^g$, dann gibt es nach Lemma 4.3 einen Morphismus t mit $f = tg$. Da $\Delta \subset \Gamma(t)^*\,\Gamma(t)$ für jeden Morphismus $t \in \mathsf{R}$ gilt ([4], Formel 4.6. (e)), folgt

$$\varrho_g = \Gamma(g)^*\,\Gamma(g) \subset \Gamma(g)^*\,\Gamma(t)^*\,\Gamma(t)\,\Gamma(g) = \Gamma(f)^*\,\Gamma(f) = \varrho_f\,. \quad (1)$$

Ist $t \in \mathsf{R}$ ein Monomorphismus, dann gilt $\Delta = \Gamma(t)^*\,\Gamma(t)$ ([4], Formel 4.6.(c)), und (1) wird eine Gleichheit, d. h. $\varrho_f = \varrho_g$.

Ist umgekehrt $\varrho_f = \varrho_g$ in R, dann existieren laut Satz 4.4 und Lemma 4.3 Morphismen t und s mit $f = tg$ und $g = sf$. Sind also f und g Epimorphismen in R und gilt $\varrho_f = \varrho_g$, dann ist t offenbar ein Isomorphismus. Es folgt

Satz 4.5. *Wenn in R die Morphismen $f \in [A, X]$ und $g \in [A, Y]$ Epimorphismen sind, dann sind X und Y isomorphe Quotienten-Objekte von A genau dann, wenn $\varrho_f = \varrho_g$ gilt.*

Dies gilt in G-regulären Kategorien für G-reguläre Quotienten-Objekte ([3]).

Da in R jeder Morphismus f eine G-reguläre Faktorisierung ($f = uh$) zuläßt, folgt, daß für jeden Morphismus f in R ein Epimorphismus $h \in$ R existiert mit $\varrho_f = \varrho_h$. Mit Satz 4.5 folgt daraus, daß jede Kongruenz $\varrho_f \in [A, A]_{\mathrm{Rel(R)}}$ bis auf einen Isomorphismus eindeutig ein Quotienten-Objekt von A (Ziel des Morphismus h) bestimmt, das mit A/ϱ_f bezeichnet wird. So wird die Formulierung der Isomorphie-Theoreme in R möglich wie im Falle der G-regulären Kategorien (siehe [3]).

Es ist offensichtlich, daß in R zwei beliebige mono-epi-Faktorisierungen eines Morphismus f äquivalent sind (die zwei von den zwei Faktorisierungen bestimmten Unterobjekte sind isomorph), und somit gilt

Theorem 4.6. *Für jeden Morphismus f in R gilt*

$$A/\varrho_f = \mathrm{Im}(f)\,.$$

Sei $g \in [A, .]$ ein Morphismus in R und $g = uh$ eine Faktorisierung von g, wobei u ein Monomorphismus und h ein Epimorphismus sind. Zur Erleichterung der Formulierung der folgenden zwei Theoreme bezeichnen wir den Epimorphismus $h \in [A, A/\varrho_g]$ mit nat g.¹)

Theorem 4.7. *Wenn f und g zwei Morphismen in R mit derselben Quelle A sind und $\varrho_g \subset \varrho_f$ gilt, dann gibt es einen eindeutig bestimmten Morphismus f_1, so daß das Diagramm*

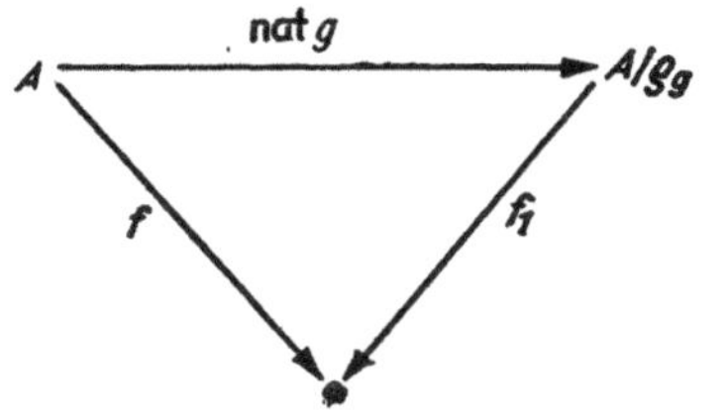

kommutativ ist.

Beweis. Mit Rücksicht auf Satz 4.4 und Lemma 4.3 folgt die Existenz eines Morphismus t mit $f = tg$. Wegen $g = u$ nat g gilt $f = (tu)$ nat $g = f_1$ nat g, und weil nat g ein Epimorphismus ist, ist das Theorem bewiesen.

Mit Hilfe von Thereom 4.7 erhält man

Theorem 4.8. *Wenn f und g zwei Morphismen in R mit derselben Quelle A sind und $\varrho_g \subset \varrho_f$ gilt, dann gibt es einen eindeutig bestimmten Morphismus $\theta \in [A/\varrho_g, A/\varrho_f]$ mit $\theta \cdot$ nat $g =$ nat f. Versteht man unter ϱ_f/ϱ_g die Kongruenz ϱ_θ, so bestimmt θ auf eindeutige Weise einen Isomorphismus $\theta_1 \in [(A/\varrho_g)/(\varrho_f/\varrho_g), A/\varrho_f]$, so daß das Diagramm*

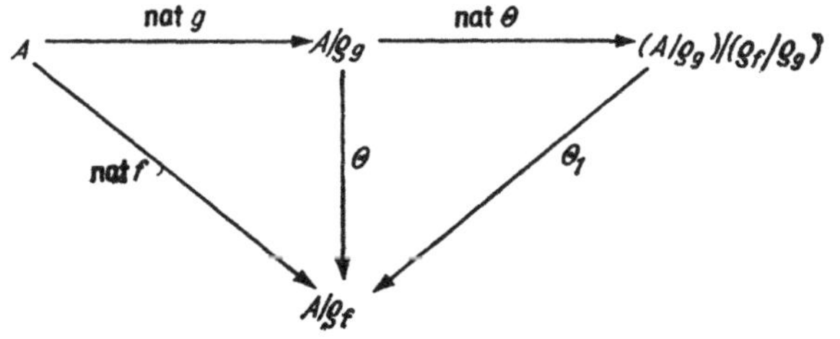

kommutativ ist.

¹) Wir betrachten sowohl die in [1] gebrauchten Bezeichnungen als auch die Darstellungsweise der Theoreme 3.3 und 3.4, Kap. I aus [1].

Literatur

[1] P. M. COHN: Universal algebra, New York, 1965.
[2] D. L. DAVIS; D. W. ROBINSON: Generalized inverses of morphisms, Linear Alg. and its Appl. 5 (1972), 319–328.
[3] P. A. GRILLET: Regular categories, Lecture Notes in Mathematics **236** (1971), 121–222.
[4] Y. KAWAHARA: Relations in categories with pullbacks, Memoirs of the Fac. of Sci. Kyushu University Ser. A 27 (1973), Nr. 1, 149–173.
[5] B. MITCHELL: Theory of categories, New York, 1965.
[6] J. VON NEUMANN: On regular rings, Proc. Acad. Sci. U.S.A. 22 (1936), 707–713.
[7] D. PUPPE: Korrespondenzen in abelschen Kategorien, Math. Ann. **148** (1962), 1–30.
[8] Y. SAI: On regular categories, Osaka, J. Math. 7 (1970), 301–306.
[9] E. SCHWAB: Über das verallgemeinerte Inverse eines Morphismus, Math. Nachr. (im Druck).

Inverse categories

by J. KASTL, Prague and Tokyo

A concept of inverse category arises if we apply a basic property of inverse semigroups (i.e. the existence of a unique generalized inverse) to morphisms of a category. As well as inverse semigroups the inverse categories are fully described by (regular) categories of partial morphisms. For example, any inverse category is isomorphic with a subcategory of the category of partial injective mappings between sets, and also with a subcategory of the category of all partial homeomorphisms of topological spaces, with a subcategory of the category of partial isomorphisms between algebras of type Δ etc.

To put it more exactly, let us embed an inverse category Y into the category X, which is well-powered and has finite intersections and every morphism of which is composed of an epimorphism and a monomorphism. Using categorical methods only it is possible to restrict (in X) the morphisms from Y so as to obtain the subcategory of category of all partial monomorphisms of X such that it is isomorphic with Y.

We can generalize this theorem for the case that the inverse category is embedded into the category of partial morphisms. Thus it is obtained a generalization of the theorem relating to inverse semigroups of translations on a set which is used by E. S. LJAPIN as well as by G. B. PRESTON.

I.

Definition. A category X is called an *inverse category* provided that for each morphism $x \in \mathsf{X}$ there exists a unique morphism $\bar{x}$ in X such that both of the following expressions are defined, and hold true: $x \cdot \bar{x} \cdot x = x$, $\bar{x} \cdot x \cdot \bar{x} = \bar{x}$.

If x is a morphism $a \xrightarrow{x} b$, then $b \xrightarrow{\bar{x}} a$. Let us observe that the set of all morphisms $a \xrightarrow{x} a$ (a — an object of X) is an inverse semigroup with an unit.

An idempotent of a category X is a morphism f such that $f \cdot f = f$ is defined and holds true.

The idempotents $a \xrightarrow{f,g} a$ belong to an inverse semigroup, consequently, they commute — see [1].

Note that in any inverse category $x \cdot \bar{x}$ is an idempotent, $x = \bar{\bar{x}}$, and $\overline{x \cdot y} = \bar{y} \cdot \bar{x}$.

Definition. A category X is called a *regular category* provided that for each morphism $x \in \mathsf{X}$ there exists a morphism $y \in \mathsf{X}$ such that $x \cdot y \cdot x = x$ is defined and holds true.

Theorem 1. *A category* X *is an inverse category if and only if it is a regular category and its idempotents commute.*

Proof. We show that a regular category with the commuting idempotents is an inverse category.

Namely, if $x \cdot y \cdot x = x$, then $x \cdot \tilde{x} \cdot x = x$, $\tilde{x} \cdot x \cdot \tilde{x} = \tilde{x}$ hold for $\tilde{x} = y \cdot x \cdot y$. If $\bar{x}$ also satisfies $x\bar{x}x = x$ and $\bar{x}x\bar{x} = \bar{x}$, then $x\bar{x}$, $x\tilde{x}\colon a \to a$ and $\bar{x}x$, $\tilde{x}x\colon b \to b$ are idempotents. Consequently, $\bar{x} = \bar{x}x\bar{x} = \bar{x}x\tilde{x}x\bar{x} = \bar{x}x\bar{x}x\tilde{x} = \bar{x}x\tilde{x} = \tilde{x}$.

Note that any subcategory of an inverse category is regular if and only if it is inverse.

Theorem 2. *Every inverse category is concrete.*

Proof. We shall verify the ISBELL concreteness condition for the inverse category X ([5], [14]).

Let

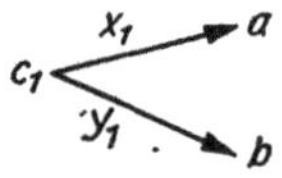

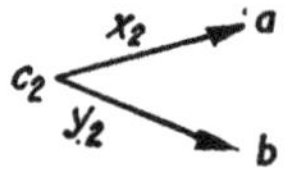

be pairs of morphisms satisfying $\bar{x}_1x_1 = \bar{x}_2x_2 \in \mathsf{X}(a, a)$, $\bar{x}_1y_1 = \bar{x}_2y_2 \in \mathsf{X}\,(a, b)$, $\bar{y}_1y_1 = \bar{y}_2y_2 \in \mathsf{X}(b, b)$. It follows that $x_1z = y_1v \Leftrightarrow x_2z = y_2v$ is true for an arbitrary pairs of morphisms

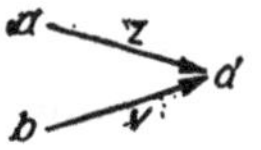

The equality $x_1z = y_1v$ implies: $\bar{y}_2y_2v = \bar{y}_1y_1v = \bar{y}_1x_1z = (\overline{\bar{x}_1y_1})\,z = \bar{y}_2x_2z$, hence $y_2v = y_2\bar{y}_2y_2v = y_2\bar{y}_2x_2z$. Equality $x_2z = x_2\bar{x}_2y_2v$ holds analogously. Consequently $x_2z = x_2\bar{x}_2y_2\bar{y}_2x_2z = y_2\bar{y}_2x_2z = y_2v$.

For any pair of objects a, b of the inverse category the relation <> from [5] always contains only a set of pairwise non-equivalent pairs

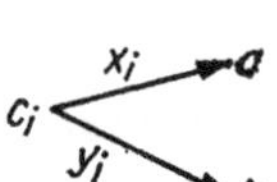

Any inverse category is concrete.

Remark 1. Some concepts of inverse semigroups can be introduced in inverse categories. For example: ordering $x \leqq y$, the GREEN's relations $\mathscr{L}$, $\mathscr{R}$ etc. A great number of propositions about inverse semigroups can be expressed for the inverse categories quite analogously ([8]).

II.

Definition. Let the class of all isomorphisms of a category X (monomorphisms, epimorphisms, respectively) be denoted as Iso X (Mono X, Epi X, respectively).

A class $\mathfrak{M}$ of morphisms of a category X is called an *M-class* provided that $\operatorname{Iso} \mathsf{X} \subseteq \mathfrak{M} \subseteq \operatorname{Mono} \mathsf{X}$ and $\mathfrak{M}$ is closed under composition.

Let $\mathfrak{M}$ be an M-class of the category X.

If the category X

1) is $\mathfrak{M}$-well-powered — i.e. there is only a set of pairwise non-isomorphic monomorphisms of $\mathfrak{M}$ with a common range,

2) has inverse $\mathfrak{M}$-images — i.e. for any $b \xrightarrow{m} a \in \mathfrak{M}$, $c \xrightarrow{x} a \in \mathsf{X}$ there exists a pullback in X

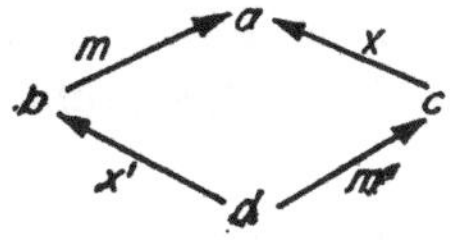

and moreover $m' \in \mathfrak{M}$,

then we define a category of partial morphisms $\mathfrak{M}$-PX.

Let us consider pairs of morphisms $a \xleftarrow{m} c \xrightarrow{x} b$, $m \in \mathfrak{M}$, $x \in \mathsf{X}$. The pairs $a \xleftarrow{m} c \xrightarrow{x} b$, $a \xleftarrow{m'} c' \xrightarrow{x'} b$ are equivalent provided that there exists an isomorphism $\varphi: c' \to c$ such that $\varphi \cdot m = m'$, $\varphi \cdot x = x'$. Let the obtained equivalence classes of pairs of morphisms be denoted by (m, x).

The category $\mathfrak{M}$-PX has the same objects as the category X. The morphisms of $\mathfrak{M}$-PX correspond to the classes of the pairs $a \xrightarrow{(m,x)} b$, where $a \xleftarrow{m} c \xrightarrow{x} b$, $m \in \mathfrak{M}$ $(x \in \mathsf{X})$.

Let $a \xrightarrow{(m,x)} b$, $b \xrightarrow{(n,y)} d \in \mathfrak{M}$-$\mathsf{PX}$ and let a pullback $n' \cdot x = x' \cdot n$ be made (in X):

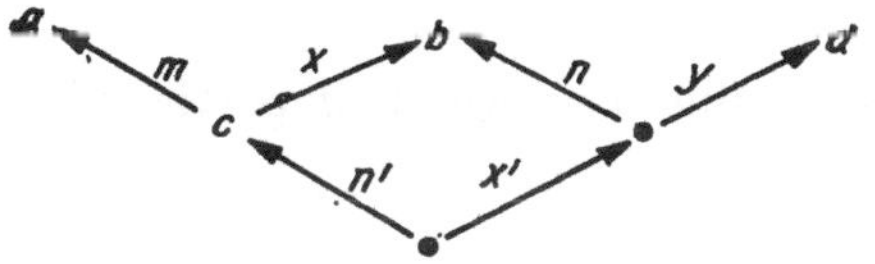

Let the composition of morphisms $(m, x) \cdot (n, y)$ be defined as $(n'm, x'y)$.

It is easy to see that $\mathfrak{M}$-PX is really a category with sets of morphisms between any two objects. (The units are $(1_a, 1_a)$.)

The category X can be embedded by an injective functor X into the category $\mathfrak{M}$-PX. For an object $a \in \mathsf{X}$ is $(a)\, X = a$; for a morphism $x: a \to b \in \mathsf{X}$ is $(x)\, X = (1_a, x)$.

Lemma 1. *The functor* $X: \mathsf{X} \to \mathfrak{M}$-$\mathsf{PX}$ *satisfies*:

1) $(\operatorname{Mono} \mathsf{X})\, X = \operatorname{Mono} \mathfrak{M}$-$\mathsf{PX}$

2) $(\operatorname{Iso} \mathsf{X})\, X = \operatorname{Iso} \mathfrak{M}$-$\mathsf{PX}$

3) *If*

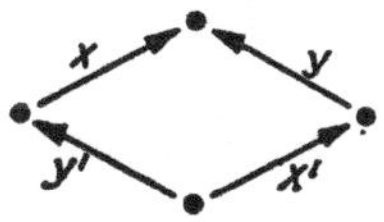

is a pullback in X, *then*

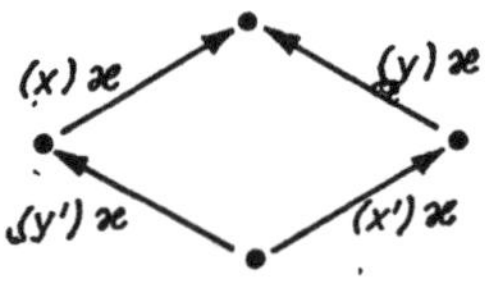

is a pullback in $\mathfrak{M}$-PX.

Proof. 1) It is easy to verify that (Mono X) $X \subseteqq$ Mono $\mathfrak{M}$-PX. Since X is injective it follows: $x \in$ Mono X $\Leftrightarrow$ $(x)\,X \in$ Mono $\mathfrak{M}$-PX .

Let $(m, x) \in$ Mono $\mathfrak{M}$-PX, $a \xleftarrow{m} c \xrightarrow{x} b$. Since $(m, m) \cdot (m, x) = (m, x) = (1_a, 1_a) \cdot$ $\cdot\, (m, x)$, then $(m, m) = (1_a, 1_a)$, and therefore m is an isomorphism. This implies $(m, x) =$ $= (1_a, m^{-1}x) = (m^{-1}x)\,X$.

2) Any $\xi \in$ Iso $\mathfrak{M}$-PX must be expressed as $\xi = (m)\,X$ where $m \in$ Mono X. Analogously $\xi^{-1} = (n)\,X$. Because $1 = (1)\,X = \xi \cdot \xi^{-1} = (nm)\,X$, then m is an isomorphism of X.

3) Let $y'x = x'y$ be a pullback in X and let us suppose that the morphisms (m, z), $(n, v) \in \mathfrak{M}$-PX satisfy $(m, z) \cdot (x)\,X = (n, v) \cdot (y)\,X$, i.e. $(m, zx) = (n, vy)$. Then there exists a $\varphi \in$ Iso X such that $\varphi zx = vy$. Hence there exists $\alpha \in$ X such that $\alpha y' = \varphi z$, $\alpha x' = v$. Because $\varphi m = n$, then for $(n, \alpha) \in \mathfrak{M}$-PX following equalities are valid: $(n, \alpha) \cdot (y')\,X = (n, \alpha y') = (m, z)$, $(n, \alpha) \cdot (x')\,X = (n, v)$.

To verify that $(y')\,X \cdot (x)\,X = (x')\,X \cdot (y)\,X$ is a pullback it suffices to show that for given (m, z), (n, v) there exists a unique (n, α) in $\mathfrak{M}$-PX If (n_1, α_1) also satisfies $(n_1, \alpha_1 y') = (m, z)$ and $(n_1, \alpha_1 x') = (n, v)$, then there exist $\psi_1, \psi_2 \in$ Iso X such that $\psi_1 n_1 = n$, $\psi_1\alpha_1 y' = \alpha y'$, $\psi_2 n_1 = n$, $\psi_2\alpha_1 x' = v = \alpha x'$. From the fact that $n_1 \in$ Mono X necessarily follows that $\psi_1 = \psi_2$. Equality $\psi_1\alpha_1 = \alpha$ follows from the property of the pullback. Then $(n_1, \alpha_1) = (n, \alpha)$.

Similarly to the definition of $\mathfrak{M}$-PX we define a category of partial bijections — $\mathfrak{M}$-PBX.

Let $\mathfrak{M}$ be an M-class of the category X, let X

1) be $\mathfrak{M}$-well-powered,

2) have (finite) $\mathfrak{M}$-intersections — i.e., for any $b \xrightarrow{m} a \in \mathfrak{M}$, $c \xrightarrow{n} a \in \mathfrak{M}$ there exists a pullback

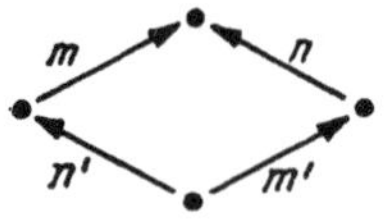

in X, and moreover $n' \in \mathfrak{M}$, $m' \in \mathfrak{M}$.

Let us consider again the equivalence classes of the pairs of morphisms (m, n).

The objects of the category $\mathfrak{M}$-PBX are equal to the objects of X, the morphisms of $\mathfrak{M}$-PBX are of the form $a \xrightarrow{(m, n)} b$ where $a \xleftarrow{m} c \xrightarrow{n} b$, $m \in \mathfrak{M}$, $n \in \mathfrak{M}$. The composition of morphisms is defined in $\mathfrak{M}$-PBX equally as in the category $\mathfrak{M}$-PX.

Theorem 3. *Let* $\mathfrak{M}$ *be an* M*-class of the category* X. *Let* X *be* $\mathfrak{M}$*-well-powered and have* $\mathfrak{M}$*-intersections.*

Then $\mathfrak{M}$-PBX *is an inverse category.*

Proof. From the equality $(m, n) \cdot (n, m) = (m, m)$ we can conclude that $\mathfrak{M}$-PBX is a regular category. We shall show now that idempotents of $\mathfrak{M}$-PBX are of the form (m, m).

Suppose that $(m, n) \cdot (m, n) = (m, n)$ holds for some morphism (m, n) of $\mathfrak{M}$-PBX. Hence

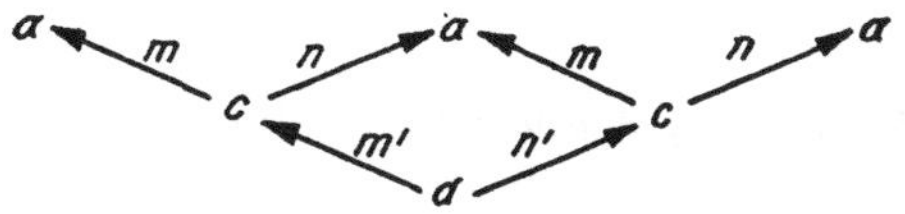

$$(m'm, n'n) = (m, n) .$$

Then there exists a $\varphi\colon c \to d \in \operatorname{Iso} X$ such that $\varphi m'm = m$, $\varphi n'n = n$, consequently $\varphi m' = 1_c = \varphi n'$. Thus $n = \varphi m'n = \varphi n'm = m$.

The following equality holds for the idempotents (m, m), (n, n): $(m, m) \cdot (n, n) = (n'm, m'n) = (n, n) \cdot (m, m)$ where m, n, n', m' again make a pullback $n'm = m'n$.

By Theorem 1 $\mathfrak{M}$-PBX is an inverse category.

Notice that $\overline{(m, n)} = (n, m)$.

Let $\mathfrak{M}$-PX be a category of partial morphisms. Then the inverse category $\mathfrak{M}$-PBX is a subcategory of $\mathfrak{M}$-PX.

Examples. (1) X is the category of all sets and mappings, all injective mappings form the M-class $\mathfrak{M}$. The category $\mathfrak{M}$-PX is isomorphic with the category of all partial mappings (of the sets), the category of all injective partial mappings corresponds to $\mathfrak{M}$-PBX.

(2) X — the arbitrary variety of algebras of type Δ, $\mathfrak{M}$ — the class of all injective homomorphisms. $\mathfrak{M}$-PX — the category morphisms of which are such partial mappings which are defined on some subalgebra, and are homomorphisms of this subalgebra. $\mathfrak{M}$-PBX — a morphism between two Δ-algebras is an arbitrary isomorphism of their subalgebras.

(3) X — the category of topological spaces and continuous mappings, $\mathfrak{M}$ — the class of all continuous injective mappings which form projectively their domain topology. $\mathfrak{M}$-PX — the morphisms are such partial mappings of topological spaces, which are continuous for all points where they are defined. $\mathfrak{M}$-PBX — partial homeomorphisms of topological spaces.

Consider that there exists a category of partial morphisms $\mathfrak{M}$-PX. It follows from Lemma 1 that $(\mathfrak{M})X$ is an M-class of the category $\mathfrak{M}$-PX, moreover $\mathfrak{M}$-PX is $(\mathfrak{M})X$-well-powered and has $(\mathfrak{M})X$-intersections. Then a category $(\mathfrak{M})X$-PB($\mathfrak{M}$-PX) can be made.

Theorem 4. *Let* $\mathfrak{M}$ *be an M-class of the category* X. *Let* X *be* $\mathfrak{M}$*-well-powered and have inverse* $\mathfrak{M}$*-images. A natural embedding of* X *into* $\mathfrak{M}$-PX *maps* $\mathfrak{M}$ *onto* $\mathfrak{M}' = (\mathfrak{M})\, X$.

The category $\mathfrak{M}'$-PB($\mathfrak{M}$-PX) *is isomorphic with the category* $\mathfrak{M}$-PBX.

Proof. Let any morphism $a \xrightarrow{(m,n)} b \in \mathfrak{M}$-PBX be in relation with the morphism $a \xrightarrow{((m)X,\,(n)X)} b$ of the category $\mathfrak{M}'$-PB($\mathfrak{M}$-PX). Because X is an injective functor, we have defined, according to Lemma 1, an injective class-mapping of $\mathfrak{M}$-PBX onto

$\mathfrak{M}'$-PB($\mathfrak{M}$-PX) correctly. Using point 3) of Lemma 1 we have a functor Φ — an isomorphism of the above mentioned categories.

Remark 2. ($\mathfrak{M}$) $X = \mathfrak{M}'$ is also an M-class for the subcategory $\mathfrak{M}$-PBX. The category $\mathfrak{M}'$-PB($\mathfrak{M}$-PBX) is identical with the category $\mathfrak{M}'$-PB($\mathfrak{M}$-PX).

III.

Theorem 5. *Let $\mathfrak{M}$ be an $\mathfrak{M}$-class of the category* X *and let the following conditions be satisfied:*

1) X *is $\mathfrak{M}$-well-powered.*

2) X *has $\mathfrak{M}$-intersections.*

3) *For any morphisms x, y of the category* X, *if $xy \in \mathfrak{M}$ holds, then $x \in \mathfrak{M}$.*

4) *Every morphism x of the category* X *is composed of an epimorphism and a monomorphism — $x = x_e \cdot x_m$ where $x_e \in$* Epi X, *$x_m \in$* Mono X.

Let Y *be a subcategory of* X, Y *be an inverse category.*

Then there exists a subcategory Z *of $\mathfrak{M}$-PBX such that* Z *is isomorphic with the category* Y.

Proof. Let us decompose a morphism $y \in$ Y and a morphism $\bar{y} \in$ Y into some epimorphism and a monomorphism.

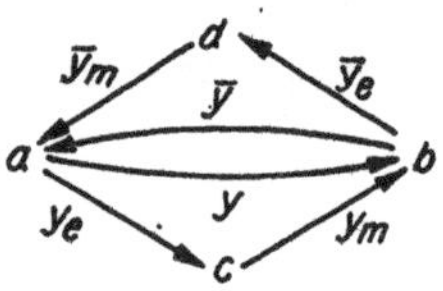

Then $y_m \cdot \bar{y}_e : c \to d$, $\bar{y}_m \cdot y_e : d \to c$ are pairwise inverse isomorphisms. $y_e \cdot y_m\bar{y}_e\bar{y}_m y_e \cdot \cdot\, y_m = y\bar{y}y = y = y_e \cdot 1_c \cdot y_m$, then $y_m\bar{y}_e \cdot \bar{y}_m y_e = 1_c$. (Then in special case $y_m\bar{y}y = y_m$.) Analogously $\bar{y}_m y_e \cdot y_m\bar{y}_e = 1_d$.

Hence $(\bar{y}_m, \bar{y}_m y) = (y_m\bar{y}_e \cdot \bar{y}_m, y_m\bar{y}_e \cdot \bar{y}_m y_e \bar{y}_m) = (y_m\bar{y}, y_m)$ which is, according to the condition 3), a morphism $a \to b$ of the category $\mathfrak{M}$-PBX. It can be immediately seen that in this way the morphism of $\mathfrak{M}$-PBX is defined for a given $y \in$ Y uniquely.

Then a functor Ψ of Y into $\mathfrak{M}$-PBX can be defined by $(y)\,\Psi = (\bar{y}_m, \bar{y}_m y) = (y_m\bar{y}, y_m)$.

Evidently $(1_a)\,\Psi = (1_a, 1_a)$. Let us verify equality $(xy)\,\Psi = (x)\,\Psi \cdot (y)\,\Psi$ $(x, y \in$ Y$)$:

Again we consider the factorization of any morphism of Y into some epimorphism and a monomorphism.

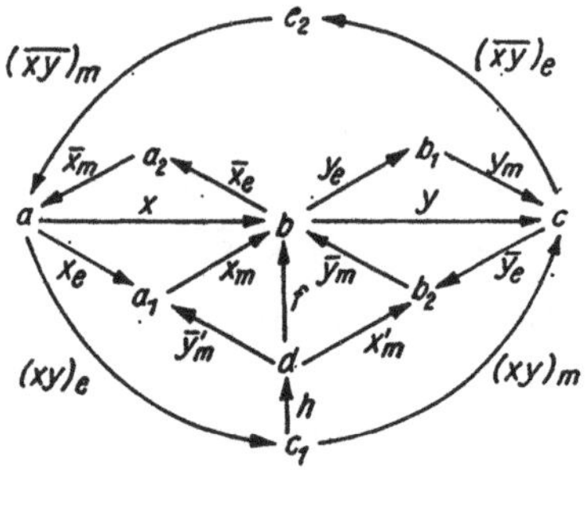

$\bar{y}'_m x_m = x'_m \bar{y}_m (= f)$ is a pullback (constituted by monomorphisms).

Then $(xy)_e \cdot (xy)_m \overline{xy} x_e x_m = xy\bar{y}\bar{x}x = x\bar{x}xy\bar{y} = (y\bar{y}, \bar{x}x$ are idempotents in $\mathsf{Y}) = xy\bar{y} = (xy)_e \cdot (xy)_m \bar{y}_e \bar{y}_m$, hence $(xy)_m \overline{xy} x_e \cdot x_m = (xy)_m \bar{y}_e \cdot \bar{y}_m$. Thus there exists an $h \in \mathsf{X}$ such that $h \cdot \bar{y}'_m = (xy)_m \overline{xy} x_e$ and $h \cdot x'_m = (xy)_m \bar{y}_e$. Then $hf = (xy)_m \overline{xy} x = (xy)_m \bar{y}$.

h is an isomorphism of X. Namely:

$$(xy)_e \cdot hf\bar{x}(xy)_e \cdot (xy)_m = (xy)_e (xy)_m \bar{y}\bar{x}xy = xy\overline{xy}xy = xy = (xy)_e \cdot 1_{c_1} \cdot (xy)_m ,$$

i.e.

$$h \cdot f\bar{x}(xy)_e = 1_{c_1} ;$$

$$f\bar{x}(xy)_e h \cdot f = f\bar{x}xy\bar{y} = \bar{y}'_m x_m \bar{x}xy\bar{y} = \bar{y}'_m x_m y\bar{y} = x'_m \bar{y}_m y\bar{y} = x'_m \bar{y}_m = f ,$$

i.e.

$$f\bar{x}(xy)_e \cdot h = 1_d .$$

Finally:

$$(x)\,\Psi \cdot (y)\,\Psi = (x_m\bar{x}, x_m) \cdot (\bar{y}_m, \bar{y}_m y) = (\bar{y}'_m x_m \bar{x}, x'_m \bar{y}_m y) = (f\bar{x}, fy) =$$
$$= (hf\bar{x}, hfy) = \big((xy)_m \bar{y}\bar{x}, (xy)_m \overline{xy}xy\big) = \big((xy)_m \overline{xy}, (xy)_m\big) = (xy)\,\Psi .$$

The functor Ψ of Y into $\mathfrak{M}$-PBX is injective. Namely, if for some $y, z \in \mathsf{Y}$ $(y)\Psi = (z)\,\Psi: a \to b$ holds, i.e. $(y_m\bar{y}, y_m) = (z_m\bar{z}, z_m)$, then there exists $\varphi \in \operatorname{Iso} \mathsf{X}$, $\varphi y_m \bar{y} = z_m\bar{z}$, $\varphi y_m = z_m$. Hence $z\bar{z} = z_e \varphi y_m \bar{y} = z_e z_m \bar{y} = z\bar{y}$.

We see that the equality $(x)\,\Psi = (y)\,\Psi$ implies that $x\bar{x} = x\bar{y}$ and $y\bar{y} = y\bar{x}$ $(x, y \in \mathsf{Y})$. Because $\mathfrak{M}$-PBX is an inverse category the functor Ψ satisfies $(\bar{y})\,\Psi = \overline{(y)\,\Psi}$. The equality implies $(\bar{x})\,\Psi = (\bar{y})\,\Psi$ and $\bar{x}x = \bar{x}y$, as well. If we summarize up, $x = x\bar{x}x = x\bar{x}y = x\bar{y}y = (y\bar{x})\,y = (y\bar{y})\,y = y$.

The constructed functor Ψ is injective and the proof is completed.

Remark 3. 1) Observe that the functor Ψ is defined by $(y)\,\Psi = (\bar{y}_m, \bar{y}_m y) = (\bar{y}_m, \bar{y}_m) \cdot (y)\,X$.

2) Let the M-class of the category X be $\mathfrak{M} = \operatorname{Mono} \mathsf{X}$, then the conditions 1)—4) are equivalent with the following three conditions:

1) X is well-powered.

2) X has (finite) intersections.

3) Every morphism of X is composed by an epimorphism and a monomorphism, i.e. X is (Epi, Mono)-factorizable.

IV.

Let a category X satisfy the assumptions of Theorem 5. If there exists an injective functor of an inverse category V into the category X, then the category V is isomorphic with the category all morphisms of which are partial bijections of X.

It can easily be obtained that there are inverse categories of no other form than inverse categories the morphisms of which are partial bijections of sets.

Theorem 6. *Any inverse category is isomorphic with a subcategory of the category of all injective partial mappings between the sets.*

Proof. Theorem 2 and Theorem 5.

Theorem 7. *Let a category* X *with its M-class* $\mathfrak{M}$ *satisfy the assumptions* 1)—4) *of Theorem* 5 *and let us suppose that there exists an injective functor of the category of sets and mappings into the category* X.

Then any inverse category V *is isomorphic with a subcategory* Z *of the category* $\mathfrak{M}$-PBX.

Proof. Theorem 2 and Theorem 5.

Examples. (1) All inverse categories are isomorphic with the inverse categories whose morphisms are partial isomorphisms of Δ-algebras of such a variety which contains at least one non-trivial algebra.

(2) Analogously, inverse subcategories of the category of all partial homeomorphisms of compact Hausdorff spaces can have an arbitrary form.

In both these cases the M-class $\mathfrak{M}$ is the class of all monomorphisms Mono X, an injective functor of sets into the category X is a left adjoint of the forgetful functor.

Definition. Analogously to the definition of M-class, an *E-class* of morphisms of a category X can be defined as the class $\mathfrak{E}$ that satisfies Iso X $\subseteqq \mathfrak{E} \subseteqq$ Epi X and that is closed under composition.

Definition. A category X with an M-class $\mathfrak{M}$ and an E-class $\mathfrak{E}$ is called ($\mathfrak{E}$, $\mathfrak{M}$)-*category* provided that every morphism $x \in$ X has a factorization $x = e_x \cdot m_x$, $e_x \in \mathfrak{E}$, $m_x \in \mathfrak{M}$, and this factorization is unique up to an isomorphism ([7]).

Lemma 2. *The following implication holds true in any* ($\mathfrak{E}$, $\mathfrak{M}$)*-category*: $x \cdot y \in \mathfrak{M} \Rightarrow$ $\Rightarrow x \in \mathfrak{M}$ (x, y — *morphisms*).

Proof. Let us use the ($\mathfrak{E}$, $\mathfrak{M}$)-factorization.

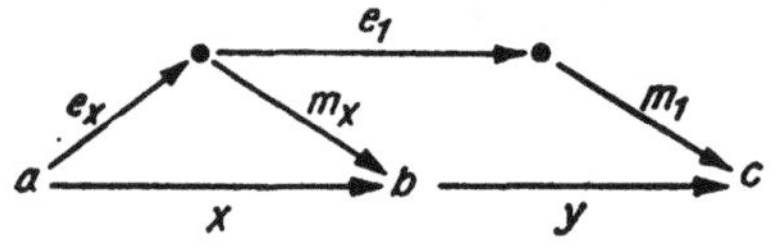

$$x = e_x \cdot m_x\,, \qquad m_x \cdot y = e_1 \cdot m_1 \quad \text{where} \quad e_x, e_1, e_x e_1 \in \mathfrak{E}, m_x, m_1, xy \in \mathfrak{M}.$$

So $xy \in \mathfrak{M}$ has the factorizations $e_x e_1 \cdot m_1 = xy = 1_a \cdot xy$. The uniqueness of the ($\mathfrak{E}$, $\mathfrak{M}$)-factorization implies that $e_x e_1 \in$ Iso X. Hence $e_1 \in$ Iso X ($\mathfrak{E} \subseteqq$ Epi X), $e_x \in$ Iso X and therefore $x \in \mathfrak{M}$.

Theorem 8. *Suppose that*:

1) X *is an* ($\mathfrak{E}$, $\mathfrak{M}$), *category* —

2) X *is* $\mathfrak{M}$*-well-powered,*

3) X *has inverse* $\mathfrak{M}$*-images,*

4) *inverse* $\mathfrak{M}$*-images preserve morphisms of* $\mathfrak{E}$ — *i.e., if there is a pullback in* X

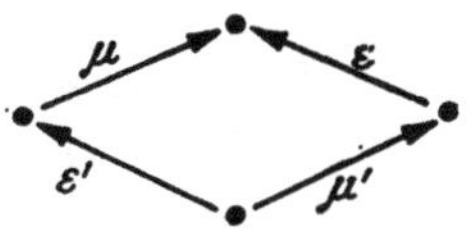

where $\mu \in \mathfrak{M}$, $\varepsilon \in \mathfrak{E}$, *then* $\varepsilon' \in \mathfrak{E}$.

Let Y *be a subcategory of* $\mathfrak{M}$-PX, *let* Y *be an inverse category.*

Then there exists a subcategory Z *such that* Z *and* Y *are isomorphic inverse categories and all morphisms of* Z *are partial bijections of* X (Z $\subseteqq$ $\mathfrak{M}$-PBX).

Proof. The embedding X: X $\to$ $\mathfrak{M}$-PX sends a morphism $x\colon a \to b \in$ X to the morphism $(x)\,X = (1_a, x)$. $\mathfrak{M}' = (\mathfrak{M})\,X$ is an M-class of the category $\mathfrak{M}$-PX. It is known that $\mathfrak{M}$-PX is $\mathfrak{M}'$-well-powered and has $\mathfrak{M}'$-intersections. Now we shall show that the category $\mathfrak{M}$-PX with the M-class $\mathfrak{M}'$ satisfies the conditions 3) and 4) of Theorem 5.

Suppose that $(m, x) \cdot (n, y) = (n'm, x'y) \in (\mathfrak{M})\,X$ ($n'x = x'n$ is a pullback). This means that $n'm \in$ Iso X, $x'y \in \mathfrak{M}$. Hence $n' \in$ Iso X, $m \in$ Iso X; by Lemma 2 $x' \in \mathfrak{M}$; therefore $x'n = n'x \in \mathfrak{M}$, $x \in \mathfrak{M}$. Thus $(m, x) \in (\mathfrak{M})\,X$.

Now let us verify the condition 4). Let (m, x) be an arbitrary morphism of $\mathfrak{M}$-PX. Let $x = en$, $e\colon c \to d \in \mathfrak{E}$, $n\colon d \to b \in \mathfrak{M}$ be the factorization of x. Then $(m, x) =$ $= (m, e) \cdot (1_d, n)$ where $(1_d, n) \in \mathfrak{M}'$. It suffices to show that $(m, e) \in$ Epi($\mathfrak{M}$-PX).

Let us suppose that $(m, e) \cdot (n_1, y_1) = (m, e) \cdot (n_2, y_2)$ holds in $\mathfrak{M}$-PX.

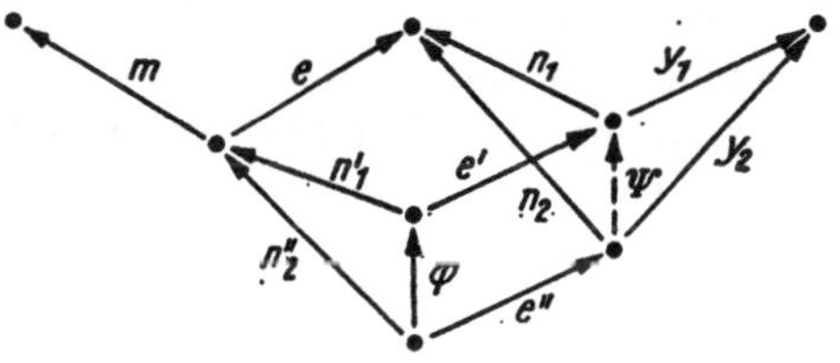

There exists a $\varphi \in$ Iso X such that $\varphi n_1' m = n_2'' m$, thus $\varphi n_1' = n_2''$; $\varphi e' y_1 = e'' y_2 \cdot n_2'' e$ can be factorized as $e'' \cdot n_2 = n_2'' e = \varphi e' \cdot n_1$ where $e', e'' \in \mathfrak{E}$ by 4). By the uniqueness of factorization, there exists a $\psi \in$ Iso X such that $e''\psi = \varphi e'$, $\psi n_1 = n_2 \cdot e'' y_2 = \varphi e' y_1 =$ $= e'' \psi y_1$ then implies that $y_2 = \psi y_1$. This shows that $(n_1, y_1) = (n_2, y_2)$.

Now, by Theorem 5, the category Y is isomorphic with a subcategory (Y) $\Psi' \subseteqq$ $\subseteqq \mathfrak{M}'$-PB($\mathfrak{M}$-PX) and by Theorem 4 it is isomorphic with Z $\subseteqq$ $\mathfrak{M}$-PBX.

Appendix. *The isomorphism* $\Omega = \Psi' \cdot \Phi^{-1}$ *of the inverse subcategory* Y *onto the subcategory of partial bijections* Z (*from the previous Theorem*) *sends every morphism* $(n, y) \in$ Y *to* $(m_{\tilde{y}}, m_{\tilde{y}}) \cdot (n, y)$ *where* $\overline{(n, y)} = (\tilde{n}, e_{\tilde{y}} \cdot m_{\tilde{y}})$ ($\tilde{n} \in \mathfrak{M}$; $e_{\tilde{y}} \in \mathfrak{E}$, $m_{\tilde{y}} \in \mathfrak{M}$).

Proof.

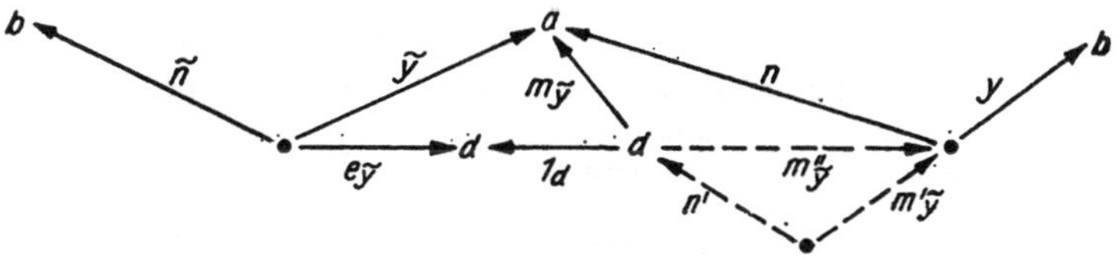

$\overline{(n, y)} = (\tilde{n}, e_{\tilde{y}}) \cdot (m_{\tilde{y}})\, X$. According to Remark 3 $(n, y)\, \Psi' = ((m_{\tilde{y}})\, X, (m_{\tilde{y}})\, X \cdot (n, y)) = ((m_{\tilde{y}})\, X, (n' \cdot 1_d, m'_{\tilde{y}} \cdot y))$ ($n' m_{\tilde{y}} = m'_{\tilde{y}} n$ is a pullback).

Since $(n', m'_{\tilde{y}} \cdot y) \in \mathfrak{M}' = (\mathfrak{M})\, X$ must hold, thus $n' \in \operatorname{Iso} \mathsf{X}$. $1_d \cdot m_{\tilde{y}} = m''_{\tilde{y}} \cdot n$ is also a pullback (in X). Hence $((n, y)\, \Psi')\, \Phi^{-1} = ((m_{\tilde{y}})\, X, (1_d \cdot 1_d, m''_{\tilde{y}} \cdot y))\, \Phi^{-1} = (m_{\tilde{y}}, m''_{\tilde{y}} \cdot y) = (m_{\tilde{y}}, m_{\tilde{y}}) \cdot (n, y)$.

Observe that Theorem 8 is an extension of the validity of Theorem 5 (from the subcategory $(\mathsf{X})\, X \subseteqq \mathfrak{M}\text{-}\mathsf{PX}$) for any inverse subcategory of the category of partial morphisms $\mathfrak{M}\text{-}\mathsf{PX}$.

For an inverse subcategory $\mathsf{Y}_0 \subseteqq \mathsf{X}$, namely $\mathsf{Y}_0 \mid X \cdot \Omega = \Psi\colon \mathsf{Y}_0 \to \mathfrak{M}\text{-}\mathsf{PBX}$ (if X satisfies the assumptions of Theorem 8). Notice that in $(\mathsf{Y}_0)X$ $\overline{(y)\, X} = (\bar{y})\, X$, $m_{\tilde{y}} = m_{\bar{y}}$.

In general, an inverse subcategory Y of the category of partical morphisms $\mathfrak{M}\text{-}\mathsf{PX}$ can be transformed to an isomorphic inverse subcategory of the category of partial bijections $\mathfrak{M}\text{-}\mathsf{PBX}$, if every morphism $y \in \mathsf{Y}$ is restricted (in the category $\mathfrak{M}\text{-}\mathsf{PX}$) to the image of its associated morphism $\bar{y} \in \mathsf{Y}$.

Examples. All examples of categories which have been mentioned until now have satisfied the conditions 1)—4) of Theorem 8. The E-classes $\mathfrak{E}$ have always been the classes of such morphisms whose underlying mappings are onto. Requirement 4) follows immediately from continuity of forgetful functor.

It is the same situation for the category X of all continuous mappings between HAUSDORFF spaces, when $\mathfrak{M}$ is the class of all injective continuous mappings which form projectively their domain topology.

Theorem 8 also holds for any well-powered abelian category X, $\mathfrak{M} = \operatorname{Mono} \mathsf{X}$.

Special case of Theorem 8 for the category of sets and mappings does not generalize only the main theorem for inverse semigroups by [12], but also contains its similar formulation for partial mappings of sets by [10].

References

[1] A. H. CLIFFORD; G. B. PRESTON: The algebraic theory of semigroups, Vol. 1, Math. Surveys No. 7; Amer. Math. Soc., Providence, R.I. (1961).

[2] A. H. CLIFFORD; G. B. PRESTON: The algebraic theory of semigroups, Vol. 2, Math. Surveys No. 7, Amer. Math. Soc., Providence, R.I. (1967).

[3] D. L. DAVIS; D. W. ROBINSON: Generalized inverse of morphisms, Linear Algebra Appl. 5 (1972), 319—328.

[4] P. J. FREYD: Aspects of topoi, Bull. Austral. Math. Soc. 7 (1972), 1—76.

[5] P. J. FREYD: Concreteness, J. Pure Appl. Alg. 3 (1973), 171—191.

[6] P. A. GRILLET: Regular categories, (M. BARR and P. A. GRILLET and D. H. VAN OSDOL, Exact categories and categories of sheaves), Lecture Notes in Mathematics 236 (1971), 121—222.

[7] H. HERRLICH; G. E. STRECKER: Category theory, Boston, 1973.

[8] J. KASTL: Inverse categories and their representations (Czech), Thesis, Charles University, Prague, 1975.

[9] Y. KAWAHARA: Relations in categories with pullbacks, Memoirs Fac. Sci. Kyushu University, Ser. A, 27 (1973), 149—173.

[10] E. S. LJAPIN: Semigroups (Russ.), Moscow, 1960.

[11] S. MACLANE: Categories for the working mathematician, Berlin-Heidelberg-New York, 1972.

[12] G. B. PRESTON: Representations of inverse semigroups by one-to-one partial transformations of a set, Semigroup Forum 6 (1973), 240—245.

[13] E. SCHWAB: Images dans les catégories régulières, Analele Universităţii din Timişoara, Ser. Şti. mat. 9 (1973), 169—174.

[14] J. VINÁREK: A new proof of the FREYD's theorem, J. Pure Appl. Alg. 8 (1976), 1—4.

Über die Vollständigkeit algebraischer induktiver Gruppoide

von Lothar Michler und Jürgen Schreckenberger, Köthen (Anhalt)

Die Menge $\mathfrak{F}$ aller eineindeutigen Funktionen zwischen den Untermengen einer Menge A bildet bezüglich der Hintereinanderausführung dieser Funktionen und bezüglich $(\subseteqq) \cap (\mathfrak{F} \times \mathfrak{F})$ ein induktives Gruppoid im Sinne von C. Ehresmann [1]), das induktive Gruppoid $\mathcal{J}(A)$ der eineindeutigen Funktionen aus A in sich. Genauso bildet die Menge $\mathfrak{A}$ aller Isomorphismen zwischen Unteralgebren einer Algebra A bezüglich der Hintereinanderausführung dieser Funktionen und bezüglich $(\subseteqq) \cap (\mathfrak{A} \times \mathfrak{A})$ ein induktives Gruppoid, das induktive Gruppoid $\mathcal{A}(\mathsf{A})$ der partiellen Automorphismen von A. In der vorliegenden Arbeit wird ein induktives Gruppoid $\mathcal{E}$ genau dann ein algebraisches induktives Gruppoid genannt, wenn eine Algebra A mit $\mathcal{E} \cong \mathcal{A}(\mathsf{A})$ existiert. Nun ist $\mathcal{J}(A)$ für jede Menge A sogar ein vollständiges induktives Gruppoid [2]), d. h., für jede kompatible Untermenge $\mathfrak{K}$ von $\mathcal{J}(A)$, d. h. jede Menge $\mathfrak{K}$ eineindeutiger Funktionen aus A in sich mit

$$\mathrm{Vb}(\varphi \cap \psi) = \mathrm{Vb}\,\varphi \cap \mathrm{Vb}\,\psi\,, \qquad \mathrm{Nb}(\varphi \cap \psi) = \mathrm{Nb}\,\varphi \cap \mathrm{Nb}\,\psi$$ [3])

für beliebige Elemente φ, ψ von $\mathfrak{K}$, gibt es eine eineindeutige Funktion χ aus A in sich mit $\varphi \subseteqq \chi$ für jedes Element φ von $\mathfrak{K}$. Weiter ist auch $\mathcal{A}(\mathsf{A})$ für jede Algebra A, von der jede definierende Operation entweder eine nulläre oder unäre Operation ist, ein vollständiges induktives Gruppoid. Das gilt aber i. a. nicht, wenn A eine Algebra ist, von der mindestens eine definierende Operation eine Arität größer als 1 hat. In [4] ist insbesondere ein Charakteristikum für die Gruppen G mit endlichem Träger angegeben worden, für die $\mathcal{A}(\mathsf{G})$ ein vollständiges induktives Gruppoid ist. In der vorliegenden Arbeit wird nun u. a. bewiesen, daß für eine Algebra A schon dann $\mathcal{A}(\mathsf{A})$ ein vollständiges induktives Gruppoid ist, wenn für jede kompatible Untermenge $\mathfrak{K}$ von $\mathcal{A}(\mathsf{A})$ mit höchstens drei und im Falle eines distributiven Unteralgebrenverbands von A sogar mit höchstens zwei partiellen Automorphismen von A als Elemente, für die $\mathrm{Vb}\,\varphi$ für jedes Element φ von $\mathfrak{K}$ der Träger einer endlich erzeugbaren Unteralgebra von A ist, ein partieller Automorphismus χ von A mit $\varphi \subseteqq \chi$ für jedes Element φ von $\mathfrak{K}$ existiert. Weiter wird bewiesen, daß für jede Algebra A gilt: Genau dann ist $\mathcal{A}(\mathsf{A})$ ein vollständiges induktives Gruppoid, wenn $\mathcal{A}(\mathsf{B})$ für jede endlich erzeugbare Unteralgebra B von A ein vollständiges induktives Gruppoid ist. Abschließend wird noch ein Zusammenhang zwischen der Vollständigkeit spezieller

[1]) Man siehe hierzu [2], [3] und § 32 von [5].

[2]) Man siehe hierzu [3] und § 32 von [5].

[3]) Ist φ eine Korrespondenz, d. h. eine Klasse geordneter Paare von Elementen, so bezeichnet „Vb φ" den Vorbereich (Definitionsbereich) und „Nb φ" den Nachbereich (Wertebereich) von φ.

induktiver Gruppoide und der Isomorphie einer bestimmten Untergruppe mit der LOEWYschen Gruppe eines bestimmten Untergruppoids[1]) herausgestellt.

Für Kategorien verwenden wird die folgenden Bezeichnungen[2]):

Definition 1. Ist $\mathscr{C}$ eine Kategorie, so bezeichnet mor $\mathscr{C}$ die Klasse aller Morphismen von $\mathscr{C}$, ob $\mathscr{C}$ die Klasse aller Einselemente (identischen Morphismen) von $\mathscr{C}$, iso $\mathscr{C}$ die Klasse aller Isomorphismen von $\mathscr{C}$, $\cdot_{\mathscr{C}}$ die Komposition von $\mathscr{C}$, $\mathrm{D}(\mathscr{C})$ den Vorbereich von $\cdot_{\mathscr{C}}$, $\alpha_{\mathscr{C}}$ die Funktion α von mor $\mathscr{C}$ in ob $\mathscr{C}$ mit $[f, \alpha(f)] \in \mathrm{D}(\mathscr{C})$ für jeden Morphismus f von $\mathscr{C}$, $\beta_{\mathscr{C}}$ die Funktion β von mor $\mathscr{C}$ in ob $\mathscr{C}$ mit $[\beta(f), f] \in \mathrm{D}(\mathscr{C})$ für jeden Morphismus f von $\mathscr{C}$ und „$\bullet_{\mathscr{C}}$" die binäre Operation $\bullet$ in $\mathfrak{P}(\text{mor } \mathscr{C})$ mit $B \bullet A = \{g \cdot_{\mathscr{C}} f \mid f \in A, g \in B, [g, f] \in \mathrm{D}(\mathscr{C})\}$ für beliebige Untermengen A, B von mor $\mathscr{C}$[3]). (Wenn keine Verwechslungen zu befürchten sind, so wird der Index an $\cdot, \alpha, \beta$ und $\bullet$ (hier also $\mathscr{C}$) auch weggelassen.) Ist $\mathscr{C}$ eine Kategorie und ist f ein Isomorphismus von $\mathscr{C}$, so bezeichnet f^{-1} (wenn keine Mißverständnisse zu befürchten sind) den bezüglich $\mathscr{C}$ zu f inversen Isomorphismus von f. ∎

Für halbgeordnete Klassen werden von uns die folgenden Bezeichnungen verwendet:

Definition 2. Es sei $\mathscr{C}$ eine halbgeordnete Klasse. Dann bezeichnet $|\mathscr{C}|$ die Klasse $\mathfrak{C}$, für die eine Halbordnungsrelation $\leqq$ in $\mathfrak{C}$ mit $\mathscr{C} = [\mathfrak{C}, \leqq]$ existiert, und $\leqq_{\mathscr{C}}$ die Halbordnungsrelation $\leqq$ in $|\mathscr{C}|$ mit $\mathscr{C} = [|\mathscr{C}|, \leqq]$. Ist A eine Unterklasse von $|\mathscr{C}|$ bzw. I eine Klasse und $(f_i)_{i \in I}$ eine Familie von Elementen von $|\mathscr{C}|$, so bezeichnet $\bigwedge_{\mathscr{C}} A$ bzw. $\bigwedge_{\mathscr{C}\, i \in I} f_i$ im Falle seiner Existenz das Infimum bezüglich $\mathscr{C}$ von A bzw. von $\{f_i \mid i \in I\}$ sowie $\bigvee_{\mathscr{C}} A$ bzw. $\bigvee_{\mathscr{C}\, i \in I} f_i$ im Falle seiner Existenz das Supremum bezüglich $\mathscr{C}$ von A bzw. von $\{f_i \mid i \in I\}$. Ist n eine natürliche Zahl, gilt $n \geqq 2$ und sind $f_1, f_2, \ldots, f_n$ Elemente von $|\mathscr{C}|$, so bezeichnet $f_1 \wedge_{\mathscr{C}} f_2 \wedge_{\mathscr{C}} \cdots \wedge_{\mathscr{C}} f_n$ im Falle seiner Existenz das Infimum bezüglich $\mathscr{C}$ von $\{f_1, f_2, \ldots, f_n\}$ und $f_1 \wedge_{\mathscr{C}} f_2 \wedge_{\mathscr{C}} \cdot\cdot \wedge_{\mathscr{C}} f_n$ im Falle seiner Existenz das Supremum bezüglich $\mathscr{C}^{\bullet}$ von $\{f_1, f_2, \ldots, f_n\}$. (Wenn keine Verwechslungen zu befürchten sind, so wird der Index an $\leqq, \bigwedge, \bigvee, \wedge$ und $\vee$ (hier also $\mathscr{C}$) auch weggelassen.) ∎

Nachstehend stellen wir noch die Definitionen einiger im folgenden gebrauchter Begriffe und weitere im folgenden verwendete Bezeichnungen zusammen.

Definition 3. Genau dann ist $\mathscr{C}$ ein *Gruppoid*, wenn $\mathscr{C}$ eine Kategorie ist und iso $\mathscr{C}$ = mor $\mathscr{C}$ gilt. Genau dann ist $\mathscr{C}$ ein *kleines Gruppoid*, wenn $\mathscr{C}$ ein Gruppoid und mor $\mathscr{C}$ eine Menge ist[4]). Ist $\mathscr{C}$ ein Gruppoid, so ist genau dann $\mathscr{D}$ ein *Untergruppoid von* $\mathscr{C}$, wenn auch $\mathscr{D}$ ein Gruppoid ist und mor $\mathscr{D} \subseteqq$ mor $\mathscr{C}$, $\mathrm{D}(\mathscr{D}) \subseteqq \mathrm{D}(\mathscr{C})$ sowie $g \cdot_{\mathscr{D}} f = g \cdot_{\mathscr{C}} f$ für beliebige Morphismen f, g von $\mathscr{D}$ mit $[g, f] \in \mathrm{D}(\mathscr{D})$ gilt. ∎

Definition 4. Ist $\mathscr{C}$ eine halbgeordnete Klasse, so ist genau dann $\mathscr{D}$ ein *Unterrelativ von* $\mathscr{C}$, wenn auch $\mathscr{D}$ eine halbgeordnete Klasse ist sowie $|\mathscr{D}| \subseteqq |\mathscr{C}|$ und für beliebige Elemente f, f' von $|\mathscr{D}|$ genau dann $f' \leqq_{\mathscr{D}} f$ gilt, wenn $f' \leqq_{\mathscr{C}} f$ gilt. ∎

[1]) Bezüglich des Begriffs der LOEWYschen Gruppe eines BRANDTschen Gruppoids siehe man etwa [6].

[2]) Wir betrachten hier die Kategorien ohne Objekte, d. h. als geordnete Paare $[\mathfrak{C}, \cdot]$ mit einer Klasse $\mathfrak{C}$ und einer speziellen Bedingungen genügenden binären partiellen Operation $\cdot$ in $\mathfrak{C}$.

[3]) Ist A eine Klasse, so bezeichnet „$\mathfrak{P}(A)$" die Klasse aller Untermengen von A.

[4]) Das ist genau dann der Fall, wenn ob $\mathscr{C}$ eine Menge ist.

Definition 5. Ist $\mathcal{C}$ eine halbgeordnete Klasse, so ist genau dann H eine *nach oben gerichtete Untermenge von* $\mathcal{C}$, wenn H eine Untermenge von $|\mathcal{C}|$ ist und für beliebige Elemente f, g von H ein Element h von H mit $f \leqq h$ und $g \leqq h$ existiert. ∎

Definition 6. Genau dann ist $\mathcal{C}$ eine *induktive Klasse*, wenn $\mathcal{C}$ eine halbgeordnete Klasse ist, $|\mathcal{C}| \neq \emptyset$ gilt, für jede nichtleere Unterklasse A von $|\mathcal{C}|$ das Infimum bezüglich $\mathcal{C}$ von A existiert und für jedes Element f von $|\mathcal{C}|$ die Klasse aller Elemente f' von $|\mathcal{C}|$ mit $f' \leqq f$ eine Menge ist.[1]) ∎

Bemerkung. Für jede halbgeordnete Klasse $\mathcal{C}$ mit $|\mathcal{C}| \neq \emptyset$, für die $\{f' \mid f' \in |\mathcal{C}|, f' \leqq f\}$ für jedes Element f von $|\mathcal{C}|$ eine Menge ist, gilt: Genau dann ist $\mathcal{C}$ eine induktive Klasse, wenn für jede bezüglich $\mathcal{C}$ nach oben beschränkte Untermenge A von $|\mathcal{C}|$ das Supremum bezüglich $\mathcal{C}$ von A existiert.

Definition 7. Ist $\mathcal{V}$ ein vollständiger Verband[2]), so ist genau dann e ein *kompaktes Element von* $\mathcal{V}$, wenn e ein Element von $|\mathcal{V}|$ ist und für jede Untermenge E von $|\mathcal{V}|$ mit $e \leqq \bigvee E$ eine endliche Untermenge E^* von E mit $e \leqq \bigvee E^*$ existiert. Genau dann ist $\mathcal{V}$ ein *algebraischer Verband*, wenn $\mathcal{V}$ ein vollständiger Verband ist und für jedes e von $|\mathcal{V}|$ eine Menge E von kompakten Elementen von $\mathcal{V}$ mit $\bigvee E = e$ existiert. ∎

Bemerkung. Für jeden vollständigen Verband $\mathcal{V}$ und beliebige kompakte Elemente e_1, e_2 von $\mathcal{V}$ ist auch $e_1 \vee e_2$ ein kompaktes Element von $\mathcal{V}$.

Definition 8. Genau dann ist $\mathcal{C}$ ein *induktives Gruppoid*, wenn eine nichtleere Klasse $\mathfrak{C}$, eine binäre partielle Operation $\cdot$ in $\mathfrak{C}$ und eine Halbordnungsrelation $\leqq$ in $\mathfrak{C}$ existieren, so daß gilt:

1. $\mathcal{C} = [\mathfrak{C}, \cdot, \leqq]$.
2. $[\mathfrak{C}, \cdot]$ ist ein Gruppoid.
3. $[\mathfrak{C}, \leqq]$ ist eine induktive Klasse.
4. Für beliebige Elemente e, f von $\mathfrak{C}$ mit $f \leqq e$ und $e \in \mathrm{ob}\,[\mathfrak{C}, \cdot]$ ist auch f ein Element von $\mathrm{ob}\,[\mathfrak{C}, \cdot]$.
5. Für beliebige Elemente f, f', g, g' von $\mathfrak{C}$ mit $f' \leqq f, g' \leqq g$ und $[g, f], [g', f'] \in \mathrm{D}([\mathfrak{C}, \cdot])$ gilt auch $g' \cdot f' \leqq g \cdot f$.
6. Für beliebige Elemente f, g, h von $\mathfrak{C}$ mit $[g, f] \in \mathrm{D}([\mathfrak{C}, \cdot])$ und $h \leqq g \cdot f$ gibt es Elemente f', g' von $\mathfrak{C}$ mit $f' \leqq f, g' \leqq g$, $[g', f'] \in \mathrm{D}([\mathfrak{C}, \cdot])$ und $g' \cdot f' = h$.[3])

Ist $\mathcal{C}$ ein induktives Gruppoid und sind $\mathfrak{C}$ die Klasse, $\cdot$ die binäre partielle Operation in $\mathfrak{C}$ und $\leqq$ die Halbordnungsrelation in $\mathfrak{C}$ mit $\mathcal{C} = [\mathfrak{C}, \cdot, \leqq]$, so bezeichnet $\mathcal{C}^{\cdot}$ das Gruppoid $[\mathfrak{C}, \cdot]$, $\mathcal{C}^{\leqq}$ die induktive Klasse $[\mathfrak{C}, \leqq]$, $\mathcal{E}_{\mathcal{C}}$ das Unterrelativ $\mathcal{E}$ von $\mathcal{C}^{\leqq}$ mit $|\mathcal{E}| = \mathrm{ob}\,\mathcal{C}^{\cdot}$, $o_{\mathcal{C}}$ das Infimum bezüglich $\mathcal{C}^{\leqq}$ von $\mathrm{ob}\,\mathcal{C}^{\cdot}$ und, falls $\mathrm{ob}\,\mathcal{C}^{\cdot}$ bezüglich $\mathcal{C}^{\leqq}$ nach oben beschränkt ist, $u_{\mathcal{C}}$ das Supremum bezüglich $\mathcal{C}^{\leqq}$ von $\mathrm{ob}\,\mathcal{C}^{\cdot}$. (Wenn keine Verwechslungen zu befürchten sind, so wird der Index an $\mathcal{E}$, o und u (hier also $\mathcal{C}$)

[1]) Man siehe hierzu [3], wo allerdings nicht verlangt wird, daß für jedes Element f von $|\mathcal{C}|$ die Klasse aller Elemente f' von $|\mathcal{C}|$ mit $f' \leqq f$ eine Menge ist.

[2]) Genau dann ist $\mathcal{V}$ ein vollständiger Verband, wenn $\mathcal{V}$ eine induktive Klasse und $|\mathcal{V}|$ bezüglich $\mathcal{V}$ nach oben beschränkt ist. Für jeden vollständigen Verband $\mathcal{V}$ ist $|\mathcal{V}|$ eine Menge.

[3]) Man siehe hierzu [2] und [3].

auch weggelassen.) Ist $\mathscr{C}$ ein induktives Gruppoid, so ist genau dann $\mathscr{D}$ ein *induktives Untergruppoid von* $\mathscr{C}$, wenn auch $\mathscr{D}$ ein induktives Gruppoid sowie $\mathscr{D}^{\cdot}$ ein Untergruppoid von $\mathscr{C}^{\cdot}$ und $\mathscr{D}^{\leqq}$ ein Unterrelativ von $\mathscr{C}^{\leqq}$ ist. Genau dann ist $\mathscr{C}$ ein *kleines induktives Gruppoid*, wenn $\mathscr{C}$ ein induktives Gruppoid und mor $\mathscr{C}^{\cdot}$ eine Menge ist.[1]) Sind $\mathscr{C}$, $\mathscr{D}$ induktive Gruppoide, so ist genau dann Φ ein *Isomorphismus von* $\mathscr{C}$ *auf* $\mathscr{D}$, wenn Φ eine eineindeutige Funktion von mor $\mathscr{C}^{\cdot}$ auf mor $\mathscr{D}^{\cdot}$ ist und für beliebige Morphismen f, g von $\mathscr{C}^{\cdot}$ gilt:

1. Wenn $[g, f] \in D(\mathscr{C}^{\cdot})$, so $[\Phi(g), \Phi(f)] \in D(\mathscr{D}^{\cdot})$ und $\Phi(g) \cdot \Phi(f) = \Phi(g \cdot f)$.
2. $f \leqq g$ genau dann, wenn $\Phi(f) \leqq \Phi(g)$.

Sind $\mathscr{C}$, $\mathscr{D}$ induktive Gruppoide, so ist genau dann $\mathscr{D}$ *zu* $\mathscr{C}$ *isomorph* (in Zeichen: $\mathscr{C} \cong \mathscr{D}$), wenn ein Isomorphismus von $\mathscr{C}$ auf $\mathscr{D}$ existiert. ▮

Ist $\mathscr{C}$ ein induktives Gruppoid, so ist für beliebige Morphismen f, g von $\mathscr{C}^{\cdot}$ die Untermenge

$$\{g' \cdot f' \mid f', g' \in \operatorname{mor} \mathscr{C}^{\cdot}, \quad f' \leqq f, \quad g' \leqq g, \quad [g', f'] \in D(\mathscr{C}^{\cdot})\}$$

von mor $\mathscr{C}^{\cdot}$ bezüglich $\mathscr{C}^{\leqq}$ nach oben beschränkt. Demnach ist die folgende Definition gerechtfertigt[2]):

Definition 9. Ist $\mathscr{C}$ ein induktives Gruppoid, so bezeichnet $\circ_{\mathscr{C}}$ die binäre Operation $\circ$ in mor $\mathscr{C}^{\cdot}$ mit für beliebige Elemente f, g von mor $\mathscr{C}^{\cdot}$

$$g \circ f = \bigvee \{g' \cdot f' \mid f', \quad g' \in \operatorname{mor} \mathscr{C}^{\cdot}, \quad f' \leqq f, \quad g' \leqq g, \quad [g', f'] \in D(\mathscr{C}^{\cdot})\},$$

und es wird unter der *Pseudomultiplikation von* $\mathscr{C}$ die binäre Operation $\circ$ in mor $\mathscr{C}^{\cdot}$ verstanden. (Wenn keine Verwechslungen zu befürchten sind, so wird der Index an $\circ$ (hier also $\mathscr{C}$) auch weggelassen.) ▮

Bemerkung. Für jedes induktive Gruppoid $\mathscr{C}$ gilt[3]):

1. Für beliebige Morphismen f, f' von $\mathscr{C}^{\cdot}$ mit $f' \leqq f$ gilt $\alpha(f') \leqq \alpha(f)$ und $\beta(f') \leqq \beta(f)$.
2. Für beliebige Morphismen f, f', g von $\mathscr{C}^{\cdot}$ mit $f \leqq g, f' \leqq g$ und a) $\alpha(f') = \alpha(f)$ oder b) $\beta(f') = \beta(f)$ gilt $f' = f$.
3. Für beliebige Morphismen f, f', g von $\mathscr{C}^{\cdot}$ mit $f \leqq g, f' \leqq g$ und a) $\alpha(f') \leqq \alpha(f)$ oder b) $\beta(f') \leqq \beta(f)$ gilt $f' \leqq f$.
4. Für jeden Morphismus f von $\mathscr{C}^{\cdot}$ gilt genau dann $\alpha(f) = o$ oder $\beta(f) = o$, wenn $f = o$ ist.
5. $\mathscr{C}$ ist eine induktive Klasse und, falls ob $\mathscr{C}^{\cdot}$ bezüglich $\mathscr{C}^{\leqq}$ nach oben beschränkt ist, sogar ein vollständiger Verband.
6. Es gilt $\bigwedge_{\mathscr{C}} E = \bigwedge_{\mathscr{C}^{\leqq}} E$ für jede nichtleere Unterklasse E von ob $\mathscr{C}^{\cdot}$ und, falls E eine bezüglich $\mathscr{C}$ nach oben beschränkte Untermenge von ob $\mathscr{C}^{\cdot}$ ist, auch $\bigvee_{\mathscr{C}} E = \bigvee_{\mathscr{C}^{\leqq}} E$.

[1]) Das ist genau dann der Fall, wenn ob $\mathscr{C}^{\cdot}$ eine Menge ist. Jedes induktive Gruppoid, für das ob $\mathscr{C}^{\cdot}$ bezüglich $\mathscr{C}^{\leqq}$ nach oben beschränkt ist, ist ein kleines induktives Gruppoid.

[2]) Man siehe hierzu [3] und § 33 von [5].

[3]) Man siehe hierzu etwa § 32 und § 33 von [5].

7. Für jede nichtleere Unterklasse A von mor $\mathcal{C}\cdot$ gilt $\alpha(\bigwedge A) \leqq \bigwedge\check{\alpha}(A)$, $\beta(\bigwedge A) \leqq \leqq \bigwedge\check{\beta}(A)$, und für jede bezüglich $\mathcal{C}^{\leqq}$ nach oben beschränkte, nichtleere Untermenge A von mor $\mathcal{C}\cdot$ gilt $\alpha(\bigwedge A) = \bigwedge\check{\alpha}(A)$ und $\beta(\bigwedge A) = \bigwedge\check{\beta}(A)$.[1])
8. Für jede bezüglich $\mathcal{C}^{\leqq}$ nach oben beschränkte Untermenge A von mor $\mathcal{C}\cdot$ sind auch $\check{\alpha}(A)$ und $\check{\beta}(A)$ bezüglich $\mathcal{C}^{\leqq}$ nach oben beschränkte Untermengen von mor $\mathcal{C}\cdot$, und es gilt $\alpha(\bigvee A) = \bigvee\check{\alpha}(A)$ und $\beta(\bigvee A) = \bigvee\check{\beta}(A)$.
9. Für beliebige Morphismen f, g von $\mathcal{C}\cdot$ mit $[g, f] \in \mathrm{D}(\mathcal{C}\cdot)$ gilt $g \circ f = g \cdot f$.
10. Für beliebige Morphismen f, g, h von $\mathcal{C}$: gilt $(h \circ g) \circ f = h \circ (g \circ f)$.
11. Für beliebige Morphismen f, g von $\mathcal{C}\cdot$ gilt $(g \circ f)^{-1} = f^{-1} \circ g^{-1}$.
12. Es gilt $o \in$ ob $\mathcal{C}\cdot$ und $f \circ o = o \circ f = o$ für jeden Morphismus f von $\mathcal{C}\cdot$.
13. Ist ob $\mathcal{C}\cdot$ bezüglich $\mathcal{C}^{\leqq}$ nach oben beschränkt, so gilt $u \in$ ob $\mathcal{C}\cdot$ und $f \circ u = = u \circ f = f$ für jeden Morphismus f von $\mathcal{C}\cdot$.
14. Für jeden Morphismus f von $\mathcal{C}\cdot$ und jedes Einselement e von $\mathcal{C}\cdot$ gilt $f \circ e \leqq f$ und $e \circ f \leqq f$.
15. Für jeden Morphismus f von $\mathcal{C}\cdot$ und jedes Einselement e von $\mathcal{C}\cdot$ gilt:
    a) Ist $e \leqq \alpha(f)$, so gilt $\alpha(f \circ e) = e$.
    b) Ist $e \leqq \beta(f)$, so gilt $\beta(e \circ f) = e$.
16. Es gilt $e \circ e' = e \wedge e'$ für beliebige Einselemente e, e' von $\mathcal{C}\cdot$.
17. Für beliebige Morphismen f, f', g, g' von $\mathcal{C}\cdot$ mit $f' \leqq f$ und $g' \leqq g$ gilt auch $g' \circ f' \leqq \leqq g \circ f$.

Für das folgende benötigen wir den nachstehenden Hilfssatz.

Hilfssatz. *Für jedes induktive Gruppoid $\mathcal{C}$, jeden Morphismus f von $\mathcal{C}\cdot$ und jedes Einselement e von $\mathcal{C}\cdot$ mit $e \leqq \beta(f)$ gilt*

$$f \circ \beta(f^{-1} \circ e) = e \circ f\,.$$

Beweis. Es seien $\mathcal{C}$ ein induktives Gruppoid, f ein Morphismus von $\mathcal{C}\cdot$ und e ein Einselement von $\mathcal{C}\cdot$, und es gelte $e \leqq \beta(f)$. Dann gilt

$$f \circ \beta(f^{-1} \circ e) = f \circ \big((f^{-1} \circ e) \cdot (f^{-1} \circ e)^{-1}\big) = f \circ \big((f^{-1} \circ e) \circ (e \circ f)\big) =$$
$$= (f \circ f^{-1}) \circ \big((e \circ e) \circ f\big) = \big(\beta(f) \circ e\big) \circ f = e \circ f\,. \blacksquare$$

Für Algebren verwenden wir die folgenden Begriffe und Bezeichnungen.

Definition 10. Es sei A eine Algebra. Dann bezeichnet $|\mathsf{A}|$ die Menge A, für die eine Familie $\mathfrak{o}$ von Operationen in A mit $\mathsf{A} = [A, \mathfrak{o}]$ existiert und $\mathfrak{o}_{\mathsf{A}}$ die Familie $\mathfrak{o}$ von Operationen in $|\mathsf{A}|$ mit $\mathsf{A} = [|\mathsf{A}|, \mathfrak{o}]$, und es wird dann unter dem *Träger von* A die Menge $|\mathsf{A}|$, unter der *Familie der definierenden Operationen von* A die Familie $\mathfrak{o}_{\mathsf{A}}$ von Operationen in $|\mathsf{A}|$ und unter einer *definierenden Operation von* A ein Element von Nb $\mathfrak{o}_{\mathsf{A}}$ verstanden. Ist B eine Algebra, so ist genau dann B *zu* A *typgleich*, wenn Vb $\mathfrak{o}_{\mathsf{B}}$ = Vb $\mathfrak{o}_{\mathsf{A}}$ und für jedes Element i von Vb $\mathfrak{o}_{\mathsf{A}}$ und jede natürliche Zahl n gilt:

[1]) Sind A, B Klassen und ist φ eine Funktion von A in B, so bezeichnet „$\check{\varphi}$" diejenige Funktion ψ von $\mathfrak{P}(A)$ in $\mathfrak{P}(B)$ mit $\psi(X) = \{\varphi(x) \mid x \in X\}$ für jede Untermenge X von A.

Ist $\mathfrak{o}_{\mathsf{A}}(i)$ eine n-äre Operation in $|\mathsf{A}|$, so ist $\mathfrak{o}_{\mathsf{B}}(i)$ eine n-äre Operation in $|\mathsf{B}|$. Genau dann ist B eine *abgeschlossene Untermenge von* A, wenn B eine Untermenge von $|\mathsf{A}|$ ist und für jede definierende Operation ω von A und jede natürliche Zahl n gilt: Ist ω eine n-äre Operation in $|\mathsf{A}|$, so gilt $\omega(b_1, b_2, \dots, b_n) \in B$ für beliebige Elemente $b_1, b_2, \dots, b_n$ von B.[1]) Genau dann ist B eine *Unteralgebra von* A, wenn B eine zu A typgleiche Algebra und $|\mathsf{B}|$ eine abgeschlossene Untermenge von A ist sowie für jedes Element i von Vb $\mathfrak{o}_{\mathsf{A}}$ und jede natürliche Zahl n gilt: Ist $\mathfrak{o}_{\mathsf{A}}(i)$ eine n-äre Operation in $|\mathsf{A}|$, so gilt $\mathfrak{o}_{\mathsf{B}}(i)(b_1, b_2, \dots, b_n) = \mathfrak{o}_{\mathsf{A}}(i)(b_1, b_2, \dots, b_n)$ für beliebige Elemente $b_1, b_2, \dots, b_n$ von $|\mathsf{B}|$.[2]) $\mathfrak{U}(\mathsf{A})$ bezeichnet die Menge aller abgeschlossenen Untermengen von A und $\mathcal{U}(\mathsf{A})$ den Unteralgebrenverband von A.[3]) Wenn X eine Untermenge von $|\mathsf{A}|$ ist, so bezeichnet „$\langle X\rangle_{\mathsf{A}}$" die Untermenge

$$\bigcap \{B \mid B \in \mathfrak{U}(\mathsf{A}), \quad X \subseteq B\}$$

von $|\mathsf{A}|$[4]), und es wird dann unter der *von X erzeugten Unteralgebra von* A die Unteralgebra B von A mit $|\mathsf{B}| = \langle X\rangle_{\mathsf{A}}$ verstanden. Genau dann ist B eine *endlich erzeugbare Unteralgebra von* A, wenn B eine Unteralgebra von A ist und es eine endliche Untermenge X von $|\mathsf{A}|$ mit $\langle X\rangle_{\mathsf{A}} = |\mathsf{B}|$ gibt. (Wenn keine Verwechslungen zu befürchten sind, so wird der Index an $\langle\ \rangle$ (hier also A) auch weggelassen.) Genau dann ist φ ein *partieller Automorphismus von* A, wenn φ eine eineindeutige Funktion ist und Vb φ, Nb $\varphi \in \mathfrak{U}(\mathsf{A})$ sowie für jede definierende Operation ω von A und jede natürliche Zahl n gilt: Ist ω eine n-äre Operation in $|\mathsf{A}|$, so gilt

$$\omega(\varphi(a_1), \varphi(a_2), \dots, \varphi(a_n)) = \varphi(\omega(a_1, a_2, \dots, a_n))$$

für beliebige Elemente $a_1, a_2, \dots, a_n$ von Vb φ. Ist B eine zu A typgleiche Algebra, so ist genau dann φ ein *Homomorphismus von* A *in* B bzw. ein *Isomorphismus von* A *auf* B, wenn φ eine Funktion von $|\mathsf{A}|$ in $|\mathsf{B}|$ bzw. eine eineindeutige Funktion von $|\mathsf{A}|$ auf $|\mathsf{B}|$ ist und für jedes Element i von Vb $\mathfrak{o}_{\mathsf{A}}$ und jede natürliche Zahl n gilt: Ist $\mathfrak{o}_{\mathsf{A}}(i)$ eine n-äre Operation in $|\mathsf{A}|$, so gilt

$$\mathfrak{o}_{\mathsf{B}}(i)\,(\varphi(a_1), \varphi(a_2), \dots, \varphi(a_n)) = \varphi(\mathfrak{o}_{\mathsf{A}}(i)\,(a_1, a_2, \dots, a_n))$$

für beliebige Elemente $a_1, a_2, \dots, a_n$ von $|\mathsf{A}|$. Ist B eine zu A typgleiche Algebra, so ist genau dann B *zu* A *isomorph*, wenn ein Isomorphismus von A auf B existiert. Genau dann ist φ ein *Automorphismus von* A, wenn φ ein Isomorphismus von A auf A ist.[5]) Genau dann ist A eine *idempotente Algebra*, wenn es keine nulläre definierende Operation von A gibt und für jede definierende Operation ω von A und jede natürliche Zahl n gilt: Ist ω eine n-äre Operation in $|\mathsf{A}|$, so gilt

$$\omega(\underset{1}{a}, \underset{2}{a}, \dots, \underset{n}{a}) = a \quad \text{für jedes Element } a \text{ von } |\mathsf{A}|. \blacksquare$$

[1]) Wenn keine Mißverständnisse zu befürchten sind, so wird bei $n = 0$ häufig ω mit $\omega(\emptyset)$ identifiziert.

[2]) Für jede abgeschlossene Untermenge B von A gibt es genau eine Unteralgebra B von A mit $|\mathsf{B}| = B$.

[3]) D. h. den vollständigen Verband $\mathcal{U}$, für den $|\mathcal{U}| = \mathfrak{U}(\mathsf{A})$ und für beliebige Elemente B, C von $\mathfrak{U}(\mathsf{A})$ genau dann $B \leqq_{\mathcal{U}} C$ gilt, wenn B eine Untermenge von C ist.

[4]) Man beachte hierbei, daß $|\mathsf{A}|$ eine abgeschlossene Untermenge von A ist. Für jede Untermenge X von $|\mathsf{A}|$ ist $\langle X\rangle_{\mathsf{A}}$ eine abgeschlossene Untermenge von A.

[5]) Jeder Automorphismus von A ist auch ein partieller Automorphismus von A. Genau dann ist φ ein partieller Automorphismus von A, wenn es Unteralgebren B, C von A gibt, so daß φ ein Isomorphismus von B auf C ist.

Es seien nun wieder A eine Algebra, $\mathfrak{A}$ die Menge aller partiellen Automorphismen von A, $\cdot$ die binäre partielle Operation in $\mathfrak{A}$ mit dem Vorbereich $\{[\psi, \varphi] \mid \varphi, \psi \in \mathfrak{A}, \mathrm{Nb}\,\varphi = \mathrm{Vb}\,\psi\}$ und mit $\psi \cdot \varphi = \psi\varphi$ für beliebige Elemente φ, ψ von $\mathfrak{A}$ mit $\mathrm{Nb}\,\varphi = \mathrm{Vb}\,\psi$[1]) und $\leqq$ die Halbordnungsrelation $(\subseteqq) \cap (\mathfrak{A} \times \mathfrak{A})$ in $\mathfrak{A}$. Dann ist $[\mathfrak{A}, \cdot, \leqq]$ ein induktives Gruppoid. Demnach ist die folgende Definition berechtigt.

Definition 11. Ist A eine Algebra, so bezeichnet $\mathcal{A}(\mathsf{A})$ das induktive Gruppoid $\mathcal{A}$ mit

1. mor $\mathcal{A}^{\cdot}$ ist die Menge aller partiellen Automorphismen von A, und für beliebige Morphismen φ, ψ von $\mathcal{A}^{\cdot}$ gilt:
2. $[\psi, \varphi) \in \mathrm{D}(\mathcal{A}^{\cdot})$ genau dann, wenn $\mathrm{Nb}\,\varphi = \mathrm{Vb}\,\psi$.
3. Wenn $[\psi, \varphi] \in \mathrm{D}(\mathcal{A}^{\cdot})$, so $\psi \cdot \varphi = \psi\varphi$.
4. $\varphi \leqq \psi$ genau dann, wenn $\varphi \subseteqq \psi$.

Wenn A eine Algebra ist, so versteht man unter dem *induktiven Gruppoid der partiellen Automorphismen von* A das induktive Gruppoid $\mathcal{A}(\mathsf{A})$. Ist A eine Menge, so bezeichnet $\mathcal{J}(A)$ das induktive Gruppoid $\mathcal{A}([A, \emptyset])$, und es wird unter dem *induktiven Gruppoid der eineindeutigen Funktionen aus A in sich* das induktive Gruppoid $\mathcal{J}(A)$ verstanden. ▮

Bemerkung. Für jede Algebra A ist $\mathcal{U}(\mathsf{A})$ ein algebraischer Verband, und zwar ist $\{\langle X\rangle \mid X$ endliche Untermenge von $|\mathsf{A}|\}$ die Menge aller kompakten Elemente von $\mathcal{U}(\mathsf{A})$.[2])

Nur für ganz spezielle induktive Gruppoide $\mathcal{C}$ gibt es eine Algebra A derart, daß $\mathcal{C}$ zu $\mathcal{A}(\mathsf{A})$ isomorph ist. Das gibt Anlaß zu folgender Definition.

Definition 12. Genau dann ist $\mathcal{C}$ ein *algebraisches induktives Gruppoid*, wenn $\mathcal{C}$ ein induktives Gruppoid ist und es eine Algebra A gibt, so daß $\mathcal{C}$ zu $\mathcal{A}(\mathsf{A})$ isomorph ist. ▮

Ein induktives Gruppoid $\mathcal{C}$ ist genau dann ein algebraisches induktives Gruppoid, wenn gilt:

1. $\mathcal{E}$ ist ein algebraischer Verband.
2. Jede nach oben gerichtete Untermenge von $\mathcal{C}^{\leqq}$ ist bezüglich $\mathcal{C}^{\leqq}$ nach oben beschränkt.[3])

Bemerkungen. Für jede Algebra A gilt:

1. $\mathcal{E}_{\mathcal{A}(\mathsf{A})}$ ist zu $\mathcal{U}(\mathsf{A})$ isomorph, und zwar ist die Funktion Φ über $\mathfrak{U}(\mathsf{A})$ mit $\Phi(B) = \mathrm{id}_B$ für jedes Element B von $\mathfrak{U}(\mathsf{A})$[4]) ein Isomorphismus von dem vollständigen Verband $\mathcal{U}(\mathsf{A})$ auf den vollständigen Verband $\mathcal{E}_{\mathcal{A}(\mathsf{A})}$. Insbesondere ist also ob $\mathcal{A}(\mathsf{A})^{\cdot}$ bezüglich $\mathcal{A}(\mathsf{A})^{\leqq}$ nach oben beschränkt.[5])
2. $\mathcal{A}(\mathsf{A})$ ist ein induktives Untergruppoid von $\mathcal{J}(|\mathsf{A}|)$.

[1]) Sind A, B, C Klassen und ist φ eine Funktion von A in B und ψ eine Funktion von B in C, so bezeichnet „$\psi\varphi$" die Funktion χ von A in C mit $\chi(a) = \psi(\varphi(a))$ für jedes Element a von A.

[2]) Man siehe hierzu [1] und auch [7].

[3]) Man siehe hierzu [7], Satz 1.

[4]) Ist A eine Klasse, so bezeichnet id_A die Funktion $\{[a, a] \mid a \in A\}$ über A.

[5]) Jedes algebraische induktive Gruppoid ist also ein kleines induktives Gruppoid.

3. Für beliebige Morphismen φ, ψ von $\mathcal{A}(\mathsf{A})^{\cdot}$ gilt $\varphi \wedge_{\mathcal{A}(\mathsf{A})\leqq} \psi = \varphi \cap \psi$.

Nunmehr definieren wir den Begriff des vollständigen induktiven Gruppoids.[1])

Definition 13. Ist $\mathscr{C}$ ein induktives Gruppoid, so ist genau dann K eine *kompatible Untermenge von* $\mathscr{C}$, wenn K eine Untermenge von mor $\mathscr{C}^{\cdot}$ ist und

$$\alpha(f \wedge g) = \alpha(f) \wedge \alpha(g)\,, \quad \beta(f \wedge g) = \beta(f) \wedge \beta(g)$$

für beliebige Elemente f, g von K gilt. Genau dann ist $\mathscr{C}$ ein *vollständiges induktives Gruppoid*, wenn $\mathscr{C}$ ein induktives Gruppoid ist und jede kompatible Untermenge K von $\mathscr{C}$, für die $\check{\alpha}(K)$ und $\check{\beta}(K)$ bezüglich $\mathscr{C}^{\leqq}$ nach oben beschränkt sind, bezüglich $\mathscr{C}^{\leqq}$ nach oben beschränkt ist. ▮

Bemerkungen.

1. Für jedes induktive Gruppoid $\mathscr{C}$ ist jede bezüglich $\mathscr{C}^{\leqq}$ nach oben beschränkte Untermenge von mor $\mathscr{C}^{\cdot}$ eine kompatible Untermenge von $\mathscr{C}$.
2. Für jedes induktive Gruppoid $\mathscr{C}$, für das ob $\mathscr{C}^{\cdot}$ bezüglich $\mathscr{C}^{\leqq}$ nach oben beschränkt ist, insbesondere also für jedes algebraische induktive Gruppoid $\mathscr{C}$, gilt: Genau dann ist $\mathscr{C}$ ein vollständiges induktives Gruppoid, wenn jede kompatible Untermenge von $\mathscr{C}$ bezüglich $\mathscr{C}^{\leqq}$ nach oben beschränkt ist.

Ist A eine Algebra und $\mathfrak{K}$ eine kompatible Untermenge von $\mathcal{A}(\mathsf{A})$, so ist $\bigcup \mathfrak{K}$ eine eineindeutige Funktion aus $|\mathsf{A}|$ in sich. Man beweist nämlich leicht:

Satz 1. *Für jede Algebra* A *und jede Untermenge* $\mathfrak{K}$ *von* mor $\mathcal{A}(\mathsf{A})^{\cdot}$[2]) *gilt: Genau dann ist* $\bigcup \mathfrak{K}$ *eine eineindeutige Funktion aus* $|\mathsf{A}|$ *in sich, wenn* $\mathfrak{K}$ *eine kompatible Untermenge von* $\mathcal{A}(\mathsf{A})$ *ist.*

Beweis. Es sei A eine Algebra und $\mathfrak{K}$ eine Untermenge von mor $\mathcal{A}(\mathsf{A})^{\cdot}$. Dann ist $\mathfrak{K}$ eine Menge von Untermengen von $|\mathsf{A}| \times |\mathsf{A}|$, also $\bigcup \mathfrak{K}$ eine Korrespondenz aus $|\mathsf{A}|$ in sich.

($\Rightarrow$) Ist dann $\bigcup \mathfrak{K}$ eine eineindeutige Funktion aus $|\mathsf{A}|$ in sich, so ist $\mathfrak{K}$ eine bezüglich $\mathcal{J}(|\mathsf{A}|)^{\leqq}$ nach oben beschränkte Untermenge von mor $\mathcal{J}(|\mathsf{A}|)^{\cdot}$ und damit eine kompatible Untermenge von $\mathcal{J}(|\mathsf{A}|)$ und folglich wegen $\mathfrak{K} \subseteqq$ mor $\mathcal{A}(\mathsf{A})^{\cdot}$ und $\alpha_{\mathcal{A}(\mathsf{A})^{\cdot}}(\varphi) = \alpha_{\mathcal{J}(|\mathsf{A}|)^{\cdot}}(\varphi)$, $\beta_{\mathcal{A}(\mathsf{A})^{\cdot}}(\varphi) = \beta_{\mathcal{J}(|\mathsf{A}|)^{\cdot}}(\varphi)$ und $\varphi \wedge_{\mathcal{A}(\mathsf{A})\leqq} \psi = \varphi \cap \psi = \varphi \wedge_{\mathcal{J}(|\mathsf{A}|)\leqq} \psi$ für beliebige Morphismen φ, ψ von $\mathcal{A}(\mathsf{A})^{\cdot}$ auch eine kompatible Untermenge von $\mathcal{A}(\mathsf{A})$.

($\Leftarrow$) Es sei nun $\mathfrak{K}$ eine kompatible Untermenge von $\mathcal{A}(\mathsf{A})$, und es sei φ^* die Korrespondenz $\bigcup \mathfrak{K}$ aus $|\mathsf{A}|$ in $|\mathsf{A}|$. Sind dann φ, ψ Elemente von $\mathfrak{K}$, so gilt

$$\mathrm{id}_{\mathrm{Vb}(\varphi \cap \psi)} = \alpha(\varphi \cap \psi) = \alpha(\varphi) \cap \alpha(\psi) = \mathrm{id}_{\mathrm{Vb}\,\varphi} \cap \mathrm{id}_{\mathrm{Vb}\,\psi} = \mathrm{id}_{\mathrm{Vb}\,\varphi \cap \mathrm{Vb}\,\psi}$$

sowie

$$\mathrm{id}_{\mathrm{Nb}(\varphi \cap \psi)} = \beta(\varphi \cap \psi) = \beta(\varphi) \cap \beta(\psi) = \mathrm{id}_{\mathrm{Nb}\,\varphi} \cap \mathrm{id}_{\mathrm{Nb}\,\psi} = \mathrm{id}_{\mathrm{Nb}\,\varphi \cap \mathrm{Nb}\,\psi}\,,$$

also

$$\mathrm{Vb}\,\varphi \cap \mathrm{Vb}\,\psi = \mathrm{Vb}(\varphi \cap \psi)\,, \qquad \mathrm{Nb}\,\varphi \cap \mathrm{Nb}\,\psi = \mathrm{Nb}(\varphi \cap \psi)\,.$$

1) Man siehe hierzu [3] und § 32 von [5].

2) Man beachte hier, daß K eine Menge von Untermengen von $|\mathsf{A}| \times |\mathsf{A}|$ ist.

Es seien nun x ein Element von Vb φ^* und y_1, y_2 Elemente von Nb φ^*, und es gelte $[x, y_1], [x, y_2] \in \varphi^*$. Dann gibt es Elemente φ, ψ von $\mathfrak{K}$ mit $[x, y_1] \in \varphi$ und $[x, y_2] \in \psi$ und folglich mit $x \in \mathrm{Vb}\,\varphi \cap \mathrm{Vb}\,\psi$, $\varphi(x) = y_1$ und $\psi(x) = y_2$. Es gilt dann aber $x \in \mathrm{Vb}(\varphi \cap \psi)$ und damit $\varphi(x) = (\varphi \cap \psi)(x) = \psi(x)$, also $y_1 = y_2$. Demnach gibt es für jedes Element x von Vb φ^* genau ein Element y von Nb φ^* mit $[x, y] \in \varphi^*$. Nun seien y ein Element von Nb φ^* und x_1, x_2 Elemente von Vb φ^*, und es gelte $[x_1, y], [x_2, y] \in \varphi^*$. Dann gibt es Elemente φ, ψ von $\mathfrak{K}$ mit $[x_1, y] \in \varphi$, $[x_2, y] \in \psi$ und folglich mit $y \in \mathrm{Nb}\,\varphi \cap \mathrm{Nb}\,\psi$, $\varphi(x_1) = y$ und $\psi(x_2) = y$. Es gilt dann aber $y \in \mathrm{Nb}(\varphi \cap \psi)$, $\varphi^{-1}(y) = x_1$, $\psi^{-1}(y) = x_2$ und damit $x_1 = \varphi^{-1}(y) = (\varphi \cap \psi)^{-1}(y) = \psi^{-1}(y) = x_2$. Demnach gibt es für jedes Element y von Nb φ^* genau ein Element x von Vb φ^* mit $[x, y] \in \varphi^*$. Mithin ist φ^* eine eineindeutige Funktion aus $|\mathsf{A}|$ in sich. ∎

Mit Hilfe von Satz 1 beweist man sofort:

Satz 2. *Für jede Menge A ist $\mathcal{J}(A)$ ein vollständiges induktives Gruppoid.* ∎

Der folgende Satz gibt nun eine notwendige und hinreichende Bedingung für die Algebren A an, für die $\mathcal{A}(\mathsf{A})$ ein vollständiges induktives Gruppoid ist.

Satz 3. *Für jede Algebra A gilt: Genau dann ist $\mathcal{A}(\mathsf{A})$ ein vollständiges induktives Gruppoid, wenn es für jede eineindeutige Funktion φ aus $|\mathsf{A}|$ in sich mit $\langle\{a\}\rangle \subseteq \mathrm{Vb}\,\varphi$ und $\varphi \mid \langle\{a\}\rangle \in \mathrm{mor}\,\mathcal{A}(\mathsf{A})^{\cdot}$ für jedes Element a von Vb φ*[1]*) einen partiellen Automorphismus ψ von A mit $\varphi \subseteq \psi$ gibt.*

Beweis. Es sei A eine Algebra.

($\Rightarrow$) Dann sei $\mathcal{A}(\mathsf{A})$ ein vollständiges induktives Gruppoid. Weiter sei φ eine eineindeutige Funktion aus $|\mathsf{A}|$ in sich, und es gelte $\langle\{a\}\rangle \subseteq \mathrm{Vb}\,\varphi$ und $\varphi \mid \langle\{a\}\rangle \in \mathrm{mor}\,\mathcal{A}(\mathsf{A})^{\cdot}$ für jedes Element a von Vb φ. Dann sei $\mathfrak{K}$ die Untermenge $\{\varphi|\langle\{a\}\rangle \mid a \in \mathrm{Vb}\,\varphi\}$ von mor $\mathcal{J}(|\mathsf{A}|)^{\cdot}$. Diese ist bezüglich $\mathcal{J}(|\mathsf{A}|)^{\leqq}$ nach oben beschränkt, da φ eine obere Schranke bezüglich $\mathcal{J}(|\mathsf{A}|)^{\leqq}$ von $\mathfrak{K}$ ist. Mithin ist $\mathfrak{K}$ eine kompatible Untermenge von $\mathcal{J}(|\mathsf{A}|)$ und damit wegen $\mathfrak{K} \subseteq \mathrm{mor}\,\mathcal{A}(\mathsf{A})^{\cdot}$ und $\alpha_{\mathcal{A}(\mathsf{A})^{\cdot}}(\varphi) = \alpha_{\mathcal{J}(|\mathsf{A}|)^{\cdot}}(\varphi)$, $\beta_{\mathcal{A}(\mathsf{A})^{\cdot}}(\varphi) = \beta_{\mathcal{J}(|\mathsf{A}|)^{\cdot}}(\varphi)$ und $\varphi \wedge_{\mathcal{A}(\mathsf{A})^{\leqq}} \psi = \varphi \wedge_{\mathcal{J}(|\mathsf{A}|)^{\leqq}} \psi$ für beliebige Morphismen φ, ψ von $\mathcal{A}(\mathsf{A})^{\cdot}$ auch eine kompatible Untermenge von $\mathcal{A}(\mathsf{A})$. Folglich gibt es eine obere Schranke ψ bezüglich $\mathcal{A}(\mathsf{A})^{\leqq}$ von $\mathfrak{K}$, und wegen $\varphi = \bigcup \mathfrak{K}$ gilt dann $\varphi \subseteq \psi$.

($\Leftarrow$) Nun existiere für jede eineindeutige Funktion φ aus $|\mathsf{A}|$ in sich mit $\langle\{a\}\rangle \subseteq \mathrm{Vb}\,\varphi$ und $\varphi \mid \langle\{a\}\rangle \in \mathrm{mor}\,\mathcal{A}(\mathsf{A})^{\cdot}$ für jedes Element a von Vb φ ein partieller Automorphismus ψ von A mit $\varphi \subseteq \psi$. Es sei $\mathfrak{K}$ eine kompatible Untermenge von $\mathcal{A}(\mathsf{A})$. Dann ist $\bigcup \mathfrak{K}$ eine eineindeutige Funktion aus $|\mathsf{A}|$ in sich (Satz 1). Diese sei φ. Wegen $\mathrm{Vb}\,\varphi = \bigcup\{\mathrm{Vb}\,\chi \mid \chi \in \mathfrak{K}\}$ gilt $\langle\{a\}\rangle \subseteq \mathrm{Vb}\,\varphi$ und $\varphi \mid \langle\{a\}\rangle \in \mathrm{mor}\,\mathcal{A}(\mathsf{A})^{\cdot}$ für jedes Element a von Vb φ. Mithin gibt es einen partiellen Automorphismus ψ von A mit $\varphi \subseteq \psi$, und ψ ist dann eine obere Schranke bezüglich $\mathcal{A}(\mathsf{A})^{\leqq}$ von $\mathfrak{K}$. Folglich ist $\mathcal{A}(\mathsf{A})$ ein vollständiges induktives Gruppoid. ∎

Mit Hilfe von Satz 3 beweist man leicht:

Satz 4. *Für jede Algebra A gilt: Ist jede definierende Operation von A entweder eine nulläre oder unäre Operation in $|\mathsf{A}|$, so ist $\mathcal{A}(\mathsf{A})$ ein vollständiges induktives Gruppoid.*

[1]) Ist φ eine Funktion und A eine Unterklasse von Vb φ, so bezeichnet „$\varphi \mid A$" die Funktion ψ über A mit $\psi(a) = \varphi(a)$ für jedes Element a von A.

Beweis. Es sei A eine Algebra, und jede definierende Operation von A sei entweder eine nulläre oder eine unäre Operation in $|A|$. Dann sei φ eine eineindeutige Funktion aus $|A|$ in sich, und es gelte $\langle\{a\}\rangle \subseteqq \mathrm{Vb}\,\varphi$ und $\varphi \mid \langle\{a\}\rangle \in \mathrm{mor}\,\mathcal{A}(A)^{\cdot}$ für jedes Element a von $\mathrm{Vb}\,\varphi$. Es gilt dann $\mathrm{Vb}\,\varphi = \bigcup\, \{\langle\{a\}\rangle \mid a \in \mathrm{Vb}\,\varphi\}$. Ist insbesondere $\mathrm{Vb}\,\varphi$ leer, so gilt $\varphi = \emptyset \subseteqq \mathrm{id}_A \in \mathrm{mor}\,\mathcal{A}(A)^{\cdot}$. Ist $\mathrm{Vb}\,\varphi$ nichtleer, so ist offenbar $\bigcup\, \{\langle\{a\}\rangle \mid a \in \in \mathrm{Vb}\,\varphi\}$, d. h. $\mathrm{Vb}\,\varphi$, eine abgeschlossene Untermenge von A und folglich φ selbst ein Element von $\mathrm{mor}\,\mathcal{A}(A)^{\cdot}$. Demnach ist $\mathcal{A}(A)$ ein vollständiges induktives Gruppoid (Satz 3). ∎

Ein algebraisches induktives Gruppoid $\mathcal{C}$ ist schon immer dann ein vollständiges induktives Gruppoid, wenn jede kompatible Untermenge K von $\mathcal{C}$, für die $\mathrm{card}\, K \leqq 3$ gilt[1]) und für die obendrein $\alpha(f)$ für jedes Element f von K ein kompaktes Element von $\mathcal{E}$ ist, bezüglich $\mathcal{C}^{\leqq}$ nach oben beschränkt ist. Wir beweisen nämlich:

Satz 5. *Für jedes algebraische induktive Gruppoid $\mathcal{C}$ gilt: Genau dann ist $\mathcal{C}$ ein vollständiges induktives Gruppoid, wenn für beliebige Morphismen f, g, h von $\mathcal{C}^{\cdot}$, für die $\{f, g, h\}$ eine kompatible Untermenge von $\mathcal{C}$ ist und $\alpha(f)$, $\alpha(g)$, $\alpha(h)$ kompakte Elemente von $\mathcal{E}$ sind, $\{f, g, h\}$ bezüglich $\mathcal{C}^{\leqq}$ nach oben beschränkt ist.*

Beweis. Es sei $\mathcal{C}$ ein algebraisches induktives Gruppoid.

($\Rightarrow$) Ist dann $\mathcal{C}$ insbesondere ein vollständiges induktives Gruppoid und sind f, g, h Morphismen von $\mathcal{C}^{\cdot}$, für die $\{f, g, h\}$ eine kompatible Untermenge von $\mathcal{C}$ ist und $\alpha(f)$, $\alpha(g)$, $\alpha(h)$ kompakte Elemente von $\mathcal{E}$ sind, so ist $\{f, g, h\}$ bezüglich $\mathcal{C}^{\leqq}$ nach oben beschränkt.

($\Leftarrow$) Es sei nun $\{f, g, h\}$ für beliebige Morphismen f, g, h von $\mathcal{C}^{\cdot}$, für die $\{f, g, h\}$ eine kompatible Untermenge von $\mathcal{C}$ ist und $\alpha(f)$, $\alpha(g)$, $\alpha(h)$ kompakte Elemente von $\mathcal{E}$ sind, bezüglich $\mathcal{C}^{\leqq}$ nach oben beschränkt. Dann sei $\mathfrak{K}$ die Menge aller kompatiblen Untermengen K von $\mathcal{C}$, für die $\alpha(f)$ für jedes Element f von K ein kompaktes Element von $\mathcal{E}$ ist. Um zu beweisen, daß $\mathcal{C}$ ein vollständiges induktives Gruppoid ist, wird nacheinander bewiesen:

1. Jede endliche Untermenge K von $\mathrm{mor}\,\mathcal{C}^{\cdot}$ mit $K \in \mathfrak{K}$ ist bezüglich $\mathcal{C}^{\leqq}$ nach oben beschränkt.
2. Für jede kompatible Untermenge K von $\mathcal{C}$ ist die Menge aller Morphismen f' von $\mathcal{C}^{\cdot}$, für die $\alpha(f')$ ein kompaktes Element von $\mathcal{E}$ ist und ein Element f von K mit $f' \leqq f$ existiert, ein Element von $\mathfrak{K}$.
3. Jede kompatible Untermenge von $\mathcal{C}$ ist bezüglich $\mathcal{C}^{\leqq}$ nach oben beschränkt.

Zu 1. Angenommen, es gäbe eine endliche Untermenge K von $\mathrm{mor}\,\mathcal{C}^{\cdot}$, die Element von $\mathfrak{K}$, aber nicht bezüglich $\mathcal{C}^{\leqq}$ nach oben beschränkt ist. Dann ist die Menge aller natürlichen Zahlen m, für die eine nicht bezüglich $\mathcal{C}^{\leqq}$ nach oben beschränkte Untermenge K von $\mathrm{mor}\,\mathcal{C}^{\cdot}$ mit $K \in \mathfrak{K}$ und $\mathrm{card}\, K = m$ existiert, nichtleer. Es sei dann n die kleinste natürliche Zahl aus dieser Menge. Notwendigerweise gilt dann $n > 3$. Es gibt dann also paarweise verschiedene Morphismen $f_1, f_2, \ldots, f_n$ von $\mathcal{C}^{\cdot}$, für die $\{f_1, f_2, \ldots, f_n\}$ ein Element von $\mathfrak{K}$ und $\{f_1, f_2, \ldots, f_n\}$ nicht bezüglich $\mathcal{C}^{\leqq}$ nach oben beschränkt ist. Für jede natürliche Zahl i mit $i \in \{3, 4, \ldots, n\}$ ist dann $\{f_1, f_2, f_i\}$ bezüglich $\mathcal{C}^{\leqq}$ nach oben beschränkt. Folglich existiert das Supremum bezüglich $\mathcal{C}^{\leqq}$ von $\{f_1, f_2\}$, und es

[1]) Ist A eine Menge, so bezeichnet „card A" die Kardinalzahl von A.

ist auch $\{f_1 \vee f_2, f_i\}$ für jede natürliche Zahl i aus $\{3, 4, \ldots, n\}$ bezüglich $\mathscr{C}^{\leqq}$ nach oben beschränkt. Demnach gilt

$$\alpha(f_1 \vee f_2) \wedge \alpha(f_i) = \alpha((f_1 \vee f_2) \wedge f_i)$$

und

$$\beta(f_1 \vee f_2) \wedge \beta(f_i) = \beta((f_1 \vee f_2) \wedge f_i)$$

für jede natürliche Zahl i aus $\{3, 4, \ldots, n\}$. Da $\alpha(f_1 \vee f_2) = \alpha(f_1) \vee \alpha(f_2)$ gilt und $\alpha(f_1)$, $\alpha(f_2)$ kompakte Elemente von $\mathscr{E}$ sind, so ist auch $\alpha(f_1 \vee f_2)$ ein kompaktes Element von $\mathscr{E}$. Mithin gilt $\{f_1 \vee f_2, f_3, \ldots, f_n\} \in \mathfrak{K}$ und $\operatorname{card}\{f_1 \vee f_2, f_3, \ldots, f_n\} < n$. Demnach ist $\{f_1 \vee f_2, f_3, \ldots, f_n\}$ bezüglich $\mathscr{C}^{\leqq}$ nach oben beschränkt und damit auch $\{f_1, f_2, \ldots, f_n\}$ im Widerspruch dazu, daß $\{f_1, f_2, \ldots, f_n\}$ nicht bezüglich $\mathscr{C}^{\leqq}$ nach oben beschränkt ist.

Zu 2. Es sei nun K eine kompatible Untermenge von $\mathscr{C}$, und es sei dann K' die Menge aller Morphismen f' von $\mathscr{C}^{\cdot}$, für die $\alpha(f')$ ein kompaktes Element von $\mathscr{E}$ ist und ein Element f von K mit $f' \leqq f$ existiert. Da $\mathscr{C}$ ein algebraisches induktives Gruppoid ist, so gibt es eine Algebra A, so daß $\mathscr{A}(\mathsf{A})$ zu $\mathscr{C}$ isomorph ist, und es existiert dann also ein Isomorphismus Φ von $\mathscr{C}$ auf $\mathscr{A}(\mathsf{A})$. Nun seien f', g' Elemente von K'. Dann gibt es Elemente f, g von K mit $f' \leqq f$ und $g' \leqq g$. Da nun K eine kompatible Untermenge von $\mathscr{C}$ ist, so ist auch $\{f, g\}$ eine kompatible Untermenge von $\mathscr{C}$ und folglich $\{\Phi(f), \Phi(g)\}$ eine kompatible Untermenge von $\mathscr{A}(\mathsf{A})$ und damit auch eine kompatible Untermenge von $\mathscr{J}(|\mathsf{A}|)$. Da weiter $\mathscr{J}(|\mathsf{A}|)$ ein vollständiges induktives Gruppoid (Satz 2) und ob $\mathscr{J}(|\mathsf{A}|)^{\cdot}$ bezüglich $\mathscr{J}(|\mathsf{A}|)^{\leqq}$ nach oben beschränkt ist, so ist also $\{\Phi(f), \Phi(g)\}$ bezüglich $\mathscr{J}(|\mathsf{A}|)^{\leqq}$ nach oben beschränkt, und wegen $\Phi(f') \subseteqq \Phi(f)$, $\Phi(g') \subseteqq \Phi(g)$ ist demnach $\{\Phi(f'), \Phi(g')\}$ eine bezüglich $\mathscr{J}(|\mathsf{A}|)^{\leqq}$ nach oben beschränkte Untermenge von $\operatorname{mor}\mathscr{A}(\mathsf{A})^{\cdot}$. Dann gilt aber

$$\alpha(\Phi(f') \cap \Phi(g')) = \alpha(\Phi(f')) \cap \alpha(\Phi(g')) ,$$
$$\beta(\Phi(f') \cap \Phi(g')) = \beta(\Phi(f')) \cap \beta(\Phi(g')) ,$$

und folglich ist $\{\Phi(f'), \Phi(g')\}$ eine kompatible Untermenge von $\mathscr{A}(\mathsf{A})$ und damit $\{f', g'\}$ eine kompatible Untermenge von $\mathscr{C}$. Mithin gilt

$$\alpha(f' \wedge g') = \alpha(f') \wedge \alpha(g') , \qquad \beta(f' \wedge g') = \beta(f') \wedge \beta(g') .$$

Demnach ist K' eine kompatible Untermenge von $\mathscr{C}$. Weiter ist $\alpha(f')$ für jedes Element f' von K' ein kompaktes Element von $\mathscr{E}$. Folglich ist K' ein Element von $\mathfrak{K}$.

Zu 3. Es sei K eine kompatible Untermenge von $\mathscr{C}$. Dann sei K' die Menge aller Morphismen f' von $\mathscr{C}^{\cdot}$, für die $\alpha(f')$ ein kompaktes Element von $\mathscr{E}$ ist and ein Element f von K mit $f' \leqq f$ existiert. Nach 2. ist dann K' ein Element von $\mathfrak{K}$. Dann ist aber auch jede endliche Untermenge von K' ein Element von $\mathfrak{K}$ und folglich nach 1. jede endliche Untermenge von K' bezüglich $\mathscr{C}^{\leqq}$ nach oben beschränkt. Demnach existiert für jede endliche Untermenge K^* von K' das Supremum bezüglich $\mathscr{C}^{\leqq}$ von K^*. Nun sei H die Menge

$$\{\bigvee K^* \mid K^* \text{ endliche Untermenge von } K'\} .$$

Dann ist H eine nach oben gerichtete Untermenge von $\mathscr{C}^{\leqq}$ und K' wegen $\bigvee\{f'\} = f'$ für jedes Element f' von K' eine Untermenge von H. Da $\mathscr{C}$ ein algebraisches induktives Gruppoid ist, so existiert also eine obere Schranke h bezüglich $\mathscr{C}^{\leqq}$ von H, und h ist dann auch eine obere Schranke bezüglich $\mathscr{C}^{\leqq}$ von der Untermenge K' von H. Nun sei

f ein Element von K. Da $\mathscr{E}$ ein algebraischer Verband ist, so gibt es dann eine Menge E von kompakten Elementen von $\mathscr{E}$ mit $\alpha(f) = \bigvee E$. Ist e ein Element von E, so gilt dann $e \leqq \alpha(f)$ und damit $\alpha(f \circ e) = e$ und folglich, da weiter $f \circ e \leqq f$ gilt, $f \circ e \in K'$. Mithin ist $\{f \circ e \mid e \in E\}$ eine Untermenge von K'. Weil f eine obere Schranke bezüglich $\mathscr{C}^{\leqq}$ von $\{f \circ e \mid e \in E\}$ ist, so existiert das Supremum bezüglich $\mathscr{C}^{\leqq}$ von $\{f \circ e \mid e \in E\}$, und es gilt $\bigvee\{f \circ e \mid e \in E\} \leqq f$ und folglich wegen

$$\alpha(f) = \bigvee E = \bigvee\{\alpha(f \circ e) \mid e \in E\} = \alpha(\bigvee\{f \circ e \mid e \in E\})$$

sogar $f = \bigvee\{f \circ e \mid e \in E\}$. Da weiter h eine obere Schranke bezüglich $\mathscr{C}^{\leqq}$ von K' und $\{f \circ e \mid e \in E\}$ eine Untermenge von K' sind, so gilt also $f \circ e \leqq h$ für jedes Element e von E und damit auch $f = \bigvee\{f \circ e \mid e \in E\} \leqq h$. Mithin ist h eine obere Schranke bezüglich $\mathscr{C}^{\leqq}$ von K und folglich K bezüglich $\mathscr{C}^{\leqq}$ nach oben beschränkt. ∎

Korollar zu Satz 5. *Für jede Algebra* A *gilt: Genau dann ist* $\mathscr{A}(\mathsf{A})$ *ein vollständiges induktives Gruppoid, wenn für beliebige partielle Automorphismen* $\varphi_1, \varphi_2, \varphi_3$ *von* A *mit* $\mathrm{Vb}(\varphi_i \cap \varphi_j) = \mathrm{Vb}\,\varphi_i \cap \mathrm{Vb}\,\varphi_j$ *und* $\mathrm{Nb}(\varphi_i \cap \varphi_j) = \mathrm{Nb}\,\varphi_i \cap \mathrm{Nb}\,\varphi_j$ *für beliebige Zahlen* i, j *aus* $\{1, 2, 3\}$, *für die die Unteralgebren* $\mathsf{B}_1, \mathsf{B}_2, \mathsf{B}_3$ *von* A *mit* $|\mathsf{B}_1| = \mathrm{Vb}\,\varphi_1$, $|\mathsf{B}_2| = \mathrm{Vb}\,\varphi_2$ *und* $|\mathsf{B}_3| = \mathrm{Vb}\,\varphi_3$ *endlich erzeugbare Unteralgebren von* A *sind, ein partieller Automorphismus* ψ *von* A *mit* $\varphi_1 \subseteqq \psi$, $\varphi_2 \subseteqq \psi$, $\varphi_3 \subseteqq \psi$ *existiert.* ∎

Ist $\mathscr{C}$ ein algebraisches induktives Gruppoid und ist speziell $\mathscr{E}$ ein distributiver voll ständiger Verband[1]), so ist sogar schon dann $\mathscr{C}$ ein vollständiges induktives Gruppoid, wenn für beliebige Morphismen f, g von $\mathscr{C}^{\cdot}$, für die $\alpha(f)$, $\alpha(g)$ kompakte Elemente von $\mathscr{E}$ sind und $\alpha(f \wedge g) = \alpha(f) \wedge \alpha(g)$, $\beta(f \wedge g) = \beta(f) \wedge \beta(g)$ gilt, ein Morphismus h von $\mathscr{C}^{\cdot}$ mit $f \leqq h$, $g \leqq h$ existiert. Es gilt nämlich:

Satz 6. *Für jedes algebraische induktive Gruppoid* $\mathscr{C}$ *mit* $(e_1 \vee e_2) \wedge e_3 = (e_1 \wedge e_3) \vee (e_2 \wedge e_3)$ *für beliebige Einselemente* e_1, e_2, e_3 *von* $\mathscr{C}^{\cdot}$ *gilt:*

Genau dann ist $\mathscr{C}$ *ein vollständiges induktives Gruppoid, wenn für beliebige Morphismen* f, g *von* $\mathscr{C}^{\cdot}$, *für die* $\{f, g\}$ *eine kompatible Untermenge von* $\mathscr{C}$ *ist sowie* $\alpha(f)$ *und* $\alpha(g)$ *kompakte Elemente von* $\mathscr{E}$ *sind,* $\{f, g\}$ *bezüglich* $\mathscr{C}^{\leqq}$ *nach oben beschränkt ist.*

Beweis. Es sei $\mathscr{C}$ ein algebraisches induktives Gruppoid, und es gelte

$$(e_1 \vee e_2) \wedge e_3 = (e_1 \wedge e_3) \vee (e_2 \wedge e_3)$$

für beliebige Einselemente e_1, e_2, e_3 von $\mathscr{C}^{\cdot}$.

$(\Rightarrow)$ Ist dann $\mathscr{C}$ insbesondere ein vollständiges induktives Gruppoid und sind f, g Morphismen von $\mathscr{C}^{\cdot}$, für die $\{f, g\}$ eine kompatible Untermenge von $\mathscr{C}$ ist und $\alpha(f)$, $\alpha(g)$ kompakte Elemente von $\mathscr{E}$ sind, so ist $\{f, g\}$ bezüglich $\mathscr{C}^{\leqq}$ nach oben beschränkt.

$(\Leftarrow)$ Es sei nun $\{f, g\}$ für beliebige Morphismen f, g von $\mathscr{C}^{\cdot}$, für die $\{f, g\}$ eine kompatible Untermenge von $\mathscr{C}$ ist und $\alpha(f)$, $\alpha(g)$ kompakte Elemente von $\mathscr{E}$ sind, bezüglich $\mathscr{C}^{\leqq}$ nach oben beschränkt. Dann seien f, g, h Morphismen von $\mathscr{C}^{\cdot}$, $\{f, g, h\}$ eine kompatible Untermenge von $\mathscr{C}$ und $\alpha(f)$, $\alpha(g)$, $\alpha(h)$ kompakte Elemente von $\mathscr{E}$. Es ist dann $\{f, g\}$ bezüglich $\mathscr{C}^{\leqq}$ nach oben beschränkt, und folglich existiert das Supremum bezüglich $\mathscr{C}^{\leqq}$ von $\{f, g\}$. Weiter ist dann $\{f \vee g, h\}$ eine kompatible Untermenge von $\mathscr{C}$.

[1]) D. h., es gilt $(e_1 \vee e_2) \wedge e_3 = (e_1 \wedge e_3) \vee (e_2 \wedge e_3)$ für beliebige Einselemente e_1, e_2, e_3 von $\mathscr{C}^{\cdot}$. Das ist hier schon dann der Fall, wenn $(e_1 \vee e_2) \wedge e_3 = (e_1 \wedge e_3) \vee (e_2 \wedge e_3)$ für beliebige kompakte Elemente e_1, e_2, e_3 von $\mathscr{E}$ gilt.

Zunächst ist nämlich sowohl $f \vee g$ als auch h eine obere Schranke bezüglich $\mathscr{C}^{\leq}$ von $\{f \wedge h, g \wedge h\}$, und folglich existiert das Supremum bezüglich $\mathscr{C}^{\leq}$ von $\{f \wedge h, g \wedge h\}$, und es gilt $(f \wedge h) \vee (g \wedge h) \leqq f \vee g$, $(f \wedge h) \vee (g \wedge h) \leqq h$ und damit

$$(f \wedge h) \vee (g \wedge h) \leqq (f \vee g) \wedge h\,.$$

Mithin gilt

$$\begin{aligned}\alpha(f \vee g) \wedge \alpha(h) &= \big(\alpha(f) \vee \alpha(g)\big) \wedge \alpha(h)\\ &= \big(\alpha(f) \wedge \alpha(h)\big) \vee \big(\alpha(g) \wedge \alpha(h)\big)\\ &= \alpha(f \wedge h) \vee \alpha(g \wedge h)\\ &= \alpha\big((f \wedge h) \vee (g \wedge h)\big)\\ &\leqq \alpha\big((f \vee g) \wedge h\big)\end{aligned}$$

Wegen $(f \vee g) \wedge h \leqq f \vee g$ und $(f \vee g) \wedge h \leqq h$ gilt weiter

$$\alpha\big((f \vee g) \wedge h\big) \leqq \alpha(f \vee g)\,, \qquad \alpha\big((f \vee g) \wedge h\big) \leqq \alpha(h)$$

und damit

$$\alpha\big((f \vee g) \wedge h\big) \leqq \alpha(f \vee g) \wedge \alpha(h)\,.$$

Demnach gilt

$$\alpha\big((f \vee g) \wedge h\big) = \alpha(f \vee g) \wedge \alpha(h)\,.$$

Entsprechend beweist man, daß auch

$$\beta\big((f \vee g) \wedge h\big) = \beta(f \vee g) \wedge \beta(h)$$

gilt. Da $\alpha(f \vee g) = \alpha(f) \vee \alpha(g)$ gilt und $\alpha(f)$, $\alpha(g)$ kompakte Elemente von $\mathscr{E}$ sind, so ist auch $\alpha(f \vee g)$ ein kompaktes Element von $\mathscr{E}$. Demnach ist $\{f \vee g, h\}$ eine kompatible Untermenge von $\mathscr{C}$, und $\alpha(f \vee g)$, $\alpha(h)$ sind kompakte Elemente von $\mathscr{E}$. Folglich ist $\{f \vee g, h\}$ und damit auch $\{f, g, h\}$ bezüglich $\mathscr{C}^{\leq}$ nach oben beschränkt. Mithin ist $\mathscr{C}$ ein vollständiges induktives Gruppoid (Satz 5). ▮

Korollar zu Satz 6. *Für jede Algebra* A *mit distributivem Unteralgebrenverband gilt: Genau dann ist* $\mathscr{A}(\mathsf{A})$ *ein vollständiges induktives Gruppoid, wenn für beliebige partielle Automorphismen* φ, ψ *von* A *mit* $\mathrm{Vb}(\varphi \cap \psi) = \mathrm{Vb}\,\varphi \cap \mathrm{Vb}\,\psi$, $\mathrm{Nb}(\varphi \cap \psi) = \mathrm{Nb}\,\varphi \cap \mathrm{Nb}\,\psi$, *für die die Unteralgebren* B, C *von* A *mit* $|\mathsf{B}| = \mathrm{Vb}\,\varphi$ *und* $|\mathsf{C}| = \mathrm{Vb}\,\psi$ *endlich erzeugbare Unteralgebren von* A *sind, ein partieller Automorphismus* χ *von* A *mit* $\varphi \subseteqq \chi$, $\psi \subseteqq \chi$ *existiert.* ▮

Mit Hilfe von Satz 6 beweist man z. B. leicht:

Satz 7. *Für jede zyklische Gruppe* Z *ist* $\mathscr{A}(\mathsf{Z})$ *ein vollständiges induktives Gruppoid.*[1])

Beweis. Es sei Z eine zyklische Gruppe. Dann ist $\mathscr{U}(\mathsf{Z})$ ein distributiver vollständiger Verband. Nun seien φ, ψ partielle Automorphismen von Z, und es gelte

$$\mathrm{Vb}(\varphi \cap \psi) = \mathrm{Vb}\,\varphi \cap \mathrm{Vb}\,\psi\,, \qquad \mathrm{Nb}(\varphi \cap \psi) = \mathrm{Nb}\,\varphi \cap \mathrm{Nb}\,\psi\,.$$

[1]) Gruppen werden hier als Algebren mit der Gruppenmultiplikation, dem Einselement und der Inversenbildung als definierende Operationen aufgefaßt. Für zyklische Gruppen mit endlichem Träger ist dieser Satz bereits in [4] bewiesen worden. Man siehe hierzu [4], Satz 7, Seite 325.

Dann gibt es Elemente a, b von $|\mathsf{Z}|$ mit $\mathrm{Vb}\,\varphi = \langle\{a\}\rangle$ und $\mathrm{Vb}\,\psi = \langle\{b\}\rangle$. Es sei dann V die Untermenge von $|\mathsf{Z}|$ mit

$$V = \{a^r \cdot b^s \mid r, s \text{ ganze Zahlen}\}\ .$$

Offenbar ist dann V ein Element von $\mathfrak{U}(\mathsf{Z})$. Nun sei weiter χ die Korrespondenz von V in $|\mathsf{Z}|$ mit

$$\chi = \{[a^r \cdot b^s, \varphi(a^r) \cdot \psi(b^s)] \mid r, s \text{ ganze Zahlen}\}\ .$$

Dann ist χ ein partieller Automorphismus von Z. Sind nämlich r_1, r_2, s_1, s_2 ganze Zahlen und gilt $a^{r_1} \cdot b^{s_1} = a^{r_2} \cdot b^{s_2}$, so gilt

$$a^{r_1 - r_2} = b^{s_2 - s_1} \in \mathrm{Vb}\,\varphi \cap \mathrm{Vb}\,\psi = \mathrm{Vb}(\varphi \cap \psi)$$

und damit $\varphi(a^{r_1-r_2}) = \psi(b^{s_2-s_1})$, also $\varphi(a^{r_1}) \cdot \psi(b^{s_1}) = \varphi(a^{r_2}) \cdot \psi(b^{s_2})$. Sind r_1, r_2, s_1, s_2 ganze Zahlen und gilt $\varphi(a^{r_1}) \cdot \psi(b^{s_1}) = \varphi(a^{r_2}) \cdot \psi(b^{s_2})$, so gilt $\varphi(a^{r_1-r_2}) = \psi(b^{s_2-s_1}) \in$ $\in \mathrm{Nb}\,\varphi \cap \mathrm{Nb}\,\psi = \mathrm{Nb}(\varphi \cap \psi) = \mathrm{Vb}\big((\varphi \cap \psi)^{-1}\big)$ und folglich $(\varphi \cap \psi)^{-1}\,\big(\varphi(a^{r_1-r_2})\big) =$ $= (\varphi \cap \psi)^{-1}\,\big(\psi(b^{s_2-s_1})\big)$, also $a^{r_1-r_2} = b^{s_2-s_1}$ und damit $a^{r_1} \cdot b^{s_1} = a^{r_2} \cdot b^{s_2}$. Demnach ist χ eine eineindeutige Funktion von V in $|\mathsf{Z}|$. Sind weiter v_1, v_2 Elemente von V, so gibt es ganze Zahlen r_1, r_2, s_1, s_2 mit $v_1 = a^{r_1} \cdot b^{s_1}$ und $v_2 = a^{r_2} \cdot b^{s_2}$, und es gilt dann

$$\begin{aligned}\chi(v_1 \cdot v_2) &= \chi(a^{r_1+r_2} \cdot b^{s_1+s_2}) = \varphi(a^{r_1+r_2}) \cdot \psi(b^{s_1+s_2})\\ &= \varphi(a^{r_1}) \cdot \psi(b^{s_1}) \cdot \varphi(a^{r_2}) \cdot \psi(b^{s_2})\\ &= \chi(a^{r_1} \cdot b^{s_1}) \cdot \chi(a^{r_2} \cdot b^{s_2})\\ &= \chi(v_1) \cdot \chi(v_2)\ .\end{aligned}$$

Schließlich gilt $\mathrm{Vb}\,\varphi \subseteqq \mathrm{Vb}\,\chi$, $\mathrm{Vb}\,\psi \subseteqq \mathrm{Vb}\,\chi$ sowie für beliebige ganze Zahlen r, s:

$$\chi(a^r) = \chi(a^r \cdot b^0) = \varphi(a^r) \cdot \psi(b^0) = \varphi(a^r)\ ,$$

$$\chi(b^s) = \chi(a^0 \cdot b^s) = \varphi(a^0) \cdot \psi(b^s) = \psi(b^s)$$

und damit $\varphi \subseteqq \chi$ und $\psi \subseteqq \chi$. Demnach ist $\mathcal{A}(\mathsf{Z})$ ein vollständiges induktives Gruppoid (Satz 6). ∎

Mit Hilfe von Satz 6 beweist man weiter leicht:

Satz 8. *Jedes algebraische induktive Gruppoid $\mathscr{C}$, für das speziell $\mathscr{E}$ eine Kette ist*[1]), *ist ein vollständiges induktives Gruppoid.*

Beweis. Es sei $\mathscr{C}$ ein algebraisches induktives Gruppoid, und es sei $\mathscr{E}$ speziell eine Kette. Dann ist $\mathscr{E}$ ein distributiver vollständiger Verband. Es seien nun f, g Morphismen von $\mathscr{C}^\cdot$, $\alpha(f)$ und $\alpha(g)$ kompakte Elemente von $\mathscr{E}$ und $\{f, g\}$ eine kompatible Untermenge von $\mathscr{C}$. Dann gilt $\alpha(f) \leqq \alpha(g)$ oder $\alpha(g) \leqq \alpha(f)$. Es gelte o. B. d. A. $\alpha(f) \leqq$ $\leqq \alpha(g)$. Dann gilt $f \wedge g \leqq f$ und $\alpha(f \wedge g) = \alpha(f) \wedge \alpha(g) = \alpha(f)$, also $f \wedge g = f$ und folglich $f \leqq g$. Demnach ist $\{f, g\}$ bezüglich $\mathscr{C}^{\leqq}$ nach oben beschränkt. Mithin ist $\mathscr{C}$ ein vollständiges induktives Gruppoid (Satz 6). ∎

Korollar zu Satz 8. *Für jede Algebra* A, *für die der Unteralgebrenverband von* A *eine Kette ist, ist* $\mathcal{A}(\mathsf{A})$ *ein vollständiges induktives Gruppoid.* ∎

[1]) D. h., es gilt $e_1 \leqq e_2$ oder $e_2 \leqq e_1$ für beliebige Einselemente e_1, e_2 von $\mathscr{C}^\cdot$.

Wenn $\mathscr{C}$ ein induktives Gruppoid und e ein Einselement von $\mathscr{C}^\cdot$ sind, so gibt es ein und auch nur ein induktives Untergruppoid $\mathscr{D}$ von $\mathscr{C}$ mit

$$\operatorname{mor} \mathscr{D}^\cdot = \{f \mid f \in \operatorname{mor} \mathscr{C}^\cdot, \alpha(f) \leqq e, \beta(f) \leqq e\} .$$

Demnach ist die folgende Definition berechtigt:

Definition 14. Ist $\mathscr{C}$ ein induktives Gruppoid und ist e ein Einselement von $\mathscr{C}^\cdot$, so bezeichnet $\mathscr{C}^{(e)}$ das induktive Untergruppoid $\mathscr{D}$ von $\mathscr{C}$ mit

$$\operatorname{mor} \mathscr{D}^\cdot = \{f \mid f \in \operatorname{mor} \mathscr{C}^\cdot, \alpha(f) \leqq e, \beta(f) \leqq e\} . \blacksquare$$

Es gilt nun der folgende Satz.

Satz 9. *Für jedes algebraische induktive Gruppoid $\mathscr{C}$ gilt: Genau dann ist $\mathscr{C}$ ein vollständiges induktives Gruppoid, wenn $\mathscr{C}^{(e)}$ für jedes kompakte Element e von $\mathscr{C}$ ein vollständiges induktives Gruppoid ist.*

Beweis. Es sei $\mathscr{C}$ ein algebraisches induktives Gruppoid.

($\Rightarrow$) Dann sei $\mathscr{C}$ ein vollständiges induktives Gruppoid, e ein kompaktes Element von $\mathscr{C}$ und K eine kompatible Untermenge von $\mathscr{C}^{(e)}$. Es gilt dann $f \wedge_{\mathscr{C}^{(e)\leqq}} g = f \wedge_{\mathscr{C}^{\leqq}} g$ für beliebige Morphismen f, g von $\mathscr{C}^{(e)\cdot}$ und $\alpha_{\mathscr{C}^{(e)\cdot}}(f) = \alpha_{\mathscr{C}^\cdot}(f)$, $\beta_{\mathscr{C}^{(e)\cdot}}(f) = \beta_{\mathscr{C}^\cdot}(f)$ für jeden Morphismus f von $\mathscr{C}^{(e)\cdot}$. Folglich ist K auch eine kompatible Untermenge von $\mathscr{C}$ und damit bezüglich $\mathscr{C}^{\leqq}$ nach oben beschränkt. Dann existiert das Supremum bezüglich $\mathscr{C}^{\leqq}$ von K, und es gilt

$$\alpha(\bigvee K) = \bigvee \check{\alpha}(K) \leqq e , \qquad \beta(\bigvee K) = \bigvee \check{\beta}(K) \leqq e .$$

Mithin ist $\bigvee K$ eine obere Schranke bezüglich $\mathscr{C}^{(e)\leqq}$ von K. Demnach ist $\mathscr{C}^{(e)}$ ein vollständiges induktives Gruppoid.

($\Leftarrow$) Es sei nun $\mathscr{C}^{(e)}$ für jedes kompakte Element e von $\mathscr{C}$ ein vollständiges induktives Gruppoid. Ist A eine Algebra, ist φ ein partieller Automorphismus von A, sind B und C die Unteralgebren von A mit $\alpha(\varphi) = \mathrm{id}_{|\mathsf{B}|}$ und $\beta(\varphi) = \mathrm{id}_{|\mathsf{C}|}$ und ist B eine endlich erzeugbare Unteralgebra von A, so ist auch C eine endlich erzeugbare Unteralgebra von A. Folglich ist für jeden Morphismus f von $\mathscr{C}^\cdot$, für den $\alpha(f)$ ein kompaktes Element von $\mathscr{C}$ ist, auch $\beta(f)$ ein kompaktes Element von $\mathscr{C}$. Nun seien f, g, h Morphismen von $\mathscr{C}^\cdot$, $\{f, g, h\}$ eine kompatible Untermenge von $\mathscr{C}$ und $\alpha(f)$, $\alpha(g)$, $\alpha(h)$ kompakte Elemente von $\mathscr{C}$. Dann sind auch $\beta(f)$, $\beta(g)$, $\beta(h)$ kompakte Elemente von $\mathscr{C}$. Es sei nun e das Einselement von $\mathscr{C}^\cdot$ mit

$$e = \alpha(f) \vee \alpha(g) \vee \alpha(h) \vee \beta(f) \vee \beta(g) \vee \beta(h) .$$

Dann ist auch e ein kompaktes Element von $\mathscr{C}$, und f, g, h sind Morphismen von $\mathscr{C}^{(e)\cdot}$. Mithin ist $\{f, g, h\}$ eine kompatible Untermenge von $\mathscr{C}^{(e)}$ und folglich, da $\mathscr{C}^{(e)}$ ein vollständiges induktives Gruppoid ist und ob $\mathscr{C}^{(e)\cdot}$ bezüglich $\mathscr{C}^{(e)\leqq}$ nach oben beschränkt ist, $\{f, g, h\}$ bezüglich $\mathscr{C}^{(e)\leqq}$ und damit auch bezüglich $\mathscr{C}^{\leqq}$ nach oben beschränkt. Folglich ist $\mathscr{C}$ ein vollständiges induktives Gruppoid (Satz 5). $\blacksquare$

Korollar zu Satz 9. *Für jede Algebra A gilt: Genau dann ist $\mathscr{A}(\mathsf{A})$ ein vollständiges induktives Gruppoid, wenn $\mathscr{A}(\mathsf{B})$ für jede endlich erzeugbare Unteralgebra B von A ein vollständiges induktives Gruppoid ist.* $\blacksquare$

Eine Gruppe G heißt genau dann eine verallgemeinerte zyklische Gruppe, wenn jede endlich erzeugbare Untergruppe von G[1]) eine zyklische Gruppe ist. Eine Gruppe G ist genau dann eine verallgemeinerte zyklische Gruppe, wenn $\mathcal{U}(G)$ ein distributiver vollständiger Verband ist.[2])

Mit Hilfe von Satz 7 und Satz 9 beweist man nun sofort:

Satz 10. *Für jede verallgemeinerte zyklische Gruppe* G *ist* $\mathcal{A}(G)$ *ein vollständiges induktives Gruppoid.* ▮

Im folgenden beschäftigen wir uns noch mit der Automorphismengruppe einer Algebra A, für die $\mathcal{A}(A)$ ein vollständiges induktives Gruppoid ist und die noch einigen weiteren Bedingungen genügt. Hierzu geben wir zunächst einige Definitionen.

Definition 15. Ist $\mathcal{C}$ ein induktives Gruppoid, so ist genau dann a ein *Atom von* $\mathcal{C}$, wenn a ein von o verschiedenes Einselement von $\mathcal{C}^\cdot$ ist und kein Einselement e von $\mathcal{C}^\cdot$ mit $o < e < a$ existiert. Ist $\mathcal{C}$ ein induktives Gruppoid, so bezeichnet at $\mathcal{C}$ die Klasse aller Atome von $\mathcal{C}$. Genau dann ist $\mathcal{C}$ ein *atomares induktives Gruppoid*, wenn $\mathcal{C}$ ein induktives Gruppoid ist und für jedes Einselement e von $\mathcal{C}^\cdot$ mit $e \neq o$ ein Atom a von $\mathcal{C}$ mit $a \leqq e$ existiert. ▮

Ist $\mathcal{C}$ ein induktives Gruppoid, so gibt es ein und nur ein Untergruppoid $\mathcal{Q}$ von $\mathcal{C}^\cdot$ mit

$$\operatorname{mor} \mathcal{Q} = \{f \mid f \in \operatorname{mor} \mathcal{C}^\cdot, \quad \text{und} \quad \alpha(f), \beta(f) \in \operatorname{at} \mathcal{C}\} .$$

Demnach ist die folgende Definition berechtigt:

Definition 16. Ist $\mathcal{C}$ ein induktives Gruppoid, so bezeichnet $\mathcal{Q}(\mathcal{C})$ das Untergruppoid $\mathcal{Q}$ von $\mathcal{C}^\cdot$ mit

$$\operatorname{mor} \mathcal{Q} = \{f \mid f \in \operatorname{mor} \mathcal{C}^\cdot, \text{und}\, \alpha(f), \beta(f) \in \operatorname{at} \mathcal{C}\} . ▮$$

Für das folgende benötigen wir weiter den Begriff der Loewyschen Gruppe eines kleinen Gruppoids.

Definition 17. Ist $\mathcal{C}$ ein kleines Gruppoid, so bezeichnet $\mathfrak{L}(\mathcal{C})$ die Menge aller Untermengen $\mathfrak{s}$ von mor $\mathcal{C}$ mit der Eigenschaft, daß für jedes Einselement e von $\mathcal{C}$ genau ein Element f von $\mathfrak{s}$ mit $\alpha(f) = e$ und genau ein Element g von $\mathfrak{s}$ mit $\beta(g) = e$ existiert. ▮

Für jedes kleine Gruppoid $\mathcal{C}$ ist nun offenbar eine Gruppe $\mathcal{L}$ definiert, deren Träger die Menge $\mathfrak{L}(\mathcal{C})$ ist und deren Gruppenmultiplikation $\cdot$ durch

$$\mathfrak{t} \cdot \mathfrak{s} = \mathfrak{t} \bullet_{\mathcal{C}} \mathfrak{s}$$

für beliebige Elemente $\mathfrak{s}$, $\mathfrak{t}$ von $\mathfrak{L}(\mathcal{C})$ gegeben ist. Demnach ist die folgende Definition berechtigt:

Definition 18. Ist $\mathcal{C}$ ein kleines Gruppoid, so bezeichnet $\mathcal{L}(\mathcal{C})$ die Gruppe, deren Träger $\mathfrak{L}(\mathcal{C})$ ist und für deren Gruppenmultiplikation $\cdot$

$$\mathfrak{t} \cdot \mathfrak{s} = \mathfrak{t} \bullet_{\mathcal{C}} \mathfrak{s}$$

für beliebige Elemente $\mathfrak{s}$, $\mathfrak{t}$ von $\mathfrak{L}(\mathcal{C})$ gilt. Wenn $\mathcal{C}$ ein kleines Gruppoid ist, so wird unter der Loewy*schen Gruppe von* $\mathcal{C}$ die Gruppe $\mathcal{L}(\mathcal{C})$ verstanden. ▮

[1]) D. h. jede endlich erzeugbare Unteralgebra von G.

[2]) Man siehe hierzu [8], Kap. I, Satz 2, Seite 4.

Bemerkung. Für jedes kleine Gruppoid $\mathscr{C}$ gilt:

1. ob $\mathscr{C}$ ist das Einselement von $\mathscr{L}(\mathscr{C})$.
2. Für jedes Element $\mathfrak{s}$ von $\mathfrak{L}(\mathscr{C})$ ist auch $\mathfrak{t}$ mit $\mathfrak{t} = \{f^{-1} \mid f \in \mathfrak{s}\}$ ein Element von $\mathfrak{L}(\mathscr{C})$, und es gilt $\mathfrak{t} \bullet \mathfrak{s} = \mathfrak{s} \bullet \mathfrak{t} =$ ob $\mathscr{C}$.

Ist nun $\mathscr{C}$ ein induktives Gruppoid und ist ob $\mathscr{C}^\cdot$ bezüglich $\mathscr{C}^\leqq$ nach oben beschränkt[1]), so ist eine Gruppe $\mathscr{G}$ offenbar wie folgt definiert: Der Träger von $\mathscr{G}$ ist die Menge aller Morphismen f von $\mathscr{C}^\cdot$ mit $\alpha(f) = \beta(f) = u$, und die Gruppenmultiplikation $\cdot$ von $\mathscr{G}$ ist durch $g \cdot f = g \cdot_{\mathscr{C}^\cdot} f$ für beliebige Morphismen f, g von $\mathscr{C}^\cdot$ mit $\alpha(f) = \beta(f) = u$ gegeben[2]). Demnach ist die folgende Definition berechtigt:

Definition 19. Ist $\mathscr{C}$ ein induktives Gruppoid und ist ob $\mathscr{C}^\cdot$ bezüglich $\mathscr{C}^\leqq$ nach oben beschränkt, so bezeichnet $\mathfrak{G}(\mathscr{C})$ die Menge aller Morphismen f von $\mathscr{C}^\cdot$ mit $\alpha(f) = \beta(f) = u$ und $\mathscr{G}(\mathscr{C})$ die Gruppe, deren Träger $\mathfrak{G}(\mathscr{C})$ ist und deren Gruppenmultiplikation $\cdot$ durch $g \cdot f = g \cdot_{\mathscr{C}^\cdot} f$ für beliebige Elemente f, g von $\mathfrak{G}(\mathscr{C})$ gegeben ist. ∎

Der folgende Satz zeigt nun für atomare induktive Gruppoide $\mathscr{C}$, für die ob $\mathscr{C}^\cdot$ bezüglich $\mathscr{C}^\leqq$ nach oben beschränkt ist[3]) und $\bigvee$ at $\mathscr{C} = u$ gilt, einen Zusammenhang zwischen den Gruppen $\mathscr{G}(\mathscr{C})$ und $\mathscr{L}(\mathscr{Q}(\mathscr{C}))$ auf.

Satz 11. *Für jedes atomare induktive Gruppoid $\mathscr{C}$, für das* ob $\mathscr{C}^\cdot$ *bezüglich $\mathscr{C}^\leqq$ nach oben beschränkt ist*[4]) *und die Funktion Φ von $\mathfrak{G}(\mathscr{C})$ in $\mathfrak{P}$(mor $\mathscr{C}^\cdot$) mit*

$$\Phi(f) = \{f \circ a \mid a \in \text{at } \mathscr{C}\} \textit{ für jedes Element } f \textit{ von } \mathfrak{G}(\mathscr{C})$$

gilt:

1. *Φ ist ein Homomorphismus von $\mathscr{G}(\mathscr{C})$ in $\mathscr{L}(\mathscr{Q}(\mathscr{C}))$.*
2. *Φ ist ein eineindeutiger Homomorphismus von $\mathscr{G}(\mathscr{C})$ in $\mathscr{L}(\mathscr{Q}(\mathscr{C}))$, wenn $\bigvee$* at $\mathscr{C} = u$ *gilt.*
3. *Φ ist ein Isomorphismus von $\mathscr{G}(\mathscr{C})$ auf $\mathscr{L}(\mathscr{Q}(\mathscr{C}))$, wenn $\bigvee$* at $\mathscr{C} = u$ *gilt und $\mathscr{C}$ ein vollständiges induktives Gruppoid ist.*

Beweis. Es sei $\mathscr{C}$ ein atomares induktives Gruppoid, es sei ob $\mathscr{C}^\cdot$ bezüglich $\mathscr{C}^\leqq$ nach oben beschränkt, und es sei Φ die Funktion von $\mathfrak{G}(\mathscr{C})$ in $\mathfrak{P}$(mor $\mathscr{C}^\cdot$) mit

$$\Phi(f) = \{f \circ a \mid a \in \text{at } \mathscr{C}\} \text{ für jedes Element } f \text{ von } \mathfrak{G}(\mathscr{C})\,.$$

1. Dann sei f ein Element von $\mathfrak{G}(\mathscr{C})$ und a ein Atom von $\mathscr{C}$. Es gilt dann $\alpha(f \circ a) = a \in$ at $\mathscr{C}$. Weiter gilt $\beta(f \circ a) \neq o$; denn gälte $\beta(f \circ a) = o$, so wäre $f \circ a = o$ und damit $a = \alpha(f \circ a) = \alpha(o) = o$ im Widerspruch zu $a \in$ at $\mathscr{C}$. Demnach gibt es ein Atom a' von $\mathscr{C}$ mit $a' \leqq \beta(f \circ a)$, und es gilt dann (Hilfssatz)

$$\beta(f^{-1} \circ a') \leqq \beta\big(f^{-1} \circ \beta(f \circ a)\big) = \beta(a \circ f^{-1}) = a\,.$$

Gälte $\beta(f^{-1} \circ a') = o$, so wäre $f^{-1} \circ a' = o$ und damit $a' = \alpha(f^{-1} \circ a') = o$ im Widerspruch zu $a' \in$ at $\mathscr{C}$. Folglich gilt

$$o < \beta(f^{-1} \circ a') \leqq a \in \text{at } \mathscr{C}\,,$$

[1]) $\mathscr{C}$ ist dann insbesondere ein kleines induktives Gruppoid.
[2]) Ist A eine Algebra und ist $\mathscr{C}$ das induktive Gruppoid $\mathscr{A}$(A), so ist $\mathfrak{G}$ die Automorphismengruppe von A.
[3]) $\mathscr{C}$ ist dann insbesondere ein kleines induktives Gruppoid.
[4]) Man beachte, daß hier $\mathscr{C}$ nicht notwendig ein algebraisches induktives Gruppoid ist.

also $\beta(f^{-1} \circ a') = a$ und damit (Hilfssatz)

$$f \circ a = f \circ \beta(f^{-1} \circ a') = a' \circ f ,$$

also $\beta(f \circ a) = \beta(a' \circ f) = a' \in \text{at } \mathscr{C}$. Mithin gilt $f \circ a \in \text{mor } \mathcal{Q}(\mathscr{C})$. Demnach gilt

$$\Phi(f) \subseteqq \text{mor } \mathcal{Q}(\mathscr{C}) \text{ für jedes Element } f \text{ von } \mathfrak{G}(\mathscr{C}) .$$

Ist weiter f ein Element von $\mathfrak{G}(\mathscr{C})$ und a ein Atom von $\mathscr{C}$, so gibt es offenbar genau ein Element f' von $\Phi(f)$ mit $\alpha(f') = a$, nämlich das Element $f \circ a$ von $\Phi(f)$. Es sei nun f ein Element von $\mathfrak{G}(\mathscr{C})$ und a ein Atom von $\mathscr{C}$. Dann ist wegen $f^{-1} \in \mathfrak{G}(\mathscr{C})$ nach dem Vorangehenden auch $\beta(f^{-1} \circ a)$ ein Atom von $\mathscr{C}$, und es gilt (Hilfssatz)

$$f \circ \beta(f^{-1} \circ a) = a \circ f$$

und damit $\beta(f \circ \beta(f^{-1} \circ a)) = \beta(a \circ f) = a$. Mithin gibt es ein Element f' von $\Phi(f)$ mit $\beta(f') = a$. Sind nun f_1', f_2' Elemente von $\Phi(f)$ und gilt $\beta(f_1') = \beta(f_2')$, so gilt wegen $f_1' \leqq f$, $f_2' \leqq f$ und $\beta(f_1') = \beta(f_2')$ notwendig $f_1' = f_2'$. Demnach gibt es auch für jedes Element f von $\mathfrak{G}(\mathscr{C})$ und jedes Atom a von $\mathscr{C}$ genau ein Element f' von $\Phi(f)$ mit $\beta(f') = a$. Damit ist bewiesen, daß $\Phi(f)$ für jedes Element f von $\mathfrak{G}(\mathscr{C})$ ein Element von $\mathfrak{L}(\mathcal{Q}(\mathscr{C}))$, also Φ eine Funktion von $\mathfrak{G}(\mathscr{C})$ in $\mathfrak{L}(\mathcal{Q}(\mathscr{C}))$ ist.

Sind weiter f, g Elemente von $\mathfrak{G}(\mathscr{C})$, so ist $g \cdot f$ eine obere Schranke bezüglich $\mathscr{C}^{\leqq}$ von $\Phi(g) \bullet \Phi(f)$ aus $\mathfrak{G}(\mathscr{C})$, und es gilt obendrein $\Phi(g) \bullet \Phi(f) \in \mathfrak{L}(\mathcal{Q}(\mathscr{C}))$. Mithin gilt offenbar

$$\Phi(g) \bullet \Phi(f) = \{(g \cdot f) \circ a \mid a \in \text{at } \mathscr{C}\} ,$$

also

$$\Phi(g) \bullet \Phi(f) = \Phi(g \cdot f) .$$

Demnach ist Φ ein Homomorphismus von $\mathscr{G}(\mathscr{C})$ in $\mathscr{L}(\mathcal{Q}(\mathscr{C}))$.

2. Es gelte nun $\bigvee \text{at } \mathscr{C} = u$. Ist dann f ein Element von $\mathfrak{G}(\mathscr{C})$ und gilt $\Phi(f) = \text{ob } \mathscr{C}^{\cdot}$, so ist f eine obere Schranke bezüglich $\mathscr{C}^{\leqq}$ von ob $\mathscr{C}^{\cdot}$, und folglich gilt $u = \bigvee \text{ob } \mathscr{C}^{\cdot} \leqq f$, also wegen $\alpha(f) = u = \alpha(u)$ notwendig $f = u$. Mithin ist der Homomorphismus Φ von der Gruppe $\mathscr{G}(\mathscr{C})$ in die Gruppe $\mathscr{L}(\mathcal{Q}(\mathscr{C}))$ (Satz 11,1.) ein eineindeutiger Homomorphismus von $\mathscr{G}(\mathscr{C})$ in $\mathscr{L}(\mathcal{Q}(\mathscr{C}))$.

3. Es gelte $\bigvee \text{at } \mathscr{C} = u$, und es sei speziell $\mathscr{C}$ ein vollständiges induktives Gruppoid. Dann ist Φ ein eineindeutiger Homomorphismus von $\mathscr{G}(\mathscr{C})$ in $\mathscr{L}(\mathcal{Q}(\mathscr{C}))$ (Satz 11,2). Nun sei $\mathfrak{z}$ ein Element von $\mathfrak{L}(\mathcal{Q}(\mathscr{C}))$. Dann gilt für beliebige Elemente f, g von $\mathfrak{z}$ mit $f \neq g$ wegen $\alpha(f), \beta(f), \alpha(g), \beta(g) \in \text{at } \mathscr{C}$ und $\alpha(f) \neq \alpha(g)$, $\beta(f) \neq \beta(g)$:

$$\alpha(f) \wedge \alpha(g) = o , \qquad \beta(f) \wedge \beta(g) = o$$

und damit wegen $\alpha(f \wedge g) \leqq \alpha(f) \wedge \alpha(g)$ und $\beta(f \wedge g) \leqq \beta(f) \wedge \beta(g)$:

$$\alpha(f \wedge g) = \alpha(f) \wedge \alpha(g) , \qquad \beta(f \wedge g) = \beta(f) \wedge \beta(g) .$$

Mithin ist $\mathfrak{z}$ eine kompatible Untermenge von $\mathscr{C}$, und folglich ist, da $\mathscr{C}$ ein vollständiges induktives Gruppoid und ob $\mathscr{C}^{\cdot}$ bezüglich $\mathscr{C}^{\leqq}$ nach oben beschränkt ist, $\mathfrak{z}$ bezüglich $\mathscr{C}^{\leqq}$ nach oben beschränkt. Demnach existiert das Supremum bezüglich $\mathscr{C}^{\leqq}$ von $\mathfrak{z}$. Wegen

$$\alpha(\bigvee \mathfrak{z}) = \bigvee \check{\alpha}(\mathfrak{z}) = \bigvee \text{at } \mathscr{C} = u$$

und

$$\beta(\bigvee \mathfrak{z}) = \bigvee \check{\beta}(\mathfrak{z}) = \bigvee \text{at } \mathscr{C} = u$$

ist dann $\bigvee \mathfrak{F}$ ein Element von $\mathfrak{G}(\mathscr{C})$. Ist nun a ein Atom von $\mathscr{C}$ und f das Element von $\mathfrak{F}$ mit $\alpha(f) = a$, so gilt

$$f \leqq \bigvee \mathfrak{F}, (\bigvee \mathfrak{F}) \circ a \leqq \bigvee \mathfrak{F}, \alpha(f) = a = \alpha((\bigvee \mathfrak{F}) \circ a)$$

und damit

$$f = (\bigvee \mathfrak{F}) \circ a .$$

Mithin gilt

$$\mathfrak{F} = \{(\bigvee \mathfrak{F}) \circ a \mid a \in \operatorname{at} \mathscr{C}\} = \Phi(\bigvee \mathfrak{F}) .$$

Demnach ist Φ eine Funktion von $\mathfrak{G}(\mathscr{C})$ *auf* $\mathfrak{L}(\mathscr{Q}(\mathscr{C}))$. Folglich ist Φ ein Isomorphismus von $\mathscr{G}(\mathscr{C})$ auf $\mathscr{L}(\mathscr{Q}(\mathscr{C}))$. ▮

Korollar zu Satz 11. *Für jedes atomare induktive Gruppoid $\mathscr{C}$, für das* ob $\mathscr{C}^{\cdot}$ *bezüglich $\mathscr{C}^{\leqq}$ nach oben beschränkt ist und $\bigvee$* at $\mathscr{C}$ *= u gilt, gilt:*

1. *$\mathscr{G}(\mathscr{C})$ ist zu einer Untergruppe von $\mathscr{L}(\mathscr{Q}(\mathscr{C}))$ isomorph.*
2. *$\mathscr{G}(\mathscr{C})$ ist zu $\mathscr{L}(\mathscr{Q}(\mathscr{C}))$ isomorph, wenn $\mathscr{C}$ ein vollständiges induktives Gruppoid ist.* ▮

Mit Hilfe von Satz 11 beweist man leicht:

Satz 12. *Für jede idempotente Algebra* A, *für die eine von* 0 *und* 1 *verschiedene natürliche Zahl n, eine natürliche Zahl k aus* $\{1, 2, \ldots, n\}$, *eine n-äre definierende Operation ω von* A *und Elemente a, b von* |A| *mit*

$$\omega(\underset{1}{a}, \underset{2}{a}, \ldots, \underset{k-1}{a}, \underset{k}{b}, \underset{k+1}{a}, \ldots, \underset{n}{a}) \neq \omega(\underset{1}{b}, \underset{2}{b}, \ldots, \underset{k-1}{b}, \underset{k}{a}, \underset{k+1}{b}, \ldots, \underset{n}{b}) \notin \{a, b\}$$

existieren, ist das induktive Gruppoid $\mathscr{A}$(A) *kein vollständiges induktives Gruppoid.*

Beweis. Es sei A eine idempotente Algebra, und es existiere eine von 0 und 1 verschiedene natürliche Zahl n, eine natürliche Zahl k aus $\{1, 2, \ldots, n\}$, eine n-äre definierende Operation ω von A und Elemente a, b von |A| mit

$$\omega(\underset{1}{a}, \underset{2}{a}, \ldots, \underset{k-1}{a}, \underset{k}{b}, \underset{k+1}{a}, \ldots, \underset{n}{a}) \neq \omega(\underset{1}{b}, \underset{2}{b}, \ldots, \underset{k-1}{b}, \underset{k}{a}, \underset{k+1}{b}, \ldots, \underset{n}{b}) \notin \{a, b\} .$$

Dann gilt

$$\operatorname{at} \mathscr{A}(\mathsf{A}) = \{\{x\} \mid x \in |\mathsf{A}|\} ,$$

und $\mathscr{A}$(A) ist ein atomares induktives Gruppoid, für das ob $\mathscr{A}(\mathsf{A})^{\cdot}$ bezüglich $\mathscr{A}(\mathsf{A})^{\leqq}$ nach oben beschränkt ist und $\bigvee$ at $\mathscr{A}(\mathsf{A}) = \mathrm{id}_{|\mathsf{A}|}$ gilt. Angenommen $\mathscr{A}$(A) wäre ein vollständiges induktives Gruppoid. Dann sei ψ die Funktion von $\{a\}$ auf $\{b\}$. Es ist dann

$$\{\psi, \psi^{-1}\} \cup \{\mathrm{id}_{\{x\}} \mid x \in |\mathsf{A}| \setminus \{a, b\}\}$$

ein Element von $\mathfrak{L}(\mathscr{Q}(\mathscr{A}(\mathsf{A})))$, und folglich gibt es einen Automorphismus φ von A mit $\varphi(a) = b$, $\varphi(b) = a$ und $\varphi(x) = x$ für jedes Element x von |A| mit $x \neq a$ und $x \neq b$ (Satz 11). Wegen

$$\omega(\underset{1}{b}, \underset{2}{b}, \ldots, \underset{k-1}{b}, \underset{k}{a}, \underset{k+1}{b}, \ldots, \underset{n}{b}) \notin \{a, b\}$$

gilt dann aber

$$\omega(\underset{1}{b}, \underset{2}{b}, \ldots, \underset{k-1}{b}, \underset{k}{a}, \underset{k+1}{b}, \ldots, \underset{n}{b}) = \varphi\big(\omega(\underset{1}{b}, \underset{2}{b}, \ldots, \underset{k-1}{b}, \underset{k}{a}, \underset{k+1}{b}, \ldots, \underset{n}{b})\big)$$
$$= \omega\big(\underset{1}{\varphi(b)}, \underset{2}{\varphi(b)}, \ldots, \underset{k-1}{\varphi(b)}, \underset{k}{\varphi(a)}, \underset{k+1}{\varphi(b)}, \ldots, \underset{n}{\varphi(b)}\big)$$
$$= \omega(\underset{1}{a}, \underset{2}{a}, \ldots, \underset{k-1}{a}, \underset{k}{b}, \underset{k+1}{a}, \ldots, \underset{n}{a})$$

im Widerspruch zu

$$\omega(\underset{1}{b}, \underset{2}{b}, \ldots, \underset{k-1}{b}, \underset{k}{a}, \underset{k+1}{b}, \ldots, \underset{n}{b}) \neq \omega(\underset{1}{a}, \underset{2}{a}, \ldots, \underset{k-1}{a}, \underset{k}{b}, \underset{k+1}{a}, \ldots, \underset{n}{a}) .$$

Mithin ist das induktive Gruppoid $\mathcal{A}(\mathsf{A})$ kein vollständiges induktives Gruppoid. ▮

Man beweist weiter leicht mit Hilfe von Satz 11, daß $\mathcal{A}(\mathsf{V})$ für jeden Verband V mit card $|\mathsf{V}| \geqq 2$[1]) kein vollständiges induktives Gruppoid ist.

Satz 13. *Für jeden Verband* V *mit* card $|\mathsf{V}| \geqq 2$ *ist das induktive Gruppoid* $\mathcal{A}(\mathsf{V})$ *kein vollständiges induktives Gruppoid.*

Beweis. Es sei V ein Verband, und es gelte card $|\mathsf{V}| \geqq 2$. Dann gilt

$$\text{at}\, \mathcal{A}(\mathsf{V}) = \{\{x\} \mid x \in |\mathsf{V}|\} ,$$

und $\mathcal{A}(\mathsf{V})$ ist ein atomares induktives Gruppoid, für das ob $\mathcal{A}(\mathsf{V})^{\cdot}$ bezüglich $\mathcal{A}(\mathsf{V})^{\leqq}$ nach oben beschränkt ist und $\bigvee$ at $\mathcal{A}(\mathsf{V}) = \text{id}_{|\mathsf{V}|}$ gilt. Weiter gibt es Elemente a, b von $|\mathsf{V}|$ mit $a \not< b$. Angenommen $\mathcal{A}(\mathsf{V})$ wäre ein vollständiges induktives Gruppoid. Dann sei ψ die Funktion von $\{a\}$ auf $\{b\}$. Es ist dann

$$\{\psi, \psi^{-1}\} \cup \{\text{id}_{\{x\}} \mid x \in |\mathsf{V}| \setminus \{a, b\}\}$$

ein Element von $\mathfrak{L}(\mathcal{Q}(\mathcal{A}(\mathsf{V})))$, und folglich gibt es einen Automorphismus φ von V mit $\varphi(a) = b$, $\varphi(b) = a$ und $\varphi(x) = x$ für jedes Element x von $|\mathsf{V}|$ mit $x \neq a$ und $x \neq b$ (Satz 11). Wegen $a < b$ gilt dann aber auch $\varphi(a) < \varphi(b)$, d. h. $b < a$ im Widerspruch zu $a < b$. Mithin ist das induktive Gruppoid $\mathcal{A}(\mathsf{V})$ kein vollständiges induktives Gruppoid. ▮

Es sei $\mathcal{C}$ ein atomares induktives Gruppoid, mor $\mathcal{C}^{\cdot}$ eine endliche Menge und ob $\mathcal{C}^{\cdot}$ bezüglich $\mathcal{C}^{\leqq}$ nach oben beschränkt, es gelte $\bigvee \{a \mid a \in \text{at}\, \mathcal{C}, a \leqq e\} = e$ für jedes Einselement e von $\mathcal{C}^{\cdot}$, also insbesondere $\bigvee$ at $\mathcal{C} = u$. Ist dann $\mathcal{C}$ speziell ein vollständiges induktives Gruppoid, so gilt card $\mathfrak{G}(\mathcal{C})$ = card $\mathfrak{L}(\mathcal{Q}(\mathcal{C}))$ (Satz 11). Wie der folgende abschließende Satz zeigt, ist dann aber auch umgekehrt $\mathcal{C}$ ein vollständiges induktives Gruppoid, wenn card $\mathfrak{G}(\mathcal{C})$ = card $\mathfrak{L}(\mathcal{Q}(\mathcal{C}))$ gilt.

Satz 14. *Für jedes atomare induktive Gruppoid* $\mathcal{C}$ *gilt:* $\mathcal{C}$ *ist ein vollständiges induktives Gruppoid, wenn* mor $\mathcal{C}^{\cdot}$ *eine endliche Menge und* ob $\mathcal{C}^{\cdot}$ *bezüglich* $\mathcal{C}^{\leqq}$ *nach oben beschränkt ist sowie*

$$\bigvee \{a \mid a \in \text{at}\, \mathcal{C},\ a \leqq e\} = e \quad \textit{für jedes Einselement } e \textit{ von } \mathcal{C}^{\cdot}.$$ [2])

[1]) Verbände V werden hier als Algebren mit der Durchschnittsbildung $\wedge$ und der Vereinigungsbildung $\vee$ als definierende Operationen aufgefaßt. Dann ist durch

$$a \leqq b \quad \text{genau dann, wenn } a \wedge b = a \qquad (a, b \in |\mathsf{V}|)$$

eine Halbordnungsrelation $\leqq$ in $|\mathsf{V}|$ definiert.

[2]) Insbesondere gilt dann $\vee$ at $\mathcal{C} = u$.

und

$$\text{card } \mathfrak{S}(\mathscr{C}) = \text{card } \mathfrak{L}(Q(\mathscr{C}))$$

gilt.

Beweis. Es seien $\mathscr{C}$ ein atomares induktives Gruppoid, mor $\mathscr{C}^{\cdot}$ eine endliche Menge und ob $\mathscr{C}^{\cdot}$ bezüglich $\mathscr{C}^{\leqq}$ nach oben beschränkt, und es gelte

$$\bigvee \{a \mid a \in \text{at } \mathscr{C}, \quad a \leqq e\} = e \quad \text{für jedes Einselement } e \text{ von } \mathscr{C}^{\cdot}$$

sowie card $\mathfrak{S}(\mathscr{C}) = \text{card } \mathfrak{L}(Q(\mathscr{C}))$. Dann sei Φ die Funktion über $\mathfrak{S}(\mathscr{C})$ mit

$$\Phi(f) = \{f \circ a \mid a \in \text{at } \mathscr{C}\} \quad \text{für jedes Element } f \text{ von } \mathfrak{S}(\mathscr{C}).$$

Es ist dann Φ ein eineindeutiger Homomorphismus von $\mathscr{S}(\mathscr{C})$ in $\mathscr{L}(Q(\mathscr{C}))$ (Satz 11) und folglich, da $\mathfrak{S}(\mathscr{C})$ eine endliche Menge ist und card $\mathfrak{S}(\mathscr{C}) = \text{card } \mathfrak{L}(Q(\mathscr{C}))$ gilt, Φ ein Isomorphismus von $\mathscr{S}(\mathscr{C})$ auf $\mathscr{L}(Q(\mathscr{C}))$.

Nun sei K eine kompatible Untermenge von $\mathscr{C}$. Dann sei $\mathfrak{s}'$ die Untermenge von mor $\mathscr{C}^{\cdot}$ mit

$$\mathfrak{s}' = \{f \circ a \mid f \in K, \quad a \in \text{at } \mathscr{C}, \quad a \leqq \alpha(f)\}.$$

Es sei nun f ein Element von K und a ein Atom von $\mathscr{C}$, und es gelte $a \leqq \alpha(f)$. Dann gilt $\alpha(f \circ a) = a \in \text{at } \mathscr{C}$. Weiter gilt $\beta(f \circ a) \neq o$; denn gälte $\beta(f \circ a) = o$, so wäre $f \circ a = o$ und damit $a = \alpha(f \circ a) = \alpha(o) = o$ im Widerspruch zu $a \in \text{at } \mathscr{C}$. Demnach gibt es ein Atom a' von $\mathscr{C}$ mit $a' \leqq \beta(f \circ a)$, und es gilt dann (Hilfssatz)

$$\beta(f^{-1} \circ a') \leqq \beta(f^{-1} \circ \beta(f \circ a)) = \beta(a \circ f^{-1}) = a.$$

Gälte $\beta(f^{-1} \circ a') = o$, so wäre $f^{-1} \circ a' = o$ und damit wegen $a' \leqq \beta(f \circ a) \leqq \beta(f)$:

$$a' = \beta(f) \circ a' = (f \cdot f^{-1}) \circ a' = f \circ (f^{-1} \circ a') = f \circ o = o$$

im Widerspruch zu $a' \in \text{at } \mathscr{C}$. Mithin gilt $o < \beta(f^{-1} \circ a') \leqq a \in \text{at } \mathscr{C}$, also $\beta(f^{-1} \circ a') = a$ und damit wegen $a' \leqq \beta(f \circ a) \leqq \beta(f) = \alpha(f^{-1})$ notwendig $f \circ a = f \circ \beta(f^{-1} \circ a') = a' \circ f$ (Hilfssatz), also $\beta(f \circ a) = \beta(a' \circ f) = a' \in \text{at } \mathscr{C}$. Folglich gilt $\mathfrak{s}' \subseteqq \text{mor } Q(\mathscr{C})$. Es seien nun f', g' Elemente von $\mathfrak{s}'$,und es gelte $\alpha(f') = \alpha(g')$ bzw. $\beta(f') = \beta(g')$. Dann existieren Elemente f, g von K mit $f' \leqq f$, $g' \leqq g$, und es gilt dann

$$\alpha(f') = \alpha(g') \leqq \alpha(f) \wedge \alpha(g) = \alpha(f \wedge g)$$

bzw.

$$\beta(f') = \beta(g') \leqq \beta(f) \wedge \beta(g) = \beta(f \wedge g),$$

also wegen $f' \leqq f, f \wedge g \leqq f$ sowie $g' \leqq g, f \wedge g \leqq g$:

$$f' \leqq f \wedge g, \quad g' \leqq f \wedge g.$$

Wegen $\alpha(f') = \alpha(g')$ bzw. $\beta(f') = \beta(g')$ gilt somit $f' = g'$. Es sei nun $\mathfrak{L}$ die Menge aller Untermengen $\bar{\mathfrak{s}}$ von mor $Q(\mathscr{C})$ mit $\mathfrak{s}' \subseteqq \bar{\mathfrak{s}}$ sowie $\alpha(\bar{f}) \neq \alpha(\bar{g})$ und $\beta(\bar{f}) \neq \beta(\bar{g})$ für beliebige verschiedene Elemente $\bar{f}, \bar{g}$ von $\bar{\mathfrak{s}}$. Dann ist $\mathfrak{s}'$ ein Element von $\mathfrak{L}$. Folglich ist $\mathfrak{L}$ eine nichtleere Menge von Untermengen der endlichen Menge mor $Q(\mathscr{C})$, und es gibt somit maximale Elemente der halbgeordneten Menge $[\mathfrak{L}, (\subseteqq) \cap (\mathfrak{L} \times \mathfrak{L})]$. Ein solches sei $\mathfrak{s}$. Dann sei $(s_a)_{a \in \text{at } \mathscr{C}}$ die wie folgt definierte Familie von Morphismen von $Q(\mathscr{C})$:

1. Ist a ein Element von $\check{\alpha}(\mathfrak{s})$, so sei s_a das Element s von $\mathfrak{s}$ mit $\alpha(s) = a$.
2. Es sei nun a ein Element von $\check{\beta}(\mathfrak{s})$, aber kein Element von $\check{\alpha}(\mathfrak{s})$.

Dann gibt es genau ein Element t_1 von $\mathfrak{S}$ mit $\beta(t_1) = a$. Wegen $\alpha(t_1) \in \check{\alpha}(\mathfrak{S})$ gilt dann notwendig $\alpha(t_1) \neq a$. Ist $\alpha(t_1)$ ein Element von $\check{\beta}(\mathfrak{S})$, so existiert genau ein Element t_2 von $\mathfrak{S}$ mit $\beta(t_2) = \alpha(t_1)$, und es gilt dann wegen $\alpha(t_2) \in \check{\alpha}(\mathfrak{S})$ auch $\alpha(t_2) \neq a$ sowie wegen $\beta(t_2) = \alpha(t_1) \neq a = \beta(t_1)$ auch $\alpha(t_2) \neq \alpha(t_1)$. Ist wieder $\alpha(t_2)$ ein Element von $\check{\beta}(\mathfrak{S})$, so gibt es genau ein Element t_3 von $\mathfrak{S}$ mit $\beta(t_3) = \alpha(t_2)$, und es gilt dann $\alpha(t_3) \neq a$, $\alpha(t_3) \neq \alpha(t_1)$, $\alpha(t_3) \neq \alpha(t_2)$ wegen $\alpha(t_3) \in \check{\alpha}(\mathfrak{S})$, $\beta(t_3) = \alpha(t_2) \neq a = \beta(t_1)$, $\beta(t_3) = \alpha(t_2) \neq \alpha(t_1) = \beta(t_2)$. Ist wieder $\alpha(t_3)$ ein Element von $\check{\beta}(\mathfrak{S})$, so gibt es genau ein Element t_4 von $\mathfrak{S}$ mit $\beta(t_4) = \alpha(t_3)$, und es gilt dann $\alpha(t_4) \neq a, \alpha(t_4) \neq \alpha(t_1), \alpha(t_4) \neq \alpha(t_2), \alpha(t_4) \neq \alpha(t_3)$ wegen $\alpha(t_4) \in \check{\alpha}(\mathfrak{S}), \beta(t_4) = \alpha(t_3) \neq a = \beta(t_1), \beta(t_4) = \alpha(t_3) \neq \alpha(t_1) = \beta(t_2), \beta(t_4) = \alpha(t_3) \neq \alpha(t_2) = \beta(t_3)$. So fährt man fort. Nun ist aber $\check{\alpha}(\mathfrak{S})$ eine endliche Menge. Demnach gibt es offenbar genau eine von 0 verschiedene natürliche Zahl n und genau eine Familie $(t_\nu)_{\nu \in \{1, 2, \ldots, n\}}$ von Elementen von $\mathfrak{S}$ mit

$$\alpha(t_n) \notin \check{\beta}(\mathfrak{S})\,, \quad \beta(t_n) = \alpha(t_{n-1})\,, \quad \beta(t_{n-1}) = \alpha(t_{n-2}), \ldots, \beta(t_2) = \alpha(t_1)\,, \quad \beta(t_1) = a,$$

und es sei dann s_a der Morphismus $(t_1 \cdot t_2 \cdot \ldots \cdot t_n)^{-1}$ von $Q(\mathscr{C})$.

Der Fall, daß ein Atom a von $\mathscr{C}$ weder ein Element von $\check{\alpha}(\mathfrak{S})$ noch ein Element von $\check{\beta}(\mathfrak{S})$ ist, kann nicht eintreten, da sonst $\mathfrak{S} \subset \mathfrak{S} \cup \{a\}$ und $\mathfrak{S} \cup \{a\} \in \mathfrak{L}$ gelten würde im Widerspruch dazu, daß $\mathfrak{S}$ ein maximales Element von $[\mathfrak{L}, (\subseteqq) \cap (\mathfrak{L} \times \mathfrak{L})]$ ist.

Nun sei $\bar{\mathfrak{S}}$ die Menge $\{s_a \mid a \in \operatorname{at} \mathscr{C}\}$ von Morphismen von $Q(\mathscr{C})$. Dann gilt $\alpha(s_a) = a$ für jedes Atom a von $\mathscr{C}$. Folglich gibt es für jedes Atom a von $\mathscr{C}$ genau ein Element s von $\bar{\mathfrak{S}}$ mit $\alpha(s) = a$. Weiter gilt $\mathfrak{S} \subseteqq \bar{\mathfrak{S}}$. Angenommen es gälte $\mathfrak{S} \subset \bar{\mathfrak{S}}$. Dann existiert ein Element s von $\bar{\mathfrak{S}} \setminus \mathfrak{S}$, und es gibt dann genau eine von 0 verschiedene natürliche Zahl n und genau eine Familie $(t_\nu)_{\nu \in \{1, 2, \ldots, n\}}$ von Elementen von $\mathfrak{S}$ mit

$$\alpha(t_n) \notin \check{\beta}(\mathfrak{S})\,, \quad \beta(t_n) = \alpha(t_{n-1})\,, \quad \beta(t_{n-1}) = \alpha(t_{n-2}), \ldots, \beta(t_2) = \alpha(t_1)\,, \quad \beta(t_1) \notin \hat{\alpha}(\mathfrak{S})$$

und $s = (t_1 \cdot t_2 \cdot \ldots \cdot t_n)^{-1}$. Dann gilt

$$\alpha(s) = \beta(t_1 \cdot t_2 \cdot \ldots \cdot t_n) = \beta(t_1) \notin \check{\alpha}(\mathfrak{S}), \quad \beta(s) = \alpha(t_1 \cdot t_2 \cdot \ldots t_n) = \alpha(t_n) \notin \check{\beta}(\mathfrak{S})\,.$$

Folglich gilt $\mathfrak{S} \subset \mathfrak{S} \cup \{s\}$ und $\mathfrak{S} \cup \{s\} \in \mathfrak{L}$ im Widerspruch dazu, daß $\mathfrak{S}$ ein maximales Element von $[\mathfrak{L}, (\subseteqq) \cap (\mathfrak{L} \times \mathfrak{L})]$ ist. Demnach stimmt $\bar{\mathfrak{S}}$ mit $\mathfrak{S}$ überein, und folglich gilt $\check{\alpha}(\mathfrak{S}) = \{\alpha(s_a) \mid a \in \operatorname{at} \mathscr{C}\} = \operatorname{at} \mathscr{C}$, $\operatorname{card} \check{\beta}(\mathfrak{S}) = \operatorname{card} \check{\alpha}(\mathfrak{S})$, $\check{\beta}(\mathfrak{S}) \subseteqq \operatorname{at} \mathscr{C} = \check{\alpha}(\mathfrak{S})$ und damit, da $\check{\alpha}(\mathfrak{S})$ eine endliche Menge ist, $\check{\beta}(\mathfrak{S}) = \check{\alpha}(\mathfrak{S}) = \operatorname{at} \mathscr{C}$. Mithin ist $\mathfrak{S}$ ein Element von $\mathfrak{L}(Q(\mathscr{C}))$.

Da nun Φ ein Isomorphismus von $\mathscr{S}(\mathscr{C})$ auf $\mathscr{L}(Q(\mathscr{C}))$ ist, so gibt es ein Element k von $\mathfrak{G}(\mathscr{C})$ mit $\Phi(k) = \mathfrak{S}$, also $\{k \circ a \mid a \in \operatorname{at} \mathscr{C}\} = \mathfrak{S}$. Dann ist aber k eine obere Schranke bezüglich $\mathscr{C}^{\leqq}$ von $\mathfrak{S}$. Nun sei f ein Element von K. Es gilt dann

$$\{f \circ a \mid a \in \operatorname{at} \mathscr{C}\,, \quad a \leqq \alpha(f)\} \subseteqq \mathfrak{S}' \subseteqq \mathfrak{S}\,,$$

also $f \circ a \leqq k$ für jedes Atom a von $\mathscr{C}$ mit $a \leqq \alpha(f)$. Folglich existiert das Supremum bezüglich $\mathscr{C}^{\leqq}$ von $\{f \circ a \mid a \in \operatorname{at} \mathscr{C}, a \leqq \alpha(f)\}$, und es gilt

$$\bigvee \{f \circ a \mid a \in \operatorname{at} \mathscr{C}\,, \; a \leqq \alpha(f)\} \leqq k\,.$$

Wegen $f \circ a \leqq f$ für jedes Element a von $\operatorname{at} \mathscr{C}$ gilt nun aber auch

$$\bigvee \{f \circ a \mid a \in \operatorname{at} \mathscr{C}\,, \quad a \leqq \alpha(f)\} \leqq f$$

und folglich wegen

$$\alpha\left(\bigvee\{f\circ a \mid a\in \operatorname{at}\mathscr{C},\ a\leqq\alpha(f)\}\right) = \bigvee\{\alpha(f\circ a)\mid a\in \operatorname{at}\mathscr{C},\ a\leqq\alpha(f)\}$$
$$= \bigvee\{a\mid a\in \operatorname{at}\mathscr{C},\ a\leqq\alpha(f)\} = \alpha(f)$$

sogar

$$\bigvee\{f\circ a\mid a\in \operatorname{at}\mathscr{C},\ a\leqq\alpha(f)\} = f\,.$$

Mithin gilt $f \leqq k$. Demnach ist k eine obere Schranke bezüglich $\mathscr{C}^{\leqq}$ von K. Folglich ist jede kompatible Untermenge von $\mathscr{C}$ bezüglich $\mathscr{C}^{\leqq}$ nach oben beschränkt, also $\mathscr{C}$ ein vollständiges induktives Gruppoid. ▮

Literatur

[1] G. Birkhoff; O. Frink: Representation of lattices by sets, Trans. Amer. math. Soc. **64** (1948), 299—316.
[2] C. Ehresmann: Gattungen von lokalen Strukturen, Jber. Deutsch. Math.-Verein. **60** (1957), 49—77.
[3] —: Catégories inductives et pseudogroupes, Ann. Inst. Fourier (Grenoble) **10** (1960), 307—332.
[4] B. Fichtner: Über die zu Gruppen gehörigen induktiven Gruppoide. I, Math. Nachr. **44** (1970), 313—339.
[5] M. Hasse; L. Michler: Theorie der Kategorien, Berlin 1966.
[6] H.-J. Hoehnke: Über die Erzeugung von monomialen Gruppendarstellungen durch Brandtsche Gruppoide, Math. Zeitschr. **77** (1961), 68—80.
[7] L. Michler; J. Schreckenberger: Über die induktiven Gruppoide der partiellen Automorphismen von Algebren, Math. Nachr. **72** (1976), 219—230.
[8] M. Suzuki: Structure of a group and the structure of its lattice of subgroups, Berlin/Göttingen/Heidelberg 1956.

Über geometrische Äquivalenz von Symmetrie-Gruppoiden

von Konrad Fichtner, Berlin

1. Einleitung

In der Mathematik und ihren Anwendungen kommt es nicht selten vor, daß man eine Definition braucht, wann gewisse mathematische Strukturen „im wesentlichen gleich", das heißt gleichwertig von einem gewissen Standpunkt aus, sind. Isomorphie ist ein Beispiel dafür: Zwei algebraische Strukturen sind im algebraisch abstrakten Sinne gleich, wenn sie isomorph sind.

Ziel der vorliegenden Betrachtungen ist es, für Symmetrie-Gruppoide verschiedene derartige „Gleichheitsbegriffe" anzugeben. Unter einem Gruppoid verstehen wir dabei ein Gruppoid im Sinne von Brandt bzw. Ehresmann, also eine Kategorie, in der jeder Morphismus invertierbar ist (vergleiche [4]). Es ist plausibel, daß die in Anwendungen benötigten Gleichheitsbegriffe für Gruppoide im allgemeinen feiner unterscheiden müssen als die Isomorphie. Denn bei Betrachtung bis auf Isomorphie ist ein Gruppoid eindeutig bestimmt durch die zu seinen Zusammenhangsbereichen gehörenden maximalen Teilgruppen und die Mächtigkeiten der Objektmengen.

Die Notwendigkeit des Vergleichs von Gruppoiden ergab sich bei der Untersuchung von Gruppoiden, die die Symmetrieverhältnisse bei aus Schichten aufgebauten Kristallen beschreiben, den von Dornberger-Schiff [1] untersuchten OD-Strukturen. Wechler [6] betrachtet Brandtsche Gruppoide, die die Symmetrieverhältnisse bei gewissen Molekülen beschreiben. Auch in diesem Fall ist die Einführung eines für die dort entwickelte Darstellungstheorie typischen „Gleichheitsbegriffes" für Gruppoide unumgänglich. In [6] wird das allerdings implizit und auf Umwegen erreicht.

Zunächst sollen Symmetriegruppen diskutiert werden, weil sich in diesem Fall die Problematik besonders leicht darlegen läßt und weil sich daraus auch der Ansatz für die Behandlung des schwierigeren Gruppoid-Falls ergibt.

2. Geometrische Äquivalenz von Symmetriegruppen

Die Symmetrie eines Objektes im n-dimensionalen Euklidischen Raum R^n wird durch die Gruppe derjenigen Bewegungen des R^n beschrieben, die das Objekt mit sich zur Deckung bringen. Das Objekt kann z. B. ein Würfel im R^3, die Idealstruktur eines Kristalls oder allgemeiner eine Funktion f mit dem R^n als Argumentbereich sein.

Daß isomorphe Gruppen dabei wesentlich verschiedene Symmetrieverhältnisse beschreiben können, sieht man an folgendem Beispiel im R^2.

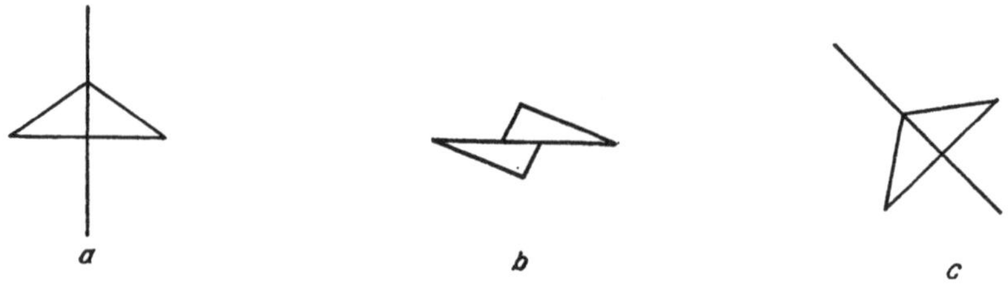

Abb. 1 Drei Objekte im R^2

Die Figur in Abb. 1a) besitzt neben der identischen Abbildung nur eine Spiegelung an der eingezeichneten Geraden als Symmetrieoperation, während die Figur in Abb. 1b) neben der identischen Abbildung nur eine Drehung um 180° als Symmetrieoperation besitzt. Die Symmetriegruppen sind also wesentlich verschieden, obwohl beide aus je zwei Elementen bestehen und damit isomorph sind. Sehr nahe stehen sich dagegen Symmetriegruppen, die die Symmetrie von zwei Objekten beschreiben, die durch eine Bewegung ineinander übergehen, z. B. die Symmetriegruppen der Figuren in den Abbildungen 1a) und 1c). Derartige Gruppen sind in der Gruppe aller Bewegungen des R^n konjugierte Untergruppen. Sei $B^{(n)}$ die Gruppe aller Bewegungen des R^n, $A^{(n)}$ die Gruppe aller affinen Transformationen des R^n. Dann legen wir fest:

Definition 1. Zwei Gruppen $G_1, G_2 \subset B^{(n)}$ heißen *geometrisch äquivalent* in einer Menge U von affinen Transformationen, $U \subseteqq A^{(n)}$, wenn es ein $\alpha \in U$ gibt, so daß $G_1 = \alpha G_2 \alpha^{-1}$ gilt.

Jetzt soll noch angegeben werden, welche Rolle dieser Begriff in der Kristallographie spielt.

Eine *kristallographische Bewegungsgruppe* im R^3 ist eine Untergruppe von $B^{(3)}$, deren Translationsuntergruppe von drei nichtkomplanaren Translationen erzeugt wird. Von praktischem Interesse bei kristallographischen Bewegungsgruppen sind zwei Fälle für U:

$$U_1 = \{\beta \mid \beta = (H, \mathfrak{t}) \in B^{(3)} \wedge \det H = +1\} ,$$
$$U_2 = \{\alpha \mid \alpha = (H, \mathfrak{t}) \in A^{(3)} \wedge \det H > 0\} ;$$

$H, \mathfrak{t}$ sind *homogener* und *translativer* Anteil einer affinen Transformation. U_1 ist also die Gruppe aller eigentlichen Bewegungen des R^3, während U_2 die Gruppe aller affinen Transformationen ist, deren homogener Anteil eine positive Determinante hat.

Die ideale Symmetrie einer dreidimensional-periodischen Kristallstruktur wird durch eine kristallographische Bewegungsgruppe beschrieben. Allerdings interessiert diese Gruppe nur bis auf geometrische Äquivalenz in U_1. Bezüglich der geometrischen Äquivalenz in U_2 zerfällt die Menge aller kristallographischen Bewegungsgruppen im R^3 in 230 Teilmengen, die sogenannten *Raumgruppen*.

Die Symmetrie einer dreidimensional-periodischen Kristallstruktur wird — bis auf geometrische Äquivalenz in U_1 — angegeben durch die Raumgruppe und die Gitterkonstanten (das sind 6 wohldefinierte Parameter).

3. Geometrische Äquivalenz von Symmetrie-Gruppoiden

Was man unter Symmetriebeschreibung durch Gruppoide verstehen muß, ist in [2] ausführlich erläutert. Kurz gesagt, es geht jetzt im Gegensatz zum obigen Fall um mehrere Objekte, und die Morphismen des Gruppoids sind die Bewegungen des R^n, die eines der Objekte mit sich oder mit einem der anderen Objekte zur Deckung bringen. Sei

$$\text{mor}\ \mathsf{B} = \{(i, j, \beta) \mid i, j \in \text{ob}\ \mathsf{B} \wedge \beta \in B^{(3)}\}\ ,$$

$$\text{mor}\ \mathsf{A} = \{(i, j, \alpha) \mid i, j \in \text{ob}\ \mathsf{A} \wedge \alpha \in A^{(3)}\}\ .$$

Die Objektmengen sollen im Moment noch nicht spezifiziert werden. Ein Symmetrie-Gruppoid G ist ein Teilgruppoid von B. Die Menge ob G ist das System von Objekten, dessen Symmetrieeigenschaften durch G beschrieben werden. $(i, j, \beta) \in \mathsf{G}$ bedeutet, daß β eine Bewegung ist, die das Objekt i mit dem Objekt j zur Deckung bringt.

Der Grundgedanke der geometrischen Äquivalenz von zwei Symmetrie-Gruppoiden G_1 und G_2 ist folgender: Geometrische Äquivalenz soll dann vorliegen, wenn das System von Objekten, dessen Symmetrie G_1 beschreibt, durch eine Transformation $\alpha \in U \subseteqq \mathsf{A}$ als Ganzes in das System von Objekten übergeht, dessen Symmetrie durch G_2 beschrieben wird. Das führt auf die folgende

Definition 2. Zwei Teilgruppoide $\mathsf{G}_1, \mathsf{G}_2 \subset \mathsf{B}$ heißen *geometrisch äquivalent* in $U \subseteqq \mathsf{A}$, wenn

(1) ein Isomorphismus $F: \mathsf{G}_1 \to \mathsf{G}_2$ existiert,

(2) (I_1, ω, I_2F) eine natürliche Transformation ist, wobei $I_1: \mathsf{G}_1 \to \mathsf{A}$ und $I_2: \mathsf{G}_2 \to \mathsf{A}$ Inklusionsfunktoren sind und $\omega(i) \in U$ für alle $i \in$ ob G_1 ist,

(3) aus $\omega(i) = (i, i', \alpha_i)$ und $\omega(j) = (j, j', \alpha_j)$ mit $i, j \in$ ob G_1 und $i', j' \in$ ob G_2 folgt $\alpha_i = \alpha_j$.

Die kategorientheoretischen Begriffe Isomorphismus, natürliche Transformation und Inklusionsfunktor sind in [5, S. 7, S. 12 bzw. S. 6] definiert. (I_1, ω, I_2F) ist eine natürliche Äquivalenz im Sinne von [4, S. 201].

Den Begriff der geometrischen Äquivalenz kann man in folgender Weise abschwächen: Geometrische Äquivalenz bis auf Translationen liegt vor, wenn die Relativlage zweier Objekte des von G_1 beschriebenen Systems von Objekten sich von der Relativlage der entsprechenden Objekte des von G_2 beschriebenen Systems von Objekten um eine Translation unterscheiden darf.

Definition 3. Zwei Teilgruppoide $\mathsf{G}_1, \mathsf{G}_2 \subset \mathsf{B}$ mit ob $\mathsf{B} = \{0, \pm 1, \pm 2, \ldots\}$ heißen *bis auf Translationen geometrisch äquivalent in* $U \subseteqq \mathsf{A}$, wenn die Bedingungen (1) und (2) von Definition 2 erfüllt sind sowie

(3′) Aus $\omega(i) = \big(i, i', (H_i, \mathfrak{t}_i)\big)$ und $\omega(j) = \big(j, j', (H_j, \mathfrak{t}_j)\big)$ mit $i, j \in$ ob G_1 und $i', j' \in$ ob G_2 folgt $H_i = H_j$,

(4) Es gilt $|F(i) - F(j)| = |i - j|$ für alle $i, j \in$ ob G_1.

4. Anwendungen des Begriffs der geometrischen Äquivalenz von Gruppoiden

Der Begriff der geometrischen Äquivalenz soll auf zwei Beispiele angewendet werden, einmal auf die in [6] betrachteten BRANDTschen Gruppoide, zum anderen auf die sogenannten OD-Gruppoide [1—3].

In [6] wird die Symmetrie eines Moleküls M, das aus mehreren gleichartigen Teilen besteht, durch ein BRANDTsches Gruppoid G beschrieben. G ist Teilgruppoid von B, wenn wir voraussetzen, daß die Menge der Teile von M (das ist ob G) in ob B enthalten ist. Wenden wir auf G die Definition 2 an, so erhalten wir für

$$U_1 = \{(i, j, \beta) \mid i, j \in \text{ob } \mathsf{B} \wedge \beta = (H, \mathfrak{t}) \in \mathsf{B} \wedge \det H = 1\}$$

$$U_2 = \{(i, j, \alpha) \mid i, j \in \text{ob } \mathsf{B} \wedge \alpha = (H, \mathfrak{t}) \in \mathsf{A} \wedge \det H > 1\}$$

zwei Begriffe für geometrische Äquivalenz, die wir hinsichtlich ihres Informationsgehalts prüfen wollen.

Sagen wir, daß G die „volle" Information über die Symmetrie von M enthält, so verlieren wir bei Betrachtung von G bis auf geometrische Äquivalenz in U_1 höchstens Informationen über die Absolutlage des Moleküls im Raum, die meist nicht interessant sein dürfte. Bei Betrachtung von G bis auf geometrische Äquivalenz in U_2 gehen außerdem gewisse, durch Parameter erfaßbare Informationen über die Relativlage der Molekülteile verloren. Das wird deutlich an dem in Abb. 2 angegebenen Beispiel von zwei ebenen Molekülen, deren Teile unterschiedlichen Abstand voneinander haben.

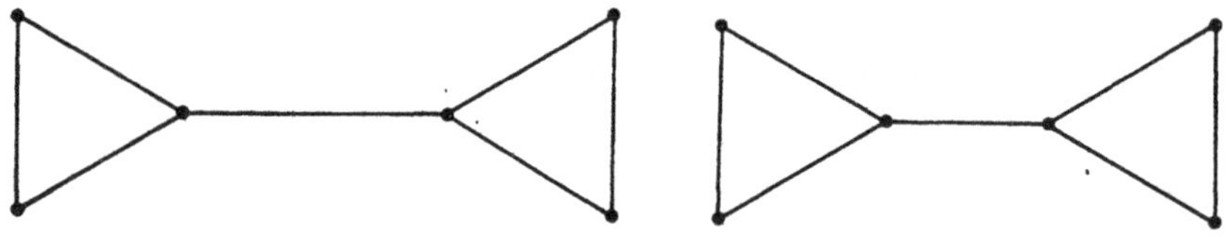

Abb. 2 Schematische Darstellung zweier ebener Moleküle, die jeweils aus zwei Teilen bestehen

Die zu diesen Molekülen gehörenden Gruppoide sind nicht geometrisch äquivalent in U_1, aber sie sind geometrisch äquivalent in U_2.

Aus dem Gesagten wird klar, daß die in [6, S. 109] vertretene Ansicht, daß das Symmetrie-Gruppoid G keine Informationen über die gegenseitige Lage der Molekülteile enthält, nicht zutrifft.

Jetzt soll noch ein Äquivalenzbegriff angegeben werden, der genau den Bedürfnissen in [6] entspricht, aber auch auf Gruppoide anwendbar ist, die nicht den Bedingungen $M1$ und $M2$ aus [6, S. 108] genügen.

Definition 4. Zwei Teilgruppoide $\mathsf{G}_1, \mathsf{G}_2 \subset \mathsf{B}$ heißen *algebraisch äquivalent hinsichtlich lokaler und globaler Symmetrie*, wenn ein Isomorphismus $F: \mathsf{G}_1 \to \mathsf{G}_2$ existiert, so daß für

$$b_1 = (i_1, j_1, \beta_1) \in \mathsf{G}_1\,, \qquad b_2 = (i_2, j_2, \beta_2) \in \mathsf{G}_1$$

sowie

$$F(b_1) = (i'_1, j'_1, \beta'_1)\,, \qquad F(b_2) = (i'_2, j'_2, \beta'_2)$$

genau dann $\beta'_1 = \beta'_2$ gilt, wenn $\beta_1 = \beta_2$.

Dieser Begriff gestattet es, das Ziel der in [6, S. 109—113] durchgeführten Untersuchungen direkt zu erreichen, d. h. ohne Zuhilfenahme der LOEWYschen Gruppe.

Der in Definition 4 eingeführte Begriff ist offensichtlich eine Verschärfung des Isomorphiebegriffes bei Gruppoiden. Algebraisch abstrakt gesehen, das heißt bis auf Isomorphie, enthält ein BRANDTches Gruppoid G nur die Information, aus wieviel Teilen das Molekül besteht und welche Symmetriegruppe die Teile haben (bis auf Isomorphie) (vgl. [6, S. 109]).

In [1] werden Kristallstrukturen betrachtet, die aus übereinandergestapelten Schichten bestehen. Dabei sind entweder alle Schichten gleich gebaut oder aber es gibt nur wenige Arten von Schichten. Relevant für die Symmetriebeschreibung ist einmal die Symmetrie der einzelnen Schichten, zum anderen die Relativlage verschiedener Schichten. Beides wird durch das Symmetrie-Gruppoid erfaßt. Die Schichten, aus denen der Kristall besteht, kann man durchnumerieren. Deshalb kann ob $\mathsf{B} = \{0, \pm 1, \pm 2, \ldots\}$ gewählt werden.

Mit Hilfe des Begriffs der geometrischen Äquivalenz aus Definition 2 werden gewisse Klassen von Untergruppoiden von B ausgezeichnet, die sogenannten OD-Gruppoide und die OD-Gruppoide von maximalem Ordnungsgrad (OD heißt order-disorder). Wichtigster Bestandteil der Bedingungen für OD-Gruppoide ist die „Nachbarschaftsbedingung":

Sei G ein Untergruppoid von B und $(i, j, \beta) \in \mathsf{G}$. Dann sind die vollen Teilgruppoide $\mathsf{V_G}(i, i-1), \mathsf{V_G}(i, i+1)$ geometrisch äquivalent in G zu den vollen Teilgruppoiden $\mathsf{V_G}(j, j-1), \mathsf{V_G}(j, j+1)$.

Bei der Definition der OD-Gruppoide von maximalem Ordnungsgrad spielt die geometrische Äquivalenz von vollen Teilgruppoiden mit mehr als zwei Objekten die entscheidende Rolle. Ausführlich sind diese Probleme in [3] dargestellt.

Für die Klasse aller OD-Gruppoide braucht man eine Einteilung in endlich viele Unterklassen. Analog wie beim Raumgruppenbegriff (vgl. Abschnitt 2) will man die Symmetrie einer OD-Kristallstruktur — bis auf geometrische Äquivalenz in U_1 — durch Angabe der Unterklasse und gewisser Parameter charakterisieren. Diese Unterklassen heißen OD-Gruppoid-Familien, und der Begriff der geometrischen Äquivalenz bis auf Translationen ist gut geeignet zu definieren, was eine OD-Gruppoid-Familie ist (vgl. [3]). Darauf aufbauend ist es möglich, einen Algorithmus aufzustellen, mit dessen Hilfe man alle OD-Gruppoid-Familien bestimmt.

5. Abschließende Bemerkungen

In den Definitionen 2 bis 4 sind drei Begriffe eingeführt worden, die für Symmetriegruppoide eine feinere Klasseneinteilung ermöglichen als es der Isomorphiebegriff zuläßt. Es wurde darauf hingewiesen, daß diese Begriffe für gewisse Anwendungen adäquat sind.

Man kann sich fragen, ob sich die Definitionen 2 und 3 als Spezialfälle einem allgemeineren Begriff unterordnen. Meiner Ansicht nach ist geometrische Äquivalenz von Gruppoiden G_1, G_2 generell durch 4 Bedingungen erfaßbar.

(i) Die Gruppoide G_1, G_2 sind isomorph.
$F\colon \mathsf{G}_1 \to \mathsf{G}_2$ sei dieser Isomorphismus.

(ii) Die Symmetriegruppen einander entsprechender Objekte sind geometrisch äquivalent im Sinne von Definition 1. Diese Eigenschaft wird erfaßt durch Bedingung (2) aus Definition 2. (I_1, ω, I_2F) ist eine natürliche Transformation des Inklusionsfunktors $I_1\colon \mathsf{G}_1 \to \mathsf{A}$ nach dem Funktor $(I_2F)\colon \mathsf{G}_1 \to \mathsf{G}_2 \to \mathsf{A}$.

(iii) Benachbarte Objekte bleiben benachbart.
In Definition 2 sichert das die Bedingung (3); in Definition 3 ist diese Forderung aufgrund von (4) erfüllt. Allgemeiner kann man diese Eigenschaft folgendermaßen erfassen:
Auf ob B ist eine geeignete Topologie definiert, die auf ob $G_1 \subseteqq$ ob B und auf ob $G_2 \subseteqq$ ob B Topologien induziert. Bezüglich dieser Topologien ist die von F bewirkte Abbildung ob $G_1 \to$ $\to$ ob G_2 ein Homöomorphismus.

(iv) Die Morphismen $\omega(i) \in A$, $i \in$ ob G_1 der natürlichen Transformation in Bedingung (ii) stehen in enger Beziehung zueinander.
In den Definitionen 2 und 3 wird diese Beziehung durch (3) bzw. (3') ausgedrückt, die sich beide auf die folgende Form bringen lassen:
Es gibt eine homomorphe Abbildung $\nu\colon A \to C$ von A in eine Gruppe C derart, daß $\nu(\omega(i)) =$ $= \nu(\omega(j))$ für alle $i, j \in$ ob G_1 gilt (homomorph heißt, daß ν mit der binären Verknüpfung der Morphismen von A bzw. der Elemente von C verträglich ist).

Es sei noch bemerkt, daß sich der in der Definition 4 eingeführte Begriff nicht in dieses Schema für die geometrische Äquivalenz einordnen läßt, weil die Bedingung (ii) nicht erfüllt ist. Der Begriff „algebraisch äquivalent hinsichtlich lokaler und globaler Symmetrie" ist feiner als der Isomorphiebegriff, aber gröber als der Begriff der geometrischen Äquivalenz.

Literatur

[1] K. Dornberger-Schiff: Grundzüge einer Theorie der OD-Strukturen aus Schichten, Abh. Deutsche Akad. d. Wiss., Kl. f. Chemie 3 (1964), Akad. Verlag Berlin 1964.
[2] K. Fichtner: Brandtsche und Ehresmannsche Gruppoide zur Symmetriebeschreibung in der Kristallographie. In „Potsdamer Forschungen", Wiss. Schriftenreihe d. Päd. Hochsch. Potsdam, Reihe B, 3 (1974), 91–95.
[3] K. Fichtner: Zur Symmetriebeschreibung von OD-Kristallstrukturen durch Brandtsche und Ehresmannsche Gruppoide. Erscheint in „Beiträge zur Algebra und Geometrie", Wiss. Beiträge d. Martin-Luther-Universität Halle-Wittenberg, Heft 6.
[4] M. Hasse und L. Michler: Theorie der Kategorien, Deutscher Verlag. d. Wiss. Berlin 1966.
[5] H. Schubert: Kategorien I, Akademie-Verlag Berlin 1970.
[6] W. Wechler: Zur Symmetriebeschreibung physikalischer Systeme. In „Studien zur Algebra und ihre Anwendungen", Schriftenreihe d. ZI f. Math. u. Mech. b. d. Akad. d. Wiss. d. DDR, 16,103 –123 (1972). Akademie-Verlag Berlin 1972.

Graphs and groups

by WALTRAUD VOSS, Ilmenau

In 1963 H. SACHS [1] proposed to investigate a certain connection between groups and systems of invariants of them consisting of orientable surfaces F_i and of graphs G_i^*, which triangulate these surfaces. Genus P_i of a surface F_i is a numerical invariant of F_i.

At first I shall formulate the problem, then I shall prove two results for special groups and state some other results. Let G be a finite group. To each ordered pair $\{x, y\}$ of elements $x, y \in G$ belongs one and only one element $z \in G$, such that $xyz = 1$. If in addition $x \neq y \neq z \neq x$ the triple (x, y, z) and the pair $\{x, y\}$ are called *regular*. To each regular triple (x, y, z) we assign the triangle $D(x, y, z)$ with the vertices x, y, z and the directed edges (x, y), (y, z), (z, x). Then the regular triples (y, z, x), (z, x, y) also correspond to the triangle $D(x, y, z)$, and we do not distinguish between them.

We consider the set $\mathfrak{M}$ of all the triangles which belong to regular triples of elements of the group G.

If (x, y, z) is regular, then evidently (y, x, z'), where $z' = (yx)^{-1}$, is regular, too.

For each regular pair $\{a, b\}$, there is in $\mathfrak{M}$ exactly one triangle D' which contains the directed edge (a, b) and there is also in $\mathfrak{M}$ exactly one triangle D'', which contains the directed edge (b, a).

To each triangle $D(x, y, z)$ belongs a sequence of edges (x, y), (y, z), (z, x). When we write down the edge sequences of all the triangles of $\mathfrak{M}$ one beneath the other we achieve a rectangular scheme each row of which contains exactly one edge sequence of a triangle. To this scheme correspond one or more surfaces (see G. RINGEL [2]). We obtain these surfaces F_i by consecutive identification of directed edges (a, b) and (b, a) from $\mathfrak{M}$. If we replace on a surface F_i the pair (a, b), (b, a) by the undirected edge $[a, b]$, we obtain a graph G_i^*, which is a triangulation of F_i. Each (regular) pair $\{x, y\}$ of elements $x, y \in G$ defines — as easily can be shown — only one graph G_i^* and also only one surface F_i, which belongs to G_i^*. So we can write: $F_i = F_i(x, y)$, $G_i^* = G_i^*(x, y)$.

The relation $F(x, y) = F(a, b)$ defines an equivalence relation $\{x, y\} \sim \{a, b\}$ and therefore the set of all regular pairs of $\mathfrak{M}$ decomposes into equivalence classes. Systematically we obtain G^* in the following way: As Figure 1 shows, the triangle $D(x, z_0 = y, z_1 = (xy)^{-1})$ is followed by the triangle $D(x, z_1, z_2 = (xz_1)^{-1})$, this triangle is followed by $D(x, z_2, z_3 = (xz_2)^{-1})$ and so on: The triangle $D(x, z_{j-1}, z_j = (xz_{j-1})^{-1})$ is followed by the triangle $D(x, z_j, z_{j+1} = (xz_j)^{-1})$.

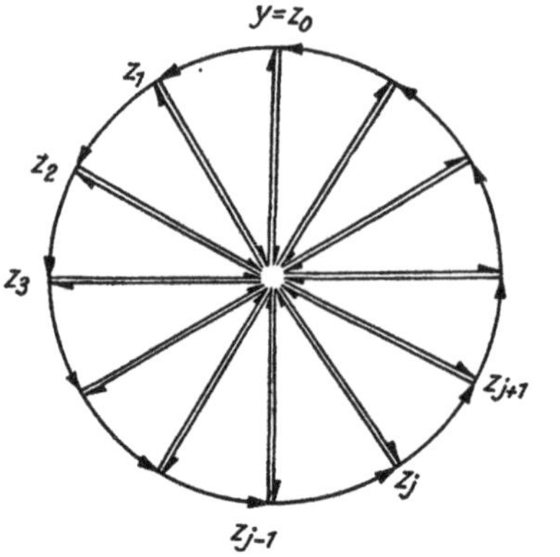

Figure 1

Obviously the sequence $\{z_\nu\}$ is periodical. The graph of Figure 1 is called a *circuit of order* 0 relative to $\{x, y\}$. We denote it by $K_0\{x, y\}$.

If $\{z_\nu\}$ has the smallest period t, then t is called the *length* of the circuit $K_0\{x, y\}$. The vertices z_ν and the directed edges which join these vertices in the graph of Figure 1 form the *boundary* of $K_0\{x, y\}$, denoted by Rd $[K_0\{x, y\}]$.

In the sequence $\{z_\nu\}$ we have $z_\nu = [xz_{\nu-1}]^{-1}$, $\nu = 1, 2, \dots$. By induction we obtain:

$$z_{2n} = x^n y x^{-n}\,, \qquad z_{2n+1} = x^n (xy)^{-1}\, x^{-n}\,, \qquad n = 0, 1, 2, \dots \tag{1}$$

Because $x^n y x^{-n} = y$ for $n = \operatorname{ord}(x)$, we obtain the relation

$$t/2 \operatorname{ord}(x)\,. \tag{2}$$

$K_0\{x, y\}$ can have odd or even length t. If the length of $K_0\{x, y\}$ is even, then let us form the $t/2$ circuits $K_0\{z_{2n-1}, z_{2n-2}\}$, $n = 1, 2, \dots, t/2$. They are called *circuits of order* 1 with respect to $\{x, y\}$ and they are denoted by $K_1^1, K_1^2, \dots, K_1^{t/2}$. With $K_0\{x, y\}$ and K_1 we form the graph $G^1(x, y)$ of Figure 2.

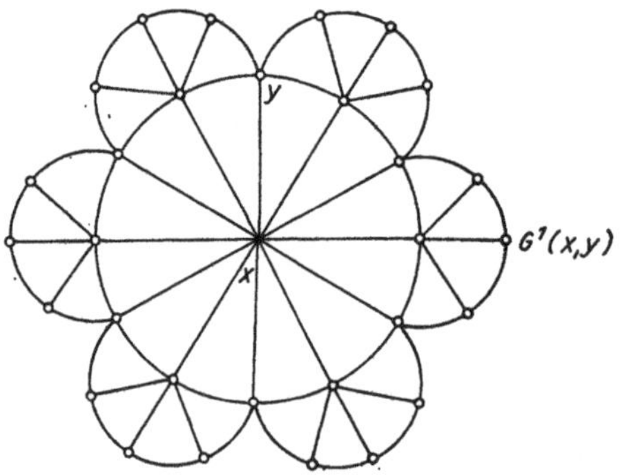

Figure 2

If to each directed edge (a, b) of $G^1(x, y)$ there exists exactly one conversely directed edge (b, a), then we obtain $G^*(x, y)$ from $G^1(x, y)$ by identification of the edges (a, b) and (b, a) for each (a, b) of $G^1(x, y)$.

The circuit K_1^1 has even or odd length. Transforming all the vertices of K_1^1 by $x^{-(\mu-1)}$ (that means transforming all the group elements, which belong to these vertices) we achieve that the circuit K_1^1 passes over to K_1^μ (see Figure 2 and cf. (1)). So all the circuits K_1^ν have the same length 1.

According to (1) and Figure 2 the following statements are obvious:

(L1) *If $x \nsim y \nsim (xy)^{-1} \nsim x$,[1] then $K_0\{x, y\}$ and K_1^1 have even length. On the boundary of $K_0\{x, y\}$ elements, conjugate to y, alternate with elements conjugate to $(xy)^{-1}$ and on the boundaries of the circuits K_1^ν elements, conjugate to x, alternate with elements conjugate to y.*

(L2) *If $xy = yx$, then the triangles $D(x, y, (xy)^{-1})$ and $D(y, x, (yx)^{-1} = (xy)^{-1})$ form already a triangulation $G^*(x, y)$ of the sphere.*

Therefore, to abelian groups there only belong surfaces F of genus $P = 0$, the corresponding triangulations of which contain exactly two triangles.

After these preliminary remarks let us deal with the nonabelian groups of order p^3, where p is a prime number $\geqq 3$. We will prove three theorems:

Theorem 1. *Let G be a nonabelian group of order p^3, where $p \neq 2$ is a prime number.*

Then it is true: If any two of the three elements $x, y, z \in G$ of the (regular) triple (x, y, z) don't commute, then x, y, z belong to three different conjugacy classes of G, and a surface of genus

$$P = 1 + p\frac{p-3}{2}$$

corresponds to the pair $\{x, y\}$.

Proof: In the proof we use some wellknown properties of the nonabelian groups of order p^3, which are stated here without proof.

There are two nonabelian groups of order p^3, denoted by G_1 and G_2.

a) G_1 is generated by the elements a and b with the relations

$$a^{p^2} = 1, \quad b^p = 1, \quad ab = ba^{p+1}.$$

$A_0 = \{1, a, a^2, \dots, a^{p^2-1}\}$ is the largest normal subgroup of G_1. The group G_1 contains the elements of A_0 and the elements of $A_i = \{b^i, b^ia, \dots, b^ia^{p^2-1}\}$ for all $i \in \{1, 2, 3, \dots, p-1\}$. The A_i are the elements of G_1 / A_0. The group G_1 / A_0 is isomorphic to the subgroup $\langle b \rangle$ of G_1 generated by $b \in G_1$. The multiplication table of G_1 / A_0 shows us that we have to consider the following kinds of triples with elements of G_1: $(A_i, A_j, A_{-(i+j)})$[2], where $i, j \in \{0, 1, 2, \dots, p-1\}$ (all the indices are to be taken modulo p).

If $b^la^i, b^ka^j \in G_1$, then the product of these elements can easily be computed:

$$b^la^ib^ka^j = b^{l+k}a^{i(1+p)^k+j} = b^{l+k}a^{ip+i+j},$$

since

$$(1+p)^k = \sum_{\nu=0}^{k} \binom{k}{\nu} p^\nu =$$

$$= \left(1 + k + \binom{k}{2} p^2 + \dots + \binom{k}{k-1} p^{k-1} + p^k\right) \equiv (1 + kp) \bmod p^2.$$

[1]) $a \sim b$ means: a and b are conjugate. $a \nsim b$ means: a and b are not conjugate.

[2]) The triple (x, y, z) belongs to the kind of triples $(A_i, A_j, A_{-(i+j)})$, when $x \in A_i$, $y \in A_j$, $z \in A_{-(i+j)}$.

The element $b^{-l}a^{-i(1+p)^{-l}}$ is the inverse of $b^l a^i$ (the exponents of a and b are to be taken mod p and mod p^2, respectively).

Centre of $G_1 : Z = \{1, a^p, a^{2p}, \dots, a^{(p-1)p}\}$. Let $K_{ij} = \{b^{i-1}a^{j-1}a^{\nu p}, 0 \leqq \nu \leqq p-1\}$, $1 \leqq i, j \leqq p$. Without taking into consideration that each element of $K_{11} = Z$ forms a conjugacy class we get $p^2 - 1$ conjugacy classes K_{ij} $(1 \leqq i, j \leqq p, (i, j) \neq (1, 1))$ of G_1, each of them containing p elements. Therefore, as it can be shown by easy computation

$$A_t = \bigcup_{j=1}^{p} K_{t+1,j}\,, \qquad t \in \{0, 1, \dots, p-1\}\,.$$

(L3) *The elements of K_{ij}, $i, j \in \{1, 2, \dots, p\}$, pairwisely commute.*

Remark. Any two elements of A_0 commute; this, however, for the elements of A_i with $i \neq 0$ is not true. Therefore it suffices to consider triples (x, y, z) only for which the following assumption V holds:

Assumption V. Non of the elements x, y, z belongs to K_{11}, but they belong to three different conjugacy classes $K_{\nu\mu}((\nu, \mu) \neq (1, 1))$, where at most one of these three classes belongs to A_0.

As a consequence of (L1) and of (2), the circuit $K_0\{x, y\}$ has the length $2p$ and the K_1^ν have the length $2p$, too. Now we prove:

(L4) *$G^*(x, y)$ consists of the triangles of $G^1(x, y)$, and the surface $F(x, y)$ has the genus*

$$P = 1 + p\frac{p-3}{2}\,.$$

Proof: From assumption V it follows $x \nsim y \nsim z = (xy)^{-1} \nsim x$ and therefore the elements x, y, z have the order p. Let x, y, z be elements of the classes K_x, K_y, K_z[1]), respectively. Each of these classes contains p elements; since $K_0\{x, y\}$ and the circuits K_1^ν have the length $2p$, on the boundary $\mathrm{Rd}[K_1^\nu]$ elements of K_x alternate with elements of K_y, and any two elements of the boundary are distinct. Therefore each of the elements of K_x and of K_y appears on the boundary of each circuit K_1^ν exactly once.

Let the elements $x_1, x_2, \dots, x_p$(conjugates of x) appear on Rd $[K_1^1]$ in this succession. By transforming x_i with x^μ we obtain $x_i \to x^{-\mu}x_i x^\mu = x_i$ for $i \in \{1, \dots, p\}$, because all elements of K_x commute (L3), so on $\mathrm{Rd}[K_1^\nu]$ the elements $x_1, x_2, \dots, x_p$ appear in the same order. On Rd $[K_1^\nu]$ each vertex y_j (let be $y = y_1$) of K_y lies between two vertices x_i. Let (x_μ, y_λ) be a directed edge of Rd $[K_1^{\nu*}]$ $(\nu^* \in \{1, 2, \dots, p\}$, ν^* fixed); on Rd $[K_1^{\nu*}]$ the vertex x_μ is followed by the vertex y_λ. Since the multiplication of group elements is uniquely determined and the centres of two circuits are distinct, there is no circuit K_1^ν between the $p-1$ circuits different to $K_1^{\nu*}$, on which x_μ is followed by y_λ. Therefore the directed edge (x_μ, y_λ) appears in $G^1(x, y)$ exactly once.

Now let us show that the conversely directed edge (y_λ, x_μ) appears in $G^1(x, y)$ exactly once. If we suppose $p \geqq 3$, then x_μ in G^* has the valency $2p \geqq 6$. Therefore

[1]) K_x, K_y, K_z denote the conjugacy classes containing x, y and z, respectively.

on Rd $[K_1^{\nu *}]$ the vertex y_λ cannot be followed by x_μ. Let x_ϱ denote the vertex by which y_λ is followed on Rd $[K_1^{\nu *}]$. Because in this case there is no circuit K_1^ν different from $K_1^{\nu *}$ such that on Rd $[K_1^\nu]$ the vertex y_λ is followed by x_ϱ because there are in K_x exactly $p - 1$ elements different from x_ϱ and because there are exactly $p - 1$ circuits different from $K_1^{\nu *}$, too, then between these $p - 1$ circuits there is exactly one circuit such that on its boundary y_λ is followed by x_μ.

$G^*(x, y)$ contains $n = 3p$ vertices, $k = 3p^2$ (undirected) edges and $f = 2p^2$ triangles.

Therefore from EULER's formula

$$2 - 2P = n - k + f$$

the genus gets the value

$$P = 1 + p\frac{p-3}{2}.$$

b) G_2 is generated by the elements a, b, c with the relations

$$a^p = 1\,, \quad b^p = 1\,, \quad c^p = 1\,, \quad ac = ca\,, \quad bc = cb\,, \quad ab = bac\,.$$

All the elements of G_2 can be written in the form $b^\nu c^\mu a^\lambda$. The product of two elements is

$$(b^m c^n a^o)\,(b^x c^y a^z) = b^{m+x} c^{n+y+xo} a^{z+o}\,. \tag{3}$$

From (3) it follows: All the elements which do not contain a b-power or an a-power commute. Obviously two elements $b^m c^n a^o$ and $b^x c^y a^z$ commute iff $x \cdot o \equiv m \cdot z \bmod p$.

Conjugacy classes are:

$$\{b^i a^j, b^i a^j c, b^i a^j c^2, b^i a^j c^3, \ldots, b^i a^j c^{p-1}\}\,,$$

where $0 \leqq i; j \leqq p - 1$, $(i, j) \neq (0, 0)$; and

$$\{1\}, \{c\}, \{c^2\}, \ldots, \{c^{p-1}\}\,.$$

Obviously:

(L5) *Any two elements of one and the same class commute.*

Again we have to consider two cases:

1) If one of the elements of the triple (x, y, z) belongs to the centre of the group G_2 or if two of the three elements belong to one and the same class, then the (regular) pair $\{x, y\}$ corresponds to a sphere.

2) No two of the elements x, y, z are in one and the same class. (It suffices to deal with the case that no two of the three elements x, y, z commut — see (L2).) The elements x, y, z have the order p. Corresponding to the group G_1 we can prove: $G^*(x, y)$ consists of the triangles of $G^1(x, y)$, and the surface $F(x, y)$ has the genus $P = 1 + p\dfrac{p-3}{2}$. This completes the proof of Theorem 1.

Now let us consider the group $G_{p,q}$ of the order pq, where p, q are primes with $p|q - 1$. $G_{p,q}$ is generated by the two elements a and b, with the relations

$$a^p = 1\,, \quad b^q = 1\,, \quad ba = ab^r\,;$$

where r is a nontrivial solution of the congruence $r^p \equiv 1(q)$. The subgroup B generated by b is normal in $G_{p,q}$; $a^i B$, $i \in \{1, 2, \dots, p-1\}$, are the elements of the group $G_{p,q}/B$.

The following Theorem can easily be proved:

Theorem 2. *In the group $G_{2,q}$ to each regular pair $\{x, y\}$, $x, y \in G_{2,q}$, there corresponds a surface $F(x, y)$ of genus 0; the triangulation $G^*(x, y)$ of $F(x, y)$ is either a "prism-graph" (see Figure 3) or $G^*(x, y)$ consists of two triangles.*

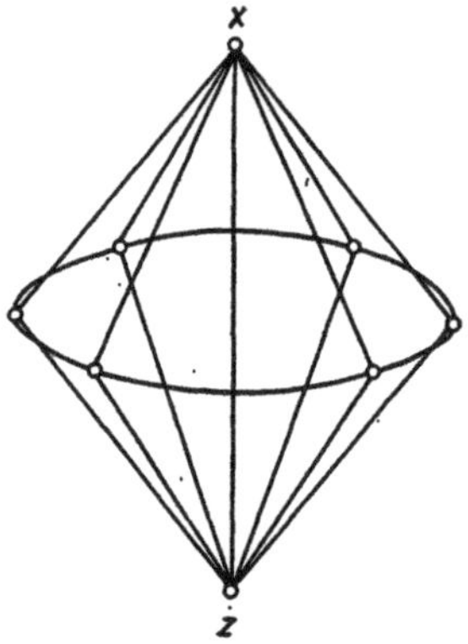

Figure 3

Proof: 1) $x, y \in B$; from (L2) follows that $G^*(x, y)$ consists of two triangles and $F(x, y)$ is the sphere.

2) $x \in B$, $y \in aB$; (from this it follows that $z \in aB$) then ord $(x) = q$, ord $(y) = 2$. With $x = b^i$ and with the definition $a_1 = y$ we form

$$\begin{aligned} a_2 &= x a_1 & \text{and} \quad b_2 &= x^{-1} a_1 \\ a_3 &= x a_2 & b_3 &= x^{-1} b_2 \\ a_4 &= x a_3 & b_4 &= x^{-1} b_3 \\ &\vdots & &\vdots \\ a_{\nu+1} &= x a_\nu & b_{\mu+1} &= x^{-1} b_\mu \end{aligned} \tag{4}$$

If $a_{\nu+1} = a_1$ or $b_{\mu+1} = a_1$, respectively, we stop. That is the case for $\nu = \mu = q + 1$.

Obviously

$$a_\nu = x^{\nu-1} a_1 , \qquad b_\mu = x^{-(\mu-1)} a_1 \qquad (\nu, \mu \geqq 2) . \tag{4'}$$

Because $a_\nu^{-1} = a_\nu$ and $b_\nu^{-1} = b_\nu$ we have $a_{\nu+1} x a_\nu = 1$ and $b_{\nu+1} x^{-1} b_\nu = 1$; therefore both

$$a_1, a_2, \dots, a_q, a_{q+1} = a_1 \quad \text{and} \quad a_1, b_2, b_3, \dots, b_q, b_{q+1} = a_1$$

form a sequence $\{z_\nu\}$ of vertices of the boundaries of the circuits $K_0\{x, a_1\}$ and $K_0\{x^{-1}, a_1\}$, respectively.

From (4) and (4') it follows that if a directed edge (v, w) is on Rd $[K_0\{x, a_1\}]$ then the conversely directed edge (w, v) is on Rd $[K_0\{x^{-1}, a_1\}]$ and vice versa.

So $G^*(x, y)$ consists of all the triangles of the circuits $K_0\{x, a_1\}$ and $K_0\{x^{-1}, a_1\}$. Obviously $G^*(x, y)$ is a triangulation of the sphere.

Note. $q - 1/2$ prism-graphs (see Figure 3) which triangulate the sphere belong to the group $G_{2,q}$.

Theorem 3. *If $p > 2$, then in $G_{p,q}$ a surface of genus*

$$P = \frac{(p-2)(q-1)}{2}$$

corresponds to each pair $\{x, y\}$, where $1 \neq x \in B$, $y \in a^i B$, $i \in \{1, 2, \dots, p-1\}$. The graph $G^*(x, y)$ consists of all the triangles of $G^1(x, y)$.

(The proof of Theorem 3 is very similar to the proof of Theorem 1).

Some further results are stated hereunder. Only surfaces of genus 0 correspond to the Dihedral groups, to the Dicyclic groups, to all the nonabelian groups of order 2^n, which have a cyclical subgroup of order 2^{n-1}. Only surfaces of genus 0 belong to the symmetric groups S_3 and S_4, too. One surface of genus 4 and one surface of genus 9 belong to the alternating group A_5 (of course, spheres belong to the group A_5, too). The triangulations $G^*(x, y)$ which correspond to these surfaces are shown in Figure 4 and Figure 5. Now we deal with the nonabelian groups of order < 32: Only spheres correspond to 41 nonabelian groups of order < 32. Surfaces of genus 0, 1 and 4

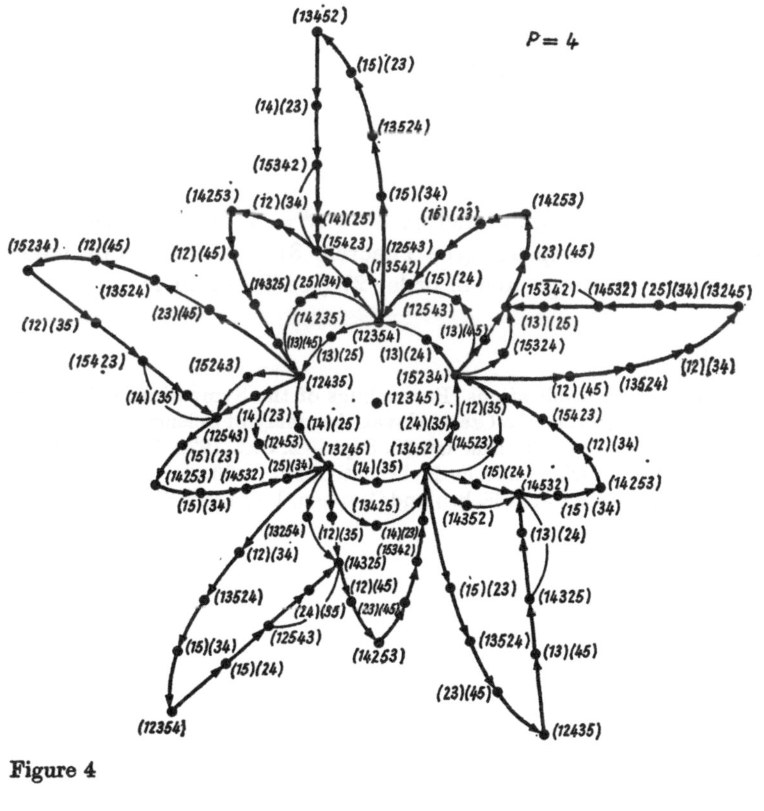

Figure 4

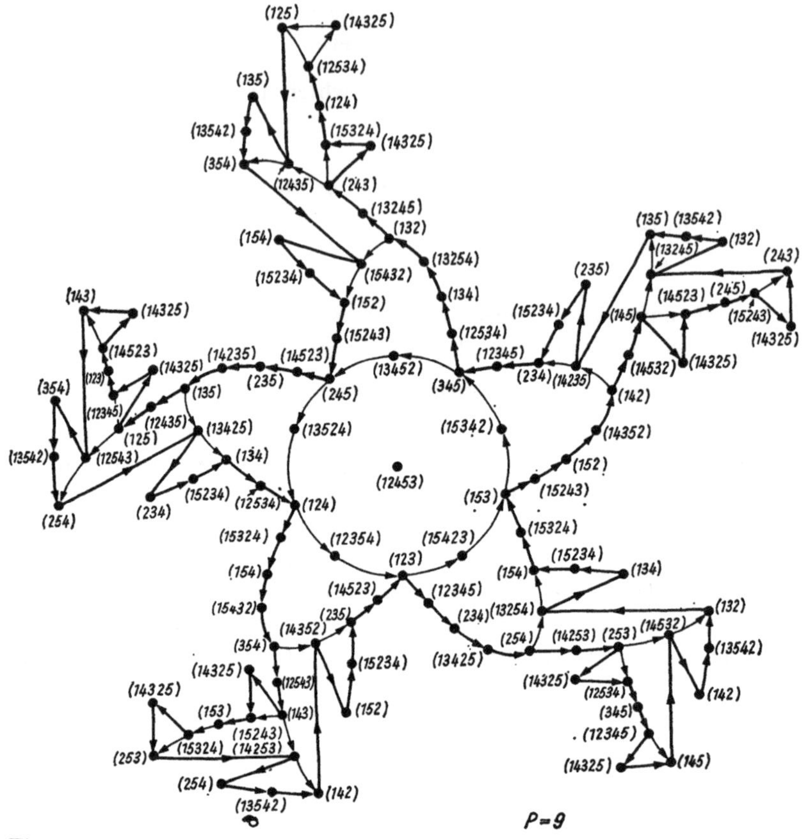

Figure 5

belong to the group of order 20, defined by $b^5 = a^4 = 1$, $ba = ab^2$. Surfaces of genus 0, 1 and 3 belong to the group $G_{3,7}$ (see Theorem 3) and surfaces of genus 0 and 1 belong to the two nonabelian groups of order 27 (see Theorem 1).

References

[1] Theory of Graphs and its Applications. Proceedings of the Symposium held in Smolenice in June 1963. Publishing House of the Czechoslovak Academy of Science; 162—163.

[2] XV. Intern. Wiss. Kolloquium TH Ilmenau 28. Sept.—2. Okt. 1970, Vortragsreihe A1 Mathematik, Rechentechnik und Ökonomische Kybernetik; 191—196.

[3] G. Ringel: Färbungsprobleme auf Flächen und Graphen, Deutscher Verlag d. Wissensch., Berlin 1959.

Semantics of monads and automata

by Andrzej Kurpiel, Toruń

In this paper we define a pair of adjoint functors Sem and Str between the category of monads [3] and the category of functors which have right Kan extension along themselves. This allows us to formulate an "algebraic form" of Beck's theorem [2]. The terminology used in this paper is the same as in [1] and [3].

§ 1. Semantics and structure

Let $\mathsf{Mon}^{\leftarrow}$ be the category of monads; objects are pairs (C, T), where C is any category and $T = (T, \mu, \eta)$ is a monad on C. Morphisms in $\mathsf{Mon}^{\leftarrow}$ are pairs $(G, \sigma)\colon (\mathsf{C}, T) \to (\mathsf{C}', T')$, where $G\colon \mathsf{C} \to \mathsf{C}'$ is a functor and $\sigma\colon T' \cdot G \to G \cdot T$ is a natural transformation such that

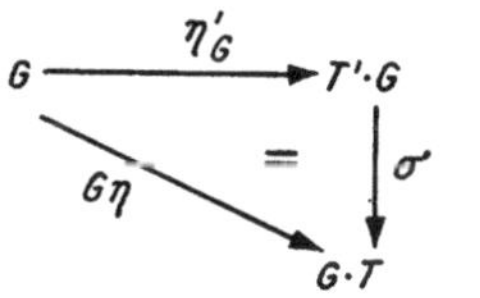

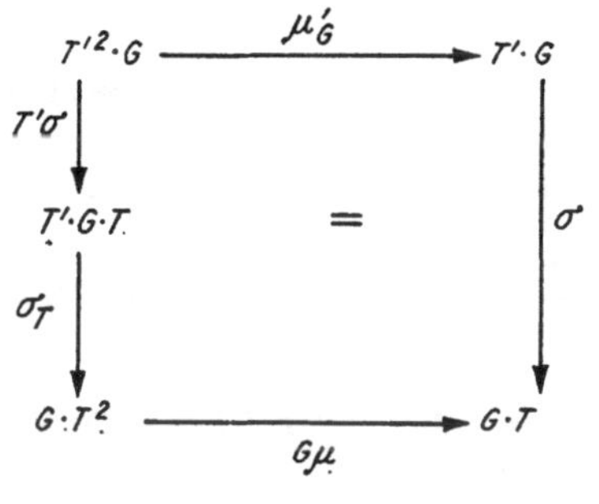

Composition of morphisms is given by the formula:

$$(F, \tau) \cdot (G, \sigma) = (F \cdot G, F\sigma \cdot \tau_G)\,.$$

Denote by $(\mathsf{Cat} \downarrow \mathsf{Cat})$ the comma-category of functors $F\colon \mathsf{A} \to \mathsf{B}$ and morphisms $(H, H')\colon (\mathsf{A} \xrightarrow{F} \mathsf{B}) \to (\mathsf{C} \xrightarrow{G} \mathsf{D})$, such that $H' \cdot F = G \cdot H$.

Fact 1. *The passage*

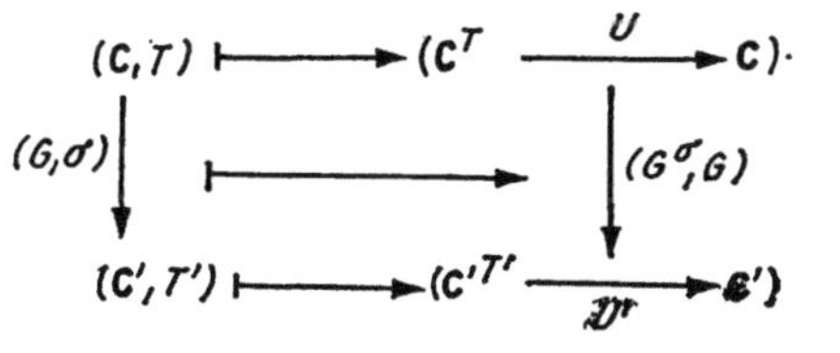

where $G^\sigma(A, a) = (GA, Ga \cdot \sigma_A)$, $G^\sigma(f) = Gf$ *defines a functor* Sem: $\mathsf{Mon}^{\leftarrow} \to$ $\to (\mathsf{Cat} \downarrow \mathsf{Cat})$.

Proof. Observe that if (G, σ) is a morphism of monads then $G: \mathsf{C}^T \to \mathsf{C}'^{T'}$ is a functor and $G \cdot U = U' \cdot G^\sigma$.

Thus G^σ is simply the comparison functor for a morphism (G, σ).

Fact 2. [2] *If the functor* $G: \mathsf{C} \to \mathsf{C}'$ *has a left adjoint* $F: \mathsf{C}' \to \mathsf{C}$ *with* $(\varepsilon, \eta): F \dashv G$, *then there exists a right* Kan *extension* $R_G(G)$ *along itself equal to* $G \cdot F$ *with a natural tranformation* $G\varepsilon: R_G(G) \cdot G = GFG \to G$ *and* $R_G(G)$ *is preserved by any functor* H: $\mathsf{C}' \to \check{\mathsf{C}}'$.

Corollary 3. *For any object* (C, T) *of the category* $\mathsf{Mon}^{\leftarrow}$ *there is a right* Kan *extension of the forgetful functor* $U: \mathsf{C}^T \to \mathsf{C}$ *along itself preserved by any functor.*

Denote by $\mathsf{R}(\mathsf{Cat} \downarrow \mathsf{Cat})$ the full subcategory of $(\mathsf{Cat} \downarrow \mathsf{Cat})$ such that $\mathsf{A} \xrightarrow{F} \mathsf{B}$ is an object of $\mathsf{R}(\mathsf{Cat} \downarrow \mathsf{Cat})$ iff $R_F(F)$ exists and is preserved by any functor.

Corollary 4. Sem *is a functor from the category* $\mathsf{Mon}^{\leftarrow}$ *into the category* $\mathsf{R}(\mathsf{Cat} \downarrow \mathsf{Cat})$.

Fact 5. [2] *If the functor* $F: \mathsf{A} \to \mathsf{B}$ *has an* $R_F(F)$, *then this is a monad (codensity monad) on* B, *that means,* $(\mathsf{B}, R_F(F))$ *is an object of the category* $\mathsf{Mon}^{\leftarrow}$.

Theorem 6. *There exists a functor* Str: $\mathsf{R}(\mathsf{Cat} \downarrow \mathsf{Cat}) \to \mathsf{Mon}^{\leftarrow}$ *defined by the rule:*

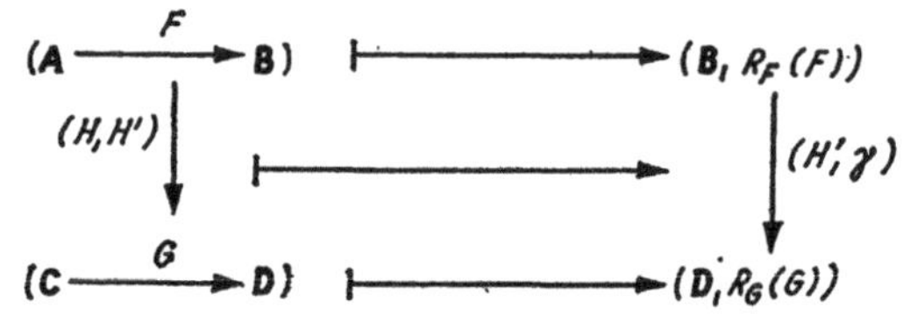

where $\gamma: R_G(G) \cdot H' \to H' \cdot R_F(F)$ *is a natural transformation given by* 1—1 *correspondence in* Kan *extension.*

Proof. At first we define the functor Str on morphisms. Let (H, H') be a morphism in $\mathsf{R}(\mathsf{Cat} \downarrow \mathsf{Cat})$ from $(\mathsf{A} \xrightarrow{F} \mathsf{B})$ into $(\mathsf{C} \xrightarrow{G} \mathsf{D})$. Denote by $(R_F(F), \tau)$ $((R_G(G), \sigma))$ the right Kan extension induce dby F(by G). Then there is an $(R_F(H' \cdot F), \alpha)$ and a natural isomorphism $z: H' \cdot R_F(F) \to R_F(H' \cdot F)$ such that

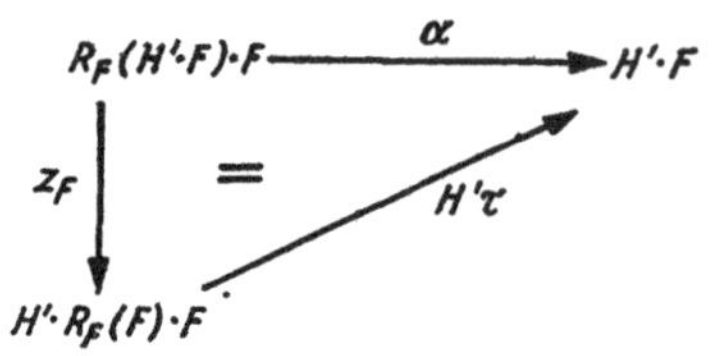

From this fact it follows that H' preserves $R_F(F)$.

From the 1—1 correspondence in the KAN extension we obtain:

$$R_G(G) \cdot H' \cdot F = R_G(G) \cdot G \cdot H \xrightarrow{\sigma_H} G \cdot H = H' \cdot F$$

$$R_G(G) \cdot H' \xrightarrow{h} R_F(H' \cdot F)$$

such that

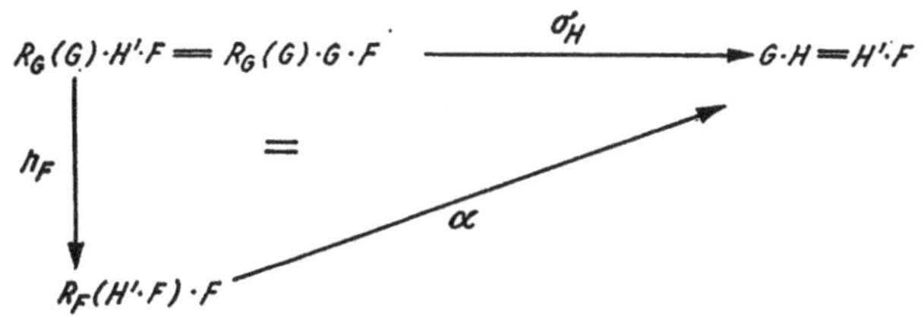

Now we define a natural transformation γ by

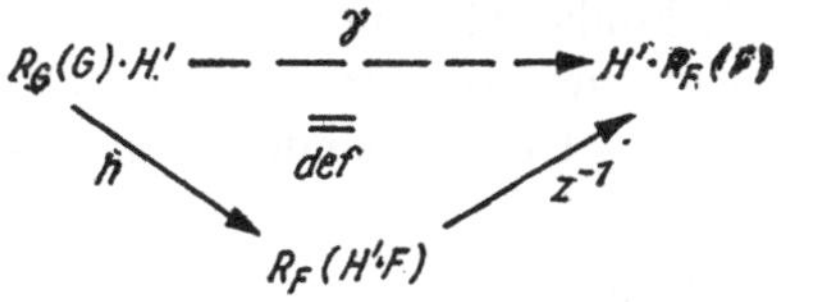

Then, of course, the pair (H', γ) is a morphism of monads and finally Str is in fact a functor.

Theorem 7. *The functor* Str *is left adjoint to* Sem.

Proof. For any object $(\mathsf{A} \xrightarrow{F} \mathsf{B})$ of $\mathsf{R}(\mathsf{Cat} \downarrow \mathsf{Cat})$ we define

$$\Psi_{(\mathsf{A} \xrightarrow{F} \mathsf{B})} : (\mathsf{A} \xrightarrow{F} \mathsf{B}) \to \mathrm{Sem} \cdot \mathrm{Str}\,(\mathsf{A} \xrightarrow{F} \mathsf{B}) = (\mathsf{B}^R \xrightarrow{U} \mathsf{B}),$$

$$\Psi_{(\mathsf{A} \xrightarrow{F} \mathsf{B})} = (F^\$, 1_\mathsf{B}),$$

where $R = R_F(F)$ is the codensity monad on B, $F^\$: \mathsf{A} \to \mathsf{B}^R$ is a functor $F^\$(X) = (FX, \tau_X)$, $F^\$(f) = F(f)$. Observe, that

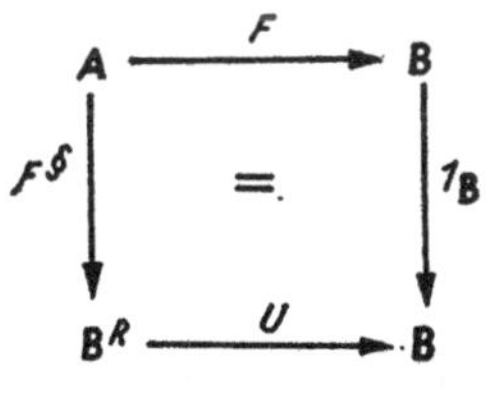

Now, Ψ is a natural transformation:

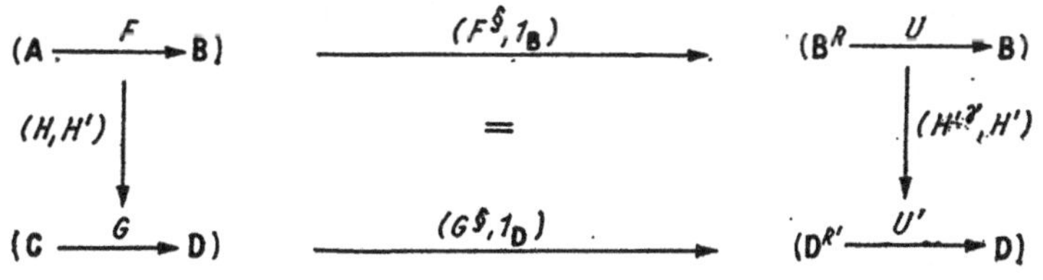

$$(H'^{\gamma}, H') \cdot (F^{\S}, 1_B) = (H'^{\gamma} \cdot F^{\S}, H'), \qquad (G^{\S}, 1_D) \cdot (H, H') = (G^{\S} \cdot H, H').$$

$$H'^{\gamma} \cdot F^{\S}(X) = H'^{\gamma}(FX, \tau_X) = (H'FX, H'\tau_X \cdot \gamma_{FX}) = (GHX, H'\tau_X \cdot z_{FX}^{-1} \cdot h_{FX}) =$$
$$= (GHX, \alpha_X \cdot h_{FX}) = (GHX, \sigma_{HX}) = G^{\S}(HX).$$

Let (C, T) be any object of $\mathsf{Mon}^{\leftarrow}$ and $(H, H')\colon (\mathsf{A} \xrightarrow{F} \mathsf{B}) \to (\mathsf{C}^T \xrightarrow{U} \mathsf{C})$ be a morphism in $\mathsf{R}(\mathsf{Cat} \downarrow \mathsf{Cat})$.

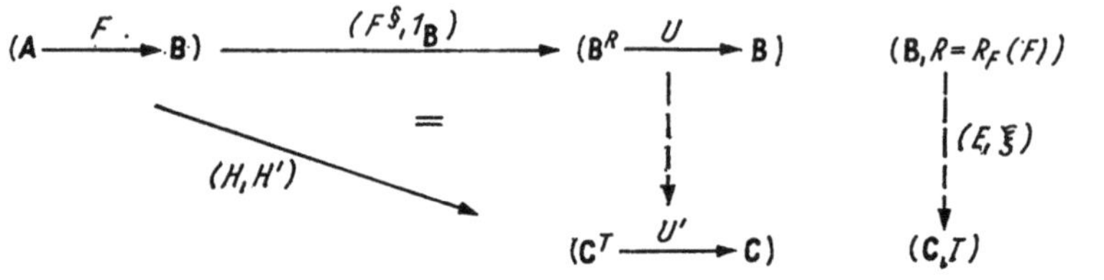

(*)

Then to prove the universality of Ψ it is of course enough to construct a morphism $(E, \xi)\colon (\mathsf{B}, R) \to (\mathsf{C}, T)$ in $\mathsf{Mon}^{\leftarrow}$ such that (*) holds. The functor E is equal to H'. For the construction of ξ let us consider the natural transformation $u\colon T \cdot U \to U$ defined by $u_{(X,x)} = x$ for $(X, x) \in (\mathsf{C}^T)^0$. From the $1-1$ correspondence in KAN extension we obtain:

$$T \cdot H' \cdot F = T \cdot U \cdot H \xrightarrow{u_H} U \cdot H = H' \cdot F$$
$$T \cdot H' \xrightarrow{w} R_F(H' \cdot F)$$

such that

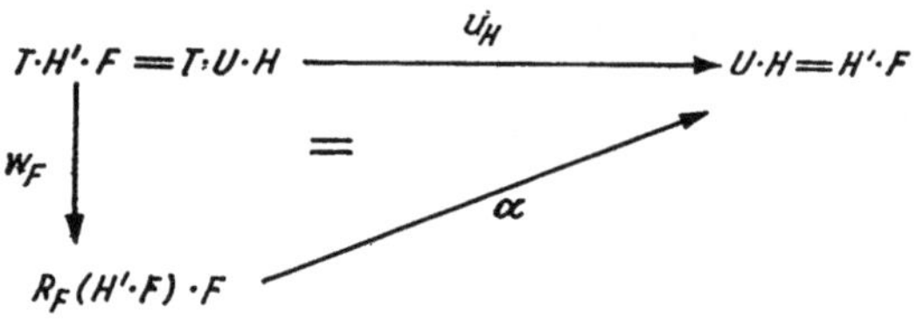

ξ is given by the rule:

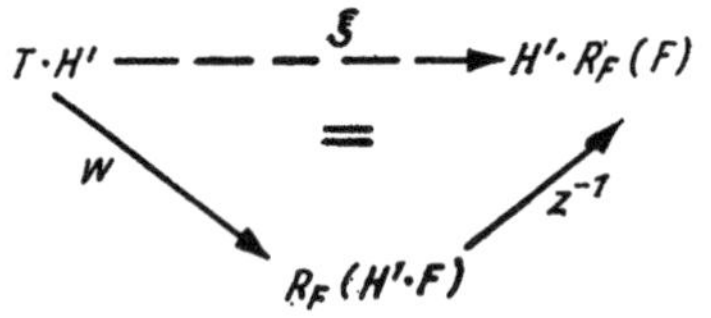

The pair (H', ξ) is a morphism of monads and ξ is unique (by the uniqueness of w). Diagram (*) is commutative: $(H'^{\xi}, H') \cdot (F^{\S}, 1_B) = (H'^{\xi} \cdot F^{\S}, H')$, $H'^{\xi}F^{\S}(X) = H'(FX, \tau_X) = (H'FX, H'\tau_X \cdot \xi_{FX}) = (H'FX, H'\tau_X \cdot z_{FX}^{-1} w_{FX}) = (UHX, \alpha_X \cdot w_{FX}) = (UHX, u_{HX}) = H(X)$ for any object of A; on morphisms: $H'^{\xi}F^{\S}(f) = H'(Ff) = H'Ff = UH(f) = H(f)$.

Theorem 8. $\mathrm{Str} \cdot \mathrm{Sem} = 1_{\mathsf{Mon}^{\leftarrow}}$.

Proof. $\mathrm{Str} \cdot \mathrm{Sem}(G, \sigma) = \mathrm{Str}(G^{\sigma}, G) = (G, \gamma)$ for any morphism from $\mathsf{Mon}^{\leftarrow}$. The equality $\sigma = \gamma$ holds by Theorem 6.

An adjunction Str $\dashv$ Sem gives us a bijection on the sets of morphisms:

$$\mathsf{Mon}^{\leftarrow}[(\mathsf{B}, R_F(F)), (\mathsf{C}, T)] \cong \mathsf{R}(\mathsf{Cat}\downarrow\mathsf{Cat})\left[(\mathsf{A} \xrightarrow{F} \mathsf{B}), (\mathsf{C}^T \xrightarrow{U} \mathsf{C})\right]$$

for any (C, T) and $(\mathsf{A} \xrightarrow{F} \mathsf{B})$. This bijection is functorial in the sense that

$$\mathsf{Mon}^{\leftarrow}[(\mathsf{B}, R_F(F)), (\mathsf{C}, T)] \quad \text{and} \quad \mathsf{R}(\mathsf{Cat}\downarrow\mathsf{Cat})\left[(\mathsf{A} \xrightarrow{F} \mathsf{B}), (\mathsf{C}^T \xrightarrow{U} \mathsf{C})\right]$$

are categories.

Let $(\mathsf{A} \xrightarrow{F} \mathsf{B})$ and $(\mathsf{C} \xrightarrow{G} \mathsf{D})$ are objects of $\mathsf{R}(\mathsf{Cat}\downarrow\mathsf{Cat})$. We define a new category $\mathsf{R}(F, G)$: objects are morphisms between F and G, i.e. a pair (H, E) of functors such that $E \cdot F = G \cdot H$, morphisms

$$(\alpha, \beta): (H, E) \to (H', E')$$

are pairs of natural transformations $\alpha: H \to H'$, $\beta: E \to E'$ such that

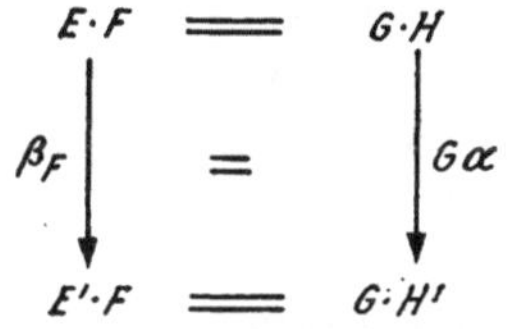

Composition of morphisms is given by composition of components. Now, for any two objects (C, T) and (C', T') of $\mathsf{Mon}^{\leftarrow}$ we define a new category $\mathsf{M}(T, T')$: objects are morphisms of monads

$$(G, \sigma): (\mathsf{C}, T) \to (\mathsf{C}', T') \quad \text{in } \mathsf{Mon}^{\leftarrow},$$

and morphisms of $\mathsf{M}(T, T')$ are natural transformations $\alpha: G \to G'$ such that

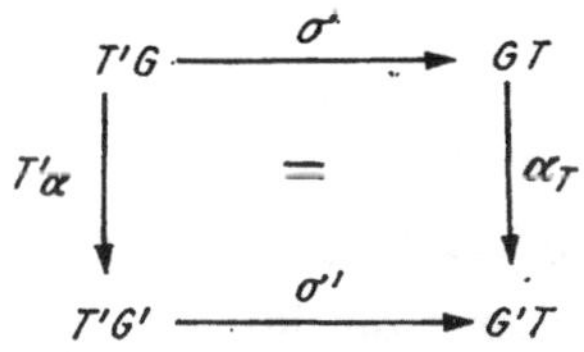

Theorem 9. *For any object* $(\mathsf{A} \xrightarrow{F} \mathsf{B})$ *and any* (C, T) *from* $\mathsf{Mon}^{\leftarrow}$, *the categories* $\mathsf{M}(R_F(F), T)$ *and* $\mathsf{R}(F, U)$ *are isomorphic.*

Proof. We define two functors b and a:

$$b: \mathsf{M}(R_F(F), T) \to \mathsf{R}(F, U),$$

$b(K, \sigma) = (K^\sigma \cdot F^\S, K)$ on objects and $b(\alpha) = (\bar{\alpha}_{F^\S}, \alpha)$ on morphisms, where K and $F^\S$ are the same as in Fact 1 and Theorem 7 and

$$\bar{\alpha}: K^\sigma \to K'^{\sigma'}\bar{\alpha}_{(X,x)} = \alpha_X .$$

$a: \mathsf{R}(F, U) \to \mathsf{M}(R_F(F), T)$ with $a(H, E) = (E, \gamma)$, $a(\alpha, \beta) = \beta$,

where γ is the same as in Theorem 6.

Of course $ab = 1$ and $ba = 1$ where 1 is the identity functor on $\mathsf{M}(R_F(F), T)$ and $\mathsf{R}(F, U)$ correspondingly.

Corollary 10. (cf. ALAGIČ [0]). *If* (C, T) *and* (C', T') *are objects of a category* $\mathsf{Mon}^{\leftarrow}$, *then the categories* $\mathsf{M}(T, T')$ *and* $\mathsf{R}(U, U')$ *are isomorphic.*

§ 2. New form of BECK's theorem

In this section we formulate and prove a new form of BECK's theorem — an "algebraical version" necessary and sufficient condition for the comparison functor [2].

A monad (C, T), where $T = (T, \mu, \eta)$ is called *idempotent* iff μ is an isomorphism.

Lemma 11. *If the monad* (T, μ, η) *is idempotent, then* η_U *is a natural isomorphism with* $u\colon T \cdot U \to U$ (cf. the proof of theorem 7) *as inverse.*

Proof. μ is an isomorphism, then $\mu = (T\eta)^{-1}$. For any object (X, x) of C^T we have $x \cdot \eta_X = 1_X$ and $Tx \cdot T\eta_X = 1_{TX}$ and $Tx = \mu_X$. Finally $1_{TX} = \mu_X \cdot \eta_{TX} = Tx \cdot \eta_{TX} = \eta_X \cdot x$. This means, that $\eta_X = \eta_{U(X,x)}$ and $x = u_{(X,x)}$ are inverse.

An adjunction Str $\dashv$ Sem gives a monad $S = \mathrm{Sem} \cdot \mathrm{Str}$ on $\mathsf{R}(\mathsf{Cat} \downarrow \mathsf{Cat})$. From Theorem 8 the monad S is idempotent.

Theorem 12 (Algebraic form of BECK's theorem). *Let the functor* $F\colon \mathsf{A} \to \mathsf{B}$ *have a right* KAN *extension* $R = R_F(F)$ *along itself, preserved by any functor.*

Then semantic comparison functor $F^{\S}\colon \mathsf{A} \to \mathsf{B}^R$ *is an isomorphism if and only if on the object* $(\mathsf{A} \xrightarrow{F} \mathsf{B})$ *there is exactly one S-algebra structure, this means, that there exists exactly one morphism*

$$(H, H')\colon S\left(\mathsf{A} \xrightarrow{F} \mathsf{B}\right) \to \left(\mathsf{A} \xrightarrow{F} \mathsf{B}\right)$$

such that the pair $(\mathsf{A} \xrightarrow{F} \mathsf{B}, (H, H'))$ *is an S-algebra.*

Proof. Let $(H, H')\colon S\left(\mathsf{A} \xrightarrow{F} \mathsf{B}\right) \to \left(\mathsf{A} \xrightarrow{F} \mathsf{B}\right)$ be an S-algebra. Then $H' = 1_{\mathsf{B}}$ and $F^{\S} \cdot H = 1_{\mathsf{A}}$. Finally $F^{\S}$ is an isomorphism (cf. Lemma 11).

If $F^{\S}\colon \mathsf{A} \to \mathsf{B}^R$ is an isomorphism with some $H\colon \mathsf{B}^R \to \mathsf{A}$ as inverse functor, then the pair $(H, 1_{\mathsf{B}})\colon S\left(\mathsf{A} \xrightarrow{F} \mathsf{B}\right) \to \left(\mathsf{A} \xrightarrow{F} \mathsf{B}\right)$ is a morphism from $\mathsf{R}(\mathsf{Cat} \downarrow \mathsf{Cat})$ and this is really an S-algebra structure. The pair $(H, 1_{\mathsf{B}})$ is unique, because $F^{\S}$ has a unique inverse.

References

[0] S. ALAGIČ: Natural state transformation, COINS Technical Report, 73 B-2 (1973).
[1] E. DUBUC: KAN Extension in Enriched Category, Lect. Notes in Math. 145 (1970).
[2] S. MAC LANE: Categories for the working mathematician, New York-Heidelberg-Berlin, 1971.
[3] J. SŁOMIŃSKI: A representation of the category of algebras over different monads by the category of algebras over one suitable monad in a category of pointed monads, Coll. Math. (in print).

An algebraic approach to Scott's flow diagrams

by Grzegorz Jarzembski, Toruń

One of the most important ideas in the theory of programs is a "lattice of flow diagram" introduced by D. Scott [1]. In this paper it will be shown, that this problem can be considered in terms of endofunctor algebra [2] and monadic algebra. The terminology used in this paper is the same as in [2] and [3].

§ 1. Category of endofunctor and monadic algebras

Let C be any category with object class |C|, and let R: C → C be any endofunctor.

Definition 1.1. A pair (A, a) where $A \in |C|$ (carrier of algebra) and $a: RA \to A$ is a morphism in a category C (structure of algebra) will be called an *R-algebra*.

Let (A, a) and (B, b) be two R-algebras.

Definition 1.2. A morphism $h \in C(A, B)$ such that the diagram

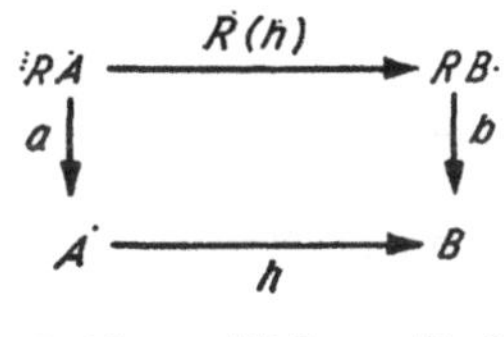

is commutative will be called an *R-morphism*.

The R-algebras together with their R-morphisms form a category which will be denoted by Alg_R C [2].

Let $T = (T, \mu, \eta)$ be a monad [3] over C.

Definition 1.3. By C^T we will denote the full subcategory of Alg_R C such that $(A, a) \in C^T$ iff the two diagrams

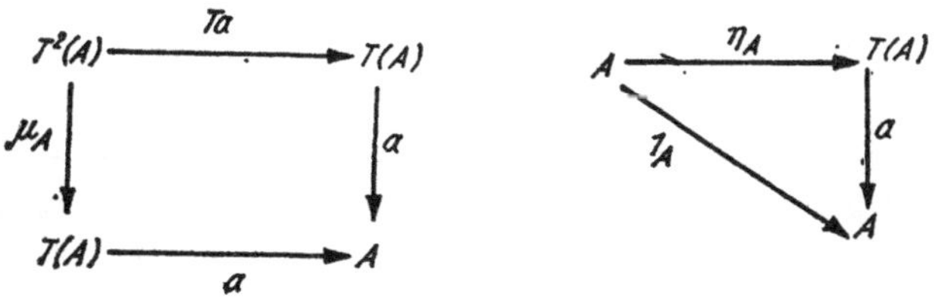

are commutative.

In natural way we can define a forgetful functor $U\colon \mathsf{Alg}_R \mathsf{C} \to \mathsf{C}$. We consider the problem of the existence of a left adjoint $V\colon \mathsf{C} \to \mathsf{Alg}_R \mathsf{C}$ to the functor U. This is equivalent to the problem of the existence of a "free R-algebra" for every object $X \in |\mathsf{C}|$.

Definition 1.4. By a *free* R-algebra over an object $X \in |\mathsf{C}|$ we understand a pair (X^*, η_X) where $X^* = (X^*, x^*)$ is an R-algebra and $\eta_X\colon X \to U(X^*, x^*)$ is a morphism in C such that

$$\bigwedge_{(A,a)\in \mathsf{Alg}_R\mathsf{C}} \ \bigwedge_{f\colon (X,x)\to U(A,a)} \ \overset{1}{\bigvee_{\tilde{f}\colon (X,x)\to(A,a)}} \quad U(\tilde{f}) \cdot \eta_X = f\,.$$

We present a special construction of a free R-algebra which will be used later.

Let C be a category with finite coproducts. Then we define a functor (for a given object $X \in |\mathsf{C}|$) $R_X\colon \mathsf{C} \to \mathsf{C}$ by the rules $R_X A = X \sqcup RA$ and $R_X f = 1_X \sqcup Rf$.

Thus we have

Lemma 1. [4] *$((X^*, x^*), \eta_X)$ is a free R-algebra over X iff $(X^*, \tilde{x})$ is an initial object in the category $\mathsf{Alg}_{R_X} \mathsf{C}$ (where $\tilde{x}$ is naturally determined by the pair (η_X, x^*)).*

Lemma 2. [4] *If (A, a) is an initial object in the category of R_X-algebras then a is an isomorphism.*

(We omit the easy proofs of those lemmas.)

Let ω be the category of natural numbers, and let $H\colon \omega \to \mathsf{C}$ be the functor defined as follows:

$$H(0) = X\,, \qquad H(n+1) = R_X H(n)\,,$$

$$H(0 \to 1) = i_X\colon X \to R_X X = X \sqcup RX \text{ where } i_X \text{ is a coproduct inclusion,}$$

$$H(n+1 \to n+2) = R_X H(n \to n+1)\,.$$

Theorem 1.3. *Suppose that* colim H *exists in a category* C.

Then colim H *is the carrier of a free R-algebra over X iff R_X preserves this colimit.*

Proof. It holds $\operatorname{colim} R_X H = \operatorname{colim} H$ because of $R_X H = H \cdot \operatorname{succ}$ where $\operatorname{succ}\colon \omega \to \omega$ is a cofinal functor defined as follows

$$\operatorname{succ}(n) = n+1\,, \qquad \operatorname{succ}(n \to n+1) = (n+1 \to n+2)\,.$$

Then the necessity of our condition follows directly from Lemma 1.1 and Lemma 1.2.

We show that the condition is sufficient. We have

$$i\colon \operatorname{colim} H \xrightarrow{\cong} R_X \operatorname{colim} H\,,$$

where $i \cdot f_n = R_X(f_{n-1})$ $(f_n\colon H(n) \to \operatorname{colim} H$ is the colimit inclusion$)$. And we define $j = i^{-1}$.

We show, that $(\operatorname{colim} H, j)$ is an initial object in $\mathsf{Alg}_{R_X} \mathsf{C}$. Let $(A, a) \in |\mathsf{Alg}_{R_X} \mathsf{C}|$. The morphism $a\colon R_X A \to A$ gives us a pair of morphisms $a_1\colon X \to A$ and $a_2\colon RA \to A$ by definition of R_X. We define a morphism $h\colon (\operatorname{colim} H, j) \to (A, a)$, i.e., a family

$$h_n\colon H(n) = R_X{}^n X \to A$$

where

$$h \cdot f_0 = h_0 = a_1 , \qquad h \cdot f_n = h_n = a \cdot R_X(h_{n-1}) .$$

We have to show, that h is an R_X-morphism, i.e.,

$$h = a \cdot R_X h \cdot i ,$$

$$h \cdot f_n = h_n = a \cdot R_X(h \cdot f_{n-1}) = a \cdot R_X h \cdot i \cdot f_n \quad \text{for every } n .$$

The proof that h is unique is obvious.

§ 2. Endofunctor algebras over the category of lattices

In this section we will consider endofunctors and endofunctor algebras over the category Lat defined as follows:

Definition 2.1. By Lat we denote the category where objects are the complete lattices and morphisms are all functions, which preserve least upper bounds of any (may be empty) subset.

Lemma 2.1. *The category* Lat *has finite products and coproducts and*

$$A \times B = A \sqcup B \quad \text{for any} \quad A, B \in |\mathsf{Lat}| .$$

Lemma 2.2. *Let* $E\colon \omega \to \mathsf{Lat}$ *be a functor such that for each natural number* n $E(n \to n+1)$ *is an s-split mono,* i.e., $E(n \to n+1) = m_n$ *and there exists* $e_n\colon E(n+1) \to En$ *such that* $e_n m_n = 1$ *and* $m_n e_n \sqsubseteq 1$.

Then colim E *exists and*

$$\operatorname{colim} E = \{(x_1, x_2, x_3, \dots, x_n, \dots) \colon \bigwedge_n x_n \in En, x_n = e_n(x_{n+1})\} ,$$

where $(x_1, x_2, \dots, x_n, \dots) \sqsubseteq (y_1, y_2, \dots, y_n, \dots) \Leftrightarrow \bigwedge_n x_n \sqsubseteq y_n$.

(Proof is obvious).

We will denote

$$f_{n\infty}\colon En \to \operatorname{colim} E \text{ (colimit inclusion)} ,$$

where

$$f_{n\infty}(x) = (\dots , e_{n-2}x, e_{n-1}x, x, m_n x, m_{n+1}x , \dots)$$

and

$$g_{n\infty}\colon \operatorname{colim} E \to En ,$$

where

$$g_{n\infty}(x_1, x_2, \dots , x_n, \dots) = x_n .$$

Then, of course, it holds:

Lemma 2.3.

a) $$g_{n\infty} \cdot f_{n\infty} = 1_{En} ,$$

b) $$f_{n\infty} g_{n\infty} \sqsubseteq 1_{\operatorname{colim} E} ,$$

c) $$\bigsqcup_{n \in N} f_{n\infty} g_{n\infty} = 1_{\operatorname{colim} E} .$$

Theorem 2.1. *Let R: Lat $\to$ Lat be a functor such that for each family of morphisms $\{\varphi_i : A \to B\}_{i \in I}$ we have*

$$R(\sqcup \varphi_i) = \sqcup \, (R(\varphi_i)) \, .$$

Then for each object $X \in |\mathrm{Lat}|$ the colim H exists and R_X preserves this colimit (where R_X and H are constructed as in § 1).

Proof: The inclusion $i_X : X \to R_X X = X \sqcup RX$, $i_X(x) = (x, \bot)$, is an s-split mono

$$(e_X : R_X X \to X \text{ with } e_X(x, y) = x \text{ and } R_X^n(i_X e_X) \sqsubseteq 1) \, .$$

Thus — by Lemma 2.2 — colim H exists. We prove, that R_X preserves this colimit. Let us consider the diagram

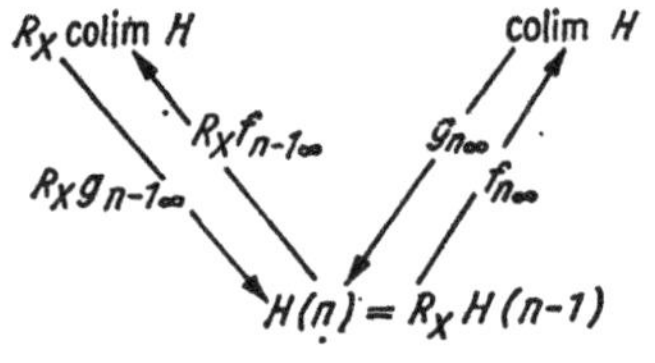

We define

$$j = \bigsqcup_{n=1}^{\infty} f_{n\infty} R_X(g_{n-1\infty}) \quad \text{and} \quad i = \bigsqcup_{n=1}^{\infty} R_X(f_{n-1\infty}) \, g_{n\infty} \, .$$

Then

$$j \cdot i = 1 \quad \text{and} \quad i \cdot j = 1 \, .$$

Example — "*Lattice of flow diagrams*".

Let B be any complete lattice. Let us consider the endofunctor

$$S : \mathrm{Lat} \to \mathrm{Lat} \, , \qquad S(X) = B \times X^2 \sqcup X^2 \, .$$

This endofunctor satisfies the condition of Theorem 2.1. From this it follows, that for each complete lattice X a free S-algebra over X exists. It is easy to see, that it is the same as a lattice of flow diagrams [1] with "alphabet" X. From this it follows that the semantics of flow diagrams may be interpreted as an S-morphism.

In a similar way we can define an endofunctor for a "*generalized flow diagram*" considered by WOJDYŁO [5].

Let V: C $\to$ Alg_R C denote a left adjoint to the forgetful functor U. It is easy to prove:

Lemma 2.4. $\mathrm{Alg}_R \, \mathrm{C} \cong \mathrm{C}^{UV}$.

(Proof is obvious by BECK's [3] theorem.)

Corollary 1. *UV is a free monad over R.*

Corollary 2. *A lattice of flow diagrams may be considered as a monadic algebra.*

References

[1] D. Scott: Lattice of flow diagrams, Lecture Notes in Math., 188 (1971), 311—366.
[2] J. Słomiński: A representation of the category of algebras over different monads by the category of algebras over one suitable monad in a category of pointed monads, Preprint No. 26, UMK Toruń.
[3] S. Mac Lane: Categories for the working mathematician, Berlin 1972.
[4] M. Barr: Coequalizers and free triples, Math. Z. 116 (1970), 307—322.
[5] B. Wojdyło: The lattices with continuous operators and their utilizing to description of generalized programs — a generalization of Scott's flow diagrams, Preprint No. 25, UMK Toruń, 1975.

Categorial realization theory. I

by JIŘÍ ADÁMEK, Prague

Contents

I. Why categories?

Let us mention three possible aspects in which category theory may turn out to be useful for automata theories.

1. Categories make it possible to study efficiently automata with a certain structure on their state sets. The most important examples are the linear automata (whose states form a vector space), and there are others: topological automata, group automata, etc. The functorial approach of ARBIB and MANES allows a unified theory of tree machines, as well: tree machines are "sequential" in a functorial sense.

2. The functorial theory presents a general background, in which "classical" sequential and tree machines are a very special case. And yet, from substantial points of view, this special case is the only one that works. As will be seen later, tree machines are the only functorial machines which allow minimal realization (!); and sequential machines are the only functorial machines for which minimal realization is universal.

Thus, category theory shows how natural the automata theory is by studying it in a more general setting. This applies also to questions such as: what is the role of finiteness? why can we restrict ourselves to outputs $\{0, 1\}$?, and others.

3. Categorial automata theory is elegant. It unifies various approaches and concepts in a natural way (the "natural" is meant both in the categorial and in the usual sense). It shows the deep relations between automata theory and algebra.

II. The model of ARBIB and MANES

In this section we recall the foundations of the categorial automata theory, investigated by ARBIB and MANES. For more details see their papers, e.g. [5]. Those who are acquainted with these papers may skip II. 1—5.

II.1. Motivation. Let us mention three basic types of automata.

A) *Sequential Σ-machine* (initial, MOORE), with Σ = set of inputs, consists of

a set Q (of states),
a map $\delta: Q \times \Sigma \to Q$ (the next-state map),
a map $\beta: Q \to Y$ (Y is the set of outputs, β is an output map)
$q_0 \in Q$ (the initial state), this can be viewed upon as a mapping $\tau: I \to Q$ where $I = \{0\}$ and $\tau(0) = q_0$.

Denote by F the functor F: **Set** $\to$ **Set** (**Set** is the category of sets and mappings) defined by $FX = X \times \Sigma$ and, for $f: X \to Y$, by $Ff = f \times \mathrm{id}_\Sigma$. Then a sequential machine is the following diagram in **Set**:

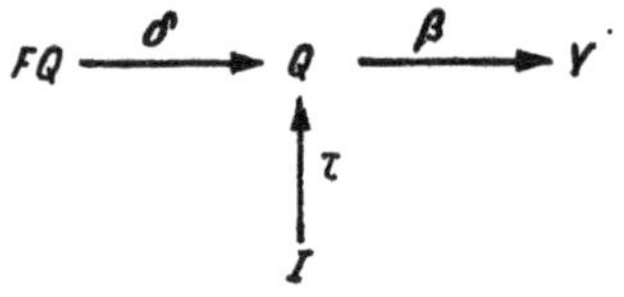

B) *Tree machine* of type $\Omega = \{n_i; i \in I\}$, where n_i are natural numbers, consists of

a set Q (of states),
operations $\delta_i: Q^{n_i} \to Q$, $i \in I$,
a map $\beta: Q \to Y$ (the output map),
$q_0 \in Q$ (or, $\tau: I \to Q$).

Considering the mappings $\delta_i: Q^{n_i} \to Q$, $i \in I$, as one mapping from the disjoint union, $\delta: \coprod_{i \in I} Q^{n_i} \to Q$, we obtain the same diagram as above, with $F = F_\Omega$: **Set** $\to$ **Set** obtained by $FX = \coprod_{i \in I} X^{n_i}$ and, for $f: X \to Y$, by $Ff = \coprod_{i \in I} f^{n_i}$.

C) *Bilinear sequential Σ-machine* (with Σ = linear space of inputs) consists of

a linear space Q (of states),
a linear map $\delta: Q \otimes \Sigma \to Q$ (the next-state map; $\otimes$ is the tensor product).

Considering here the trivial initialization $\tau: I \to Q$ where $I = \{0\}$, we obtain the same diagram as above, this time in the category **Vect** of vector spaces and linear maps, with F: **Vect** $\to$ **Vect** defined by $FX = X \otimes \Sigma$, $Ff = f \otimes \mathrm{id}_\Sigma$.

II.2. Let K be a category, let F: K → K be a functor. By an F-automaton we mean $M = (Q, \delta, Y, \beta, I, \tau)$ given by the following diagram in K:

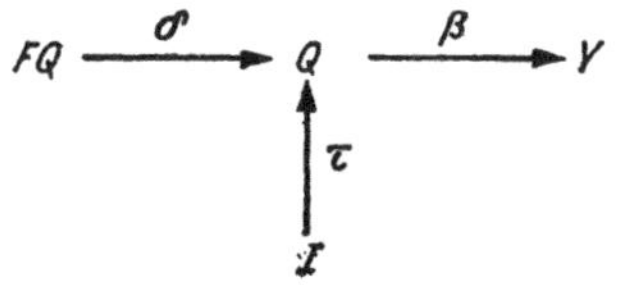

i.e., Q, Y and I are objects of K; δ, β and τ are morphisms. We write just $M = (Q, \delta, \beta, \tau)$ if I and Y are explicitely stated.

For fixed Y and I, F-automata present a new category K(F, Y, I) whose objects are F-automata and whose morphisms p: $M \to M'$, called *machine morphisms*, are such morphisms p: $Q \to Q'$, for which the following diagrams commute

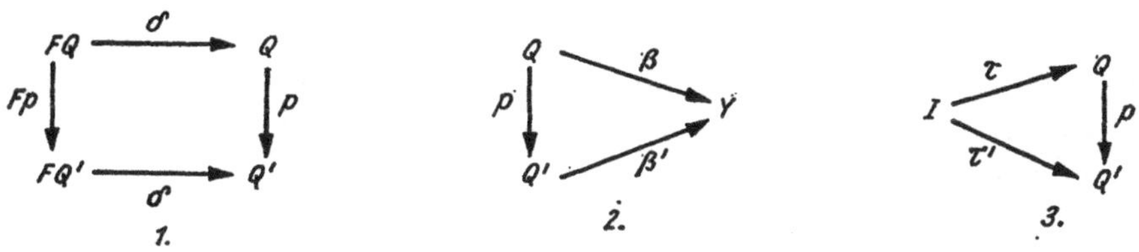

It is easy to see that, in the above examples A), B), C), these machine morphisms agree with the usual definitions (for machines with common sets of inputs and outputs!).

II.3. The algebraic theory of semigroups is a background for sequential machines theory; the background for F-automata is presented by the algebraic theory of F-algebras. For a more detailed study of functorial (triple-less) algebra see [4].

By an *F-algebra* is meant a pair (Q, δ), where Q is an object of K and δ: $FQ \to Q$ is a morphism. An *F-homomorphism* f: $(Q, \delta) \to (Q', \delta')$ is such a morphism f: $Q \to Q'$ for which $f \cdot \delta = \delta' \cdot Ff$ (i.e., the above square 1. commutes).

In agreement with universal algebra, by a *free F-algebra*, generated by an object X, we mean an F-algebra $(X^{\#}, \varphi^X)$ and a morphism $s: X \to X^{\#}$ such that every morphism f: $X \to Q$ into an F-algebra (Q, δ) can be uniquely extended to a homomorphism from $(X^{\#}, \varphi^X)$:

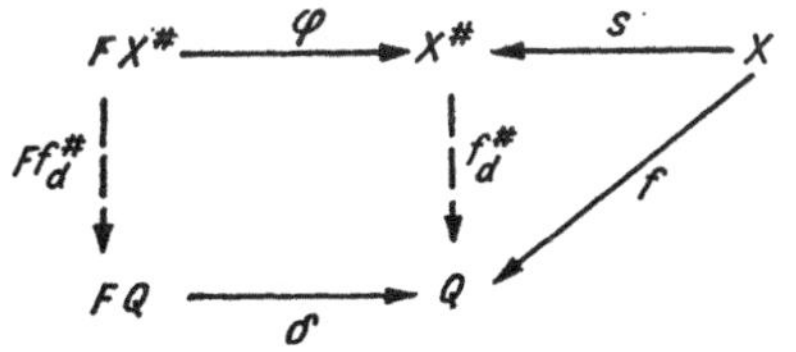

The last means that there exists a unique F-homomorphism $f_d^{\#}$: $(X^{\#}, \varphi^X) \to (Q, \delta)$ such that $f = f_d^{\#} \cdot s$.

Example. K = Set, $F = - \times \Sigma$ (see II.1. A). Then $X^{\#} = X \times \Sigma^*$ and φ: $(X \times \Sigma^*) \times \Sigma \to X \times \Sigma^*$ is the usual concatenation:

$$\varphi(x, s, \sigma) = (x, s\sigma) .$$

In particular, for a singleton set I we may identify $I^{\#}$ with the free monoid Σ^*.

The existence of free F-algebras presents a problem, to which we devote Section III. Following ARBIB and MANES we say that F is an *input process* if every object of K generates a free F-algebra.

II.4. For input processes F we can study F-machines dynamically. First we recall that in case of a sequential Σ-machine $M = (Q, \delta, \beta, \tau)$ (here $I = \{0\}$ and Y is fixed but arbitrary) we extend δ to δ^* and we define the *run map* $r \colon \Sigma^* \to Q$ recurrently by

$$\left.\begin{aligned} r(\Lambda) &= \tau(0) \\ r(s\sigma) &= \delta\big(r(s), \sigma\big) \quad \text{for} \quad s \in \Sigma^*, \sigma \in \Sigma \end{aligned}\right\} \quad r(s) = sq_0 .$$

From the algebraic point of view, $\Sigma^* = I^{\#}$ with $I = \{0\}$; and r is the extension of τ: $I \to Q$ to a homomorphism $r = \tau_\delta^{\#}$: $(I^{\#}, \varphi^I) \to (Q, \delta)$.

Analogously, for a general F-machine $M = (Q, \delta, \beta, \tau)$ with F an input process we define the run morphism $r \colon I^{\#} \to Q$ by $r = \tau_\delta^{\#}$.

The external behaviour of an initial sequential machine is the mapping $\Sigma^* \to Y$, sending a string s to $\beta(sq_0)$, i.e. to $\beta \cdot r(s)$. Generally, the *behaviour* of an F-machine M is $f_M = \beta \cdot r \colon I^{\#} \to Y$:

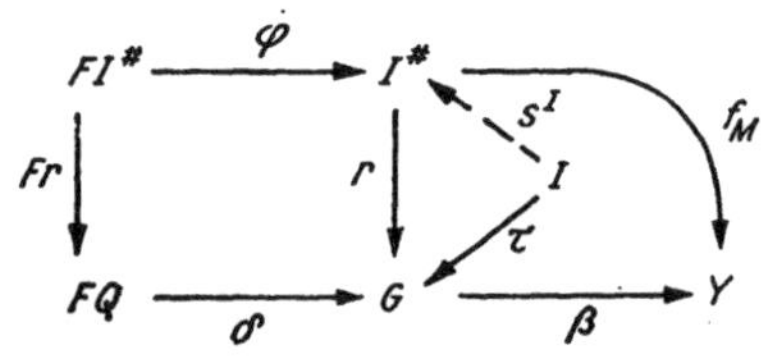

Notice that, whenever a machine morphism between F-machines M and M' exists, then $f_M = f_{M'}$. Indeed, given $p \colon M \to M'$, then $f_M = \beta \cdot r = \beta' \cdot p \cdot r$ and $p \cdot r = r'$ (clearly $p \cdot r$ is a homomorphism and, since $p \cdot r \cdot s^I = p \cdot \tau = \tau'$, necessarily $p \cdot r$ is an extension of τ' to a homomorphism. By uniqueness of extensions, $p \cdot r = (\tau')_{\delta'}^{\#} = r'$). Thus $f_M = \beta' \cdot r' = f_{M'}$.

II.5. The problem of synthesis for F-automata can be stated as follows: given a morphism $f \colon I^{\#} \to Y$, find an F-machine $M = (Q, \delta, \beta, \tau)$ whose behaviour is f; then M is said to be a *realization* of the behaviour f. Every behaviour has a "maximal", or a "free", realization, viz.

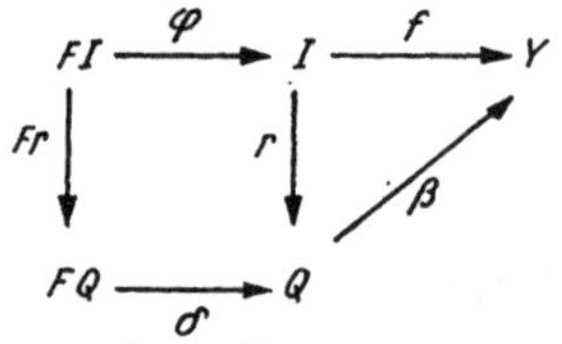

Realization

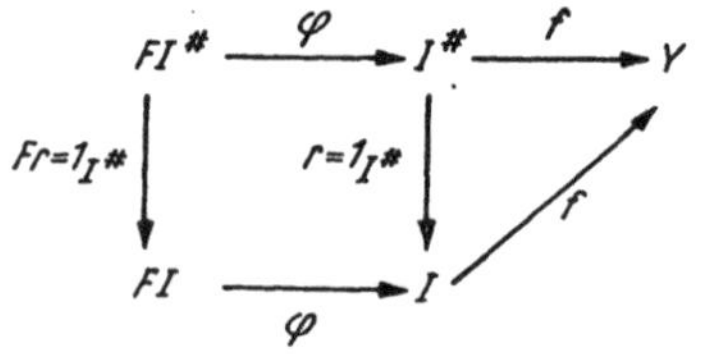

Free realization

$M = (I^{\#}, \varphi^I, s^I, f)$. Indeed, here $r = 1_{I^{\#}}$ (because 1 is a homomorphism and $r \cdot s^I = s^I$ — thus, necessarily $1 = (s^I)_{\varphi I}^{\#}$) and so the behaviour of M is $f_M = f \cdot 1_{I^{\#}} = f$. But what we are interested in is a minimal realization. Minimality has been defined originally (for finite systems) by the number of states. But, fortunately, for reachable systems, minimality can be expressed by means of morphism, too: a system, which is reachable, is minimal iff it simulates all other systems.

A sequential machine is *reachable* if it "uses all of its states", i.e. if the map $e = \delta^*(q_0, -): \Sigma^* \to Q$ is onto (every state $q \in Q$ can be reached from $q_0 = \tau(0)$, i.e. $q = r(s)$ for some string $s \in \Sigma^*$). The concept "onto" is categorially expressed by epimorphisms.

Definition. An F-machine M is *reachable* if $r(= \tau_\delta^\#)$ is an epimorphism. A *minimal realization* of a behaviour $f: I \to Y$ is its reachable realization M_0 such that the following holds: for every reachable realization M of f there exists a (necessarily unique) machine morphism from M to M_0.

II.6. The sections II,1—II,5 are a brief account of the model of ARBIB and MANES. This model does not pick up the theory of finite machines. To do this, we introduce a new parameter — a class $\mathfrak{E}$ of epimorphisms of K. The basic examples:

$\mathsf{K} = \mathsf{Set}$	is the category of sets,
$\mathfrak{E} = \mathfrak{E}_0$	is the class of all mappings onto $f: X \to Y$ (this leads to infinite machines),
$\mathfrak{E} = \mathfrak{E}_f$	is the class of all mappings onto a finite set $f: X \to Y$ with Y finite (this leads to finite machines),
$\mathsf{K} = \mathsf{Vect}$	is the category of vector spaces,
$\mathfrak{E} = \mathfrak{E}_0$	is the class of all linear mappings onto,
$\mathfrak{E} = \mathfrak{E}_f$	is the class of all linear mappings onto a finite-dimensional space.

A finite and reachable machine is then an F-machine M with $r \in \mathfrak{E}$, in case $\mathfrak{E}$ models finite quotients $f: X \to Y$ (Y is a "finite object" of K, however defined). If $\mathfrak{E}$ models all quotients then $r \in \mathfrak{E}$ means just that M is reachable. In this way we proceed simultaneously with the theory of finite machines and with the theory of arbitrary machines, depending only on the choice of the class $\mathfrak{E}$.

II.7. Given a behaviour $f: I^\# \to Y$, its $\mathfrak{E}$-reachable realization is an F-machine M, whose behaviour is f and which fulfils $r \in \mathfrak{E}$ (compare with the last definition in II,5). An *$\mathfrak{E}$-minimal realization* is an $\mathfrak{E}$-reachable realization M_0 with a machine morphism $M \to M_0 \in \mathfrak{E}$ for any other $\mathfrak{E}$-reachable realization M.

The basic difference between $\mathfrak{E}$-reachability and reachability is that not every behaviour has any $\mathfrak{E}$-reachable realization (which fact leads to the term "regular language" in the usual sequential machine theory). Indeed, we do not assume that $\mathfrak{E}$ contains identity morphisms — such an assumption would spoil finitarity, of course. Thus, the free realization of f (II.5) need not be $\mathfrak{E}$-reachable. We shall call a behaviour f *$\mathfrak{E}$-regular* if it has any $\mathfrak{E}$-reachable realization. E.g., a language $A \subseteq \Sigma^*$ is a regular event iff the corresponding behaviour $f: \Sigma^* \to \{0, 1\}$ is $\mathfrak{E}_f$-regular.

The problem of realizations has two steps, now. Given a behaviour f,

1) is f $\mathfrak{E}$-regular?

2) if so, has f an $\mathfrak{E}$-minimal realization?

The first question is very complicated even in the classical case of sequential machines. The latter has an affirmative answer there. We shall be interested only in the second one in this paper. Thus, we shall say that F *admits* $\mathfrak{E}$*-minimal realization* if every $\mathfrak{E}$-regular behaviour has an $\mathfrak{E}$-minimal realization.

In case $\mathsf{K} = \mathsf{Set}$ or Vect there is a complete descriptive answer to the question, which functors admit $\mathfrak{E}_0$-minimal realization: just the finitary ones; for $\mathfrak{E}_f$ the answer is entirely different: every functor admits $\mathfrak{E}_f$-minimal realization. See Section IV.

II.8. In what follows we shall use the notion of co-intersection of quotients very often. That is the dual concept to intersection of subobjects. Let $e_i\colon X \to A_i$, $i \in I$, be a non-void collection of quotients of X, i.e. e_i are epis. (Here I might be a proper class.) Then a quotient $e\colon X \to B$ is their *co-intersection*, $e = \bigcap\limits_{i\in I}^{*} e_i$, if

1) there exist $k_i\colon A_i \to B$ with $e = k_i \cdot e_i$ for $i \in I$;

2) given morphisms $f_i\colon A_i \to C$ and $g\colon X \to C$ such that $g = f_i \cdot e_i$ for all i, there exists a (necessarily unique) f $B \to C$ with $g = f \cdot e$ and $f_i = f \cdot k_i$ for $i \in I$.
Shortly: a co-intersection is a multiple pushout of epis.

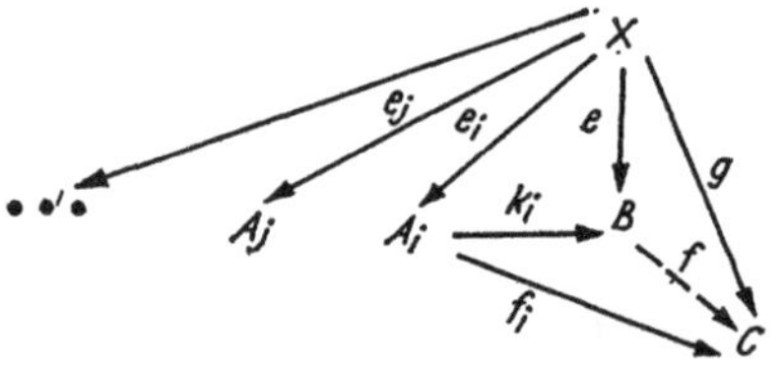

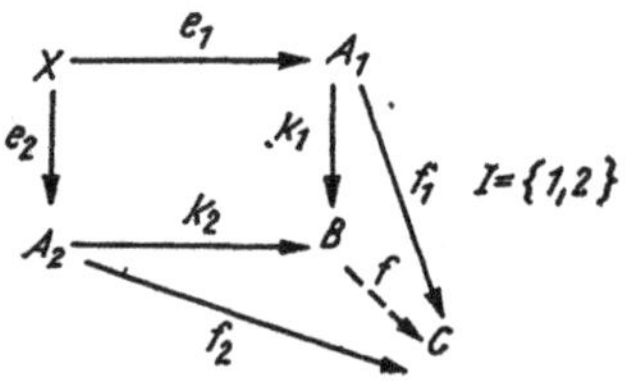

We shall say that a class of epis $\mathfrak{E}$ is *closed to co-intersections* if an arbitrary collection of $\mathfrak{E}$-epis with a common domain has a co-intersection, belonging to $\mathfrak{E}$. Given a functor $F\colon \mathsf{K} \to \mathsf{K}$ which preserves quotients (i.e., Fe is epi for $e \in \mathfrak{E}$), then we say that F *preserves* $\mathfrak{E}$*-co-intersections* if, whenever $e = \bigcap^{*} e_i$ in $\mathfrak{E}$, then $Fe = \bigcap^{*} Fe_i$.

We shall often need the following properties of $\mathfrak{E}$. Conditions S2., S3., S4. below are very natural and are fulfilled by all current classes $\mathfrak{E}$. For short, we shall say that $\mathfrak{E}$ is *standard* if

S1. $\mathfrak{E}$ is closed to co-intersections.

S2. $\mathfrak{E}$ is closed to finite sums, i.e. given $e_1\colon A_1 \to B_1$ and $e_2\colon A_2 \to B_2$ in $\mathfrak{E}$, then the natural morphism $e\colon A_1 \vee A_2 \to B_1 \vee B_2$ is in $\mathfrak{E}$, too.

S3. Any commutative square (A), where $e \in \mathfrak{E}$ and i is the summand-injection, can

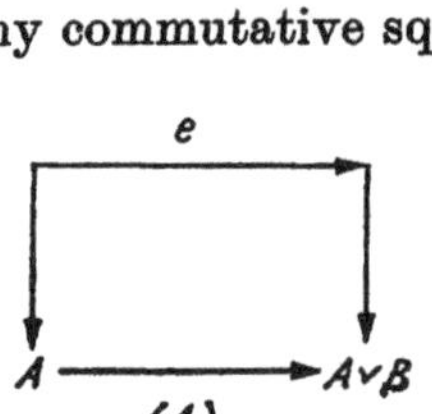

(A)

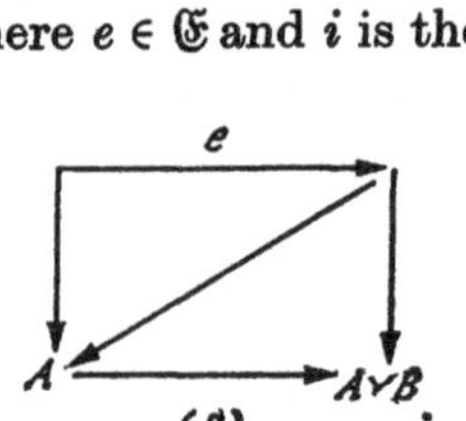

(B)

be completed to a commutative diagram (B).

S4. Given $e = e' \cdot r$ where $e' \in \mathfrak{E}$ and r is a retraction (i.e. $r \cdot r_1 = 1$ for some r_1) then also $e \in \mathfrak{E}$.

These conditions are quite natural. E.g., dropping finiteness for a while, consider $\mathfrak{E}$ as a part of a factorization system $(\mathfrak{E}, \mathfrak{M})$ for K. Then [8] proves that S2., S4. always hold and S3. says just that $i \in \mathfrak{M}$, i.e. that summands are subobjects of a sum. Finally, if K is co-complete and co-locally small, then S1. holds as well.

Furthermore, none of these conditions contradicts our intuitive concept of a finite quotient. In fact, stronger conditions (say, a quotient of a finite quotient is also finite) could be assumed. But we shall not need them.

II.9. Arbib and Manes also consider $\mathfrak{E}$-reachability, but only in case $\mathfrak{E}$ is a part of a factorization system $(\mathfrak{E}, \mathfrak{M})$; e.g., they use the class $\mathfrak{E}$ of all coequalizers. Since then $\mathfrak{E}$ contains all isomorphisms, there is no problem of $\mathfrak{E}$-regularity (the free realization is $\mathfrak{E}$-reachable).

III. Construction of free F-algebras

III. 1. This section is devoted to the existence of free F-algebras (defined in II.3). While in universal algebra free algebras always exist, this is not the case with F-algebras generally. The proof in the universal algebra is constructive: one constructs the algebra of terms and proves that it is free. We shall proceed analogously for F-algebras, defining a construction which, if it "stops", characterizes free F-algebras. Moreover, under additional assumptions, the converse is true: if a free F-algebra exists then the construction stops.

The below construction has been defined in [1] as a generalization of the one for set-functors from [12] and [11]. This construction itself is generalized to two endofunctors in [13], where lots of interesting properties are proved; see also [10].

We show first an interesting ("finitary") special case.

III.2. Let K be a co-complete category, let $F\colon \mathsf{K} \to \mathsf{K}$ be arbitrary. For a fixed object A define a sequence W_n in K as follows:

$$W_0 = A,\ W_1 = A \vee FA, \ldots, W_{n+1} = A \vee FW_n\,,$$

$s_0\colon W_0 \to W_1$ is the sum-injection ,

$s_{n+1}\colon W_{n+1} \to W_{n+2}$ is defined by s_n as

$$s_{n+1} = 1_A \vee Fs_n\ (\colon A \vee FW_n \to A \vee FW_{n+1})\,.$$

We have sum-injections from FW_n to W_{n+1}:

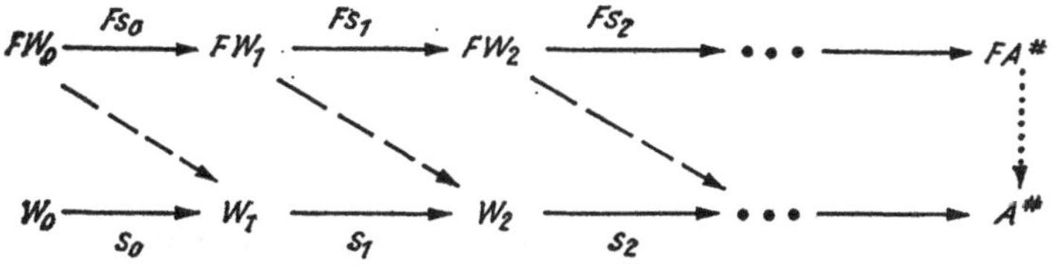

Therefore, denoting by $A^{\#}$ the colimit of the sequence (W_n, s_n) and assuming that the functor F preserves this colimit, we obtain a natural morphism $\varphi^A\colon FA^{\#} \to A^{\#}$. Then we say that the *free-algebra algorithm converges*. It can be proved rather easily that the F-algebra $(A^{\#}, \varphi^A)$ with the colimit-injection $s^A\colon A = W_0 \to A^{\#}$ is freely generated by A.

III.3. The above free-algebra algorithm generalizes the situation with finitary universal algebras (then $A = W_0$ are variables, W_1, ... are terms of order 1, ...). For infinitary algebras this algorithm does not stop and one must use a "chain construction" rather than the above "sequence construction". Let K, F be as above.

We shall define objects W_i and morphisms $s_{ij}\colon W_j \to W_i$ for arbitrary ordinals i, j with $j \leqq i$. Put $s_{ii} = 1$; $s_{ij} \cdot s_{jk} = s_{ik}$.

1) $W_0 = A$; $W_{i+1} = A \vee FW_i$ (with injections $v_i\colon A \to W_{i+1}$, $w_i\colon FW_i \to W_{i+1}$).

2) $s_0 = v_0\colon A \to A \vee FA$, $s_{i+1,i+2} = 1_A \vee Fs_{i,i+1}$ $(\colon A \vee FW_i \to A \vee FW_{i+1})$.

3) If i is a limit ordinal, then W_i is the colimit of the preceding chain $(\{W_j\}_{j<i}, \{(s_{j_1 j_2}\}_{j_2 \leqq j_1 \leqq i})$ with injections $s_{ij}\colon W_j \to W_i$.

4) For i limit define $s_{i+1,i}\colon W_i \to A \vee FW_i$ by

$$s_{i+1,i} \cdot s_{i,0} = v_i(\colon A \to A \vee FW_i)\,,$$
$$s_{i+1,i} \cdot s_{i,j+1} = 1_A \vee Fs_{ij}(\colon A \vee FW_j \to A \vee FW_i)\,.$$

We say that the *free-algebra construction stops after* α *steps* (α limit) if $s_{\alpha+1,\alpha}$ is an isomorphism. Then we have $\varphi^A = s_{\alpha+1,\alpha}^{-1} \cdot w_\alpha\colon FW_\alpha \to W_\alpha$. Put $W_\alpha = A^{\#}$.

$$FW_\alpha \xrightarrow{w_\alpha} A \vee FW_\alpha \xrightarrow{s_{\alpha+1,\alpha}^{-1}} W_\alpha\,.$$

Then the F-algebra $(A^{\#}, \varphi^A)$, with $s^A = s_{\alpha,0}\colon A \to A^{\#}$ is freely generated by A. The proof is inductive — see [1].

III.4. We say that F is a *constructive input process* if the free-algebra construction stops for every object A; analogously: an algorithmic input process. These notions are closely related to excessivity: F is said to be *non-excessive* if for any object A there exists an object B, isomorphic to the sum $A \vee FB$.

In [14] a characterization of input processes via the free-algebra construction is exhibited. Mild additional assumptions on K are needed, viz.:

M1. K is co-complete and locally small (no object has a proper class of subobjects).

M2. Summand-injections $H \to H \vee K$ are always monics.

M3. The category K^M, whose objects are the same as in K and whose morphisms are all monics of K, has colimits of chains. These colimits coincide with those in K.

Theorem. *Let* K *fulfil* M1—M3 *and let* $F\colon \mathsf{K} \to \mathsf{K}$ *be any functor, preserving monics. Then the following conditions are equivalent:*

(i) *F is an input process.*

(ii) *F is a constructive input process.*

(iii) *F is non-excessive.*

III.5. Examples. If K is the category of sets or of vector spaces, then it fulfils M1—M3. Moreover, all monics split (but the void monics $\emptyset \to X$ in **Set**, which do not play any serious role), thus every functor $F\colon \mathsf{K} \to \mathsf{K}$ preserves monics. Furthermore, the condition: "$B = A \vee FB$" in these categories means only "card $B =$ $=$ card $A +$ card FB". Thus we obtain

Corollary. Let K be Set or Vect. Then a functor F: K → K is an input process iff it is non-excessive, which means: for every cardinal α there exists a set (a space) B such that card $FB \leqq$ card $B = \alpha$.

III.6. If F is an input process, then a new functor $F^{\#}$: K → K arises, sending an object A to $A^{\#}$ and a morphism $f\colon A \to B$ to $(s_B \cdot f)^{\#}_{\varphi B}\colon A^{\#} \to B^{\#}$. This functor is a triple, analogous to the free-algebra triples in universal algebra. It is rather easy to see that $F^{\#}$ is a *free triple* in the sense of [6]. A deep result of Barr states the converse (for suitable categories K): If F generates a free triple then it is an input process. Thus, the question of input processes is closely related to triple algebra.

Barr also proves that F is an input process if it preserves colimits of chains of monics of a given length (while, from the free-algebra construction it only follows — immediately — that F is an input process if it preserves colimits of all chains of a given length). These and other conditions are investigated carefully in [13], [10].

III.7. The importance of the *algorithmic* input processes (those, for which the algorithm in III.2 converges) for the realization theory lies in the fact that, for suitable categories K, the following holds: if an input process admits minimal realizations then it is necessarily algorithmic, see [2]. These "suitable" categories include Set, Vect and other varieties of finitary algebras, categories of graphs etc.

IV. Realization theorem

IV.1. Throughout this section we assume that a co-complete category K is given, together with a class $\mathfrak{E}$ of its epimorphisms, which is standard (see II.8). Furthermore, we consider a functor F: K → K.

By an *$\mathfrak{E}$-congruence* on an F-algebra (X, δ) is meant a morphism $e\colon X \to Y$ in $\mathfrak{E}$ such that there exists an F-algebra (Y, δ_1), making e a homomorphism, $e\colon (X, \delta) \to$ $\to (Y, \delta_1)$.

IV.2. Definition. We say that F *preserves co-intersections of $\mathfrak{E}$-congruences*, if $F \bigcap^* e_i = \bigcap^* Fe_i$ holds for any non-void collection e_i of $\mathfrak{E}$-congruences of some F-algebra.

Proposition [2]. *Every input process, preserving co-intersections of $\mathfrak{E}$-congruences, admits $\mathfrak{E}$-minimal realization.*

Realization theorem I [2]. *Let F be an input process, preserving quotients* (*i.e.* $F(\mathfrak{E}) \subseteqq \mathfrak{E}$).

Then F admits minimal realization iff it preserves co-intersections of $\mathfrak{E}$-congruences.

IV.3. The condition "F preserves co-intersections of $\mathfrak{E}$-congruences" is unpleasant because the concept of a congruence engages in itself F-algebras. The natural question is if this condition cannot be strengthened to "F preserves all $\mathfrak{E}$-co-intersections". Generally, the answer is negative: there exist input processes, not preserving all $\mathfrak{E}$-co-intersections, which admit $\mathfrak{E}$-minimal realizations. To strengthen the above theorem, we shall have to assume a bit more about K.

Recall that an object D is initial, if to any other object X there leads exactly one morphism $D \to X$ (examples: $\emptyset$ in **Set**; the trivial space $\{\vec{0}\}$ in **Vect**). Dually, an object T is terminal if there always leads just one morphism $X \to T$ (examples: any singleton set in **Set**; $\{\vec{0}\}$, again, in **Vect**). We say that a category K is *connected* if for arbitrary non-initial objects X, Y there exists some morphism $X \to Y$. Examples: **Set**, **Vect**, groups, topological spaces etc. are connected categories; graphs and unary algebras form typical non-connected categories.

IV.4. Recall that we assume that K is a complete category and $\mathfrak{E}$ is a standard class of epis.

Realization theorem II [2]. *Let* K *be connected and have a terminal object; let F be an input process with $F(\mathfrak{E}) \subseteq \mathfrak{E}$ and $FT = T$.*

Then F admits $\mathfrak{E}$-minimal realization iff it preserves $\mathfrak{E}$-co-intersections.

IV.5. Examples: **Set** and **Vect**. The categories K $=$ **Set** or **Vect** fulfil all of the assumptions of IV.4 for both $\mathfrak{E}_0$ and $\mathfrak{E}_f$ (see II.6). Thus the problem of minimal realizations, respectively of finite minimal realizations for regular behaviours, reduces to the preservation of co-intersections. The following is proved in [2].

Theorem. *Every input process F*: **Set** $\to$ **Set** *or F*: **Vect** $\to$ **Vect** *admits finite minimal realization.*

The description for $\mathfrak{E} = \mathfrak{E}_0$ involves finitarity of functors. A functor F: **Set** $\to$ **Set** is *finitary* if for an arbitrary set X and a point $a \in FX$ there exists a finite subset $Y \subseteq X$ and a point $a' \in FY$ such that $Fi(a') = a$ holds for the inclusion map i: $Y \to X$. Analogously for **Vect** (here we consider a finite-dimensional space $Y \subseteq X$). The part about **Set** in the theorem below was proved by [16].

Theorem. *A functor F*: **Set** $\to$ **Set** *or F*: **Vect** $\to$ **Vect** *is an input process admitting minimal realizations iff it is finitary.*

A lot of interesting conditions are equivalent with finitarity. E.g.: F preserves filtered colimits; or (for **Set**) F is a quotient functor of some F_Ω (see II.1. B) and (for **Vect**) a normal finitary functor is equivalent to some $- \otimes \Sigma$ (see II.1. C). This has an interesting interpretation: if F is a quotient of F_Ω, then F-automata can be considered as a variety of F_Ω-automata.

Corollary. *The only F-machines in* **Set** *which have minimal realizations are varieties of tree machines. And in* **Vect**, *with F normal, these are only the classical linear sequential machines.*

V. Anti-GOGUEN: Realization is not universal

V.1. An interesting result of [9] for sequential machines states that *realization is universal*. This means that "behaviour" (as a functor from the category **Mach** of machines to the category **Beh** of behaviours) is left adjoint to "minimal realization" (as a functor from **Beh** to **Mach**). Another way, how to formulate it, is: minimal machines form a reflective subcategory of reachable machines. More in detail: for a reachable machine M denote by M^r the minimal realization of its behaviour f_M; we have a quotient $\varrho^M: M \to M^r$ (which is just the reduction of M, of course). Then universality of realization means: for every machine morphism $p: M \to \overline{M}$ there exists a unique morphism $p^r: M^r \to \overline{M}^r$ with $p^r \cdot \varrho^M = \varrho^{\overline{M}} \cdot p$:

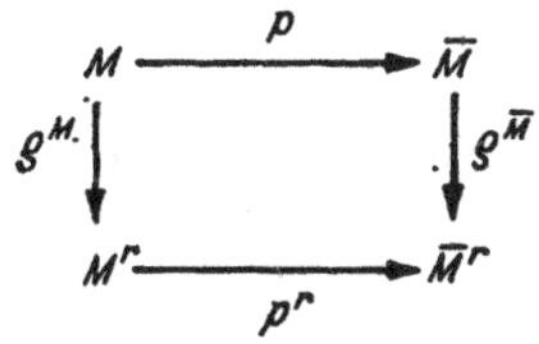

V.2. Surprisingly, GOGUEN's result cannot be extended beyond sequential machines. E.g., realization *is not universal for tree machines*. Universal realization for F-machines was defined in [18] and it is investigated in [15]; let us mention one result.

Assume that $F: \mathsf{K} \to \mathsf{K}$ is an input process which admits $\mathfrak{E}$-minimal realizations. Then, as above, for each $\mathfrak{E}$-reachable machine M there exists its reduction $\varrho^M: M \to M^r$ which is the minimal realization of the (regular!) behaviour f_M. We say that F admits *$\mathfrak{E}$-universal reduction* if for an arbitrary machine morphism $p: M \to \overline{M}$, with $M, \overline{M}$ $\mathfrak{E}$-reachable, there exists a machine morphism $p^r: M^r \to \overline{M}^r$ with $p^r \varrho^M = \varrho^{\overline{M}} \cdot p$ (notice that p^r is unique).

V.3. Assume that all of the assumptions of the Realization theorem II (IV.4) are fulfiled and that, moreover, $\mathfrak{E}$ has the property that in any pushout

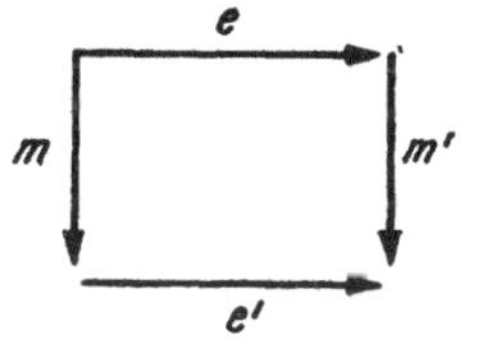

with m monic and $e \in \mathfrak{E}$ we have: also m' is monic and $e' \in \mathfrak{E}$. We call such pushouts *mixed*.

Universal realization theorem [15]. *Let F be an input process with $F(\mathfrak{E}) \subseteqq \mathfrak{E}$ and $FT = T$.*

Then F admits $\mathfrak{E}$-universal realization iff it preserves

1) *$\mathfrak{E}$-co-intersections,*
2) *mixed pushouts.*

Corollary. *If* $\mathsf{K} = \mathsf{Set}$ *or* Vect, *realization is universal only for sequential, resp. bilinear sequential machines and their varieties.*

V.4. Example of non-universality of binary tree machines (i.e., $\Omega = \{2\}$, see II.1.B). We shall show a *reduced automaton $\overline{M}$ with a non-reduced subautomaton M.* Then, of course, realization cannot be universal: for the inclusion morphism $i: M \to \overline{M}$ there exists no i^r.

Consider the binary automaton $M = (Z, \delta, Y, \beta, I, \tau)$ where (Z, δ) is the additive group of integers, $\delta(x, y) = x + y$; $I = \{0\}$ and $\tau(0) = 0$; $Y = \{0, 1\}$ and $\beta(1) = 1$, $\beta(n) = 0$ for all $n \neq 1$. This automaton is reduced. Else, $\varrho^{\overline{M}}$ would present a congruence $\equiv$ (mod k), $k > 0$. Then $\varrho^{\overline{M}}(1) = \varrho^{\overline{M}}(k + 1)$ but $\beta(1) \neq \beta(k + 1)$, a contradiction. But the submachine on the set E of even numbers is not reduced, since $\beta = 0$ on E.

References

[1] J. ADÁMEK: Free algebras and automata realizations in language of categories, CMUC[1]) 15 (1974), 589–602.
[2] J. ADÁMEK: Realization theory for automata in categories, to appear in J. Pure and Appl. Algebra.
[3] J. ADÁMEK: Automata and categories—finiteness contra minimality, Lecture Notes in Comp. Sci. 32, 160–166.
[4] J. ADÁMEK; V. KOUBEK: Algebraic theory over a functor, to appear.
[5] M. A. ARBIB; E. G. MANES: A categorist's view of automata and systems, Lecture Notes in Comp. Sci. 28.
[6] M. BARR: Coequalizers and free triples, Math. Z. 166 (1970), 307–322.
[7] M. BARR: Right exact functors, J. Pure Appl. Algebra 4 (1974), 1–8.
[8] M. BARR: Factorizations, generators and rank, a preprint.
[9] J. A. GOGUEN: Realization is universal, Math. System Theory 6 (1973), 359–374.
[10] V. KOUBEK; J. REITERMAN: Automata in categories — input processes, Lecture Notes in Comp. Sci. 32, 280–286.
[11] V. KURKOVÁ-POHLOVÁ; V. KOUBEK: When a generalized algebraic category is monadic, CMUC 15 (1974), 577–602.
[12] V. POHLOVÁ: On sums in generalized algebraic categories, Czech. Math. J. 23 (1973), 235–251.
[13] J. REITERMAN: A left adjoint construction related to free triples, to appear.
[14] V. TRNKOVÁ; J. ADÁMEK; V. KOUBEK; J. REITERMAN: Free algebras free monads and input processes, CMUC 16 (1975), 339–352.
[15] V. TRNKOVÁ; J. ADÁMEK: Realization is not universal, to appear.
[16] V. TRNKOVÁ: On minimal realizations of behaviour maps in categorial automata theory, CMUC 15 (1975), 555–566.
[17] V. TRNKOVÁ: Minimal realizations for finite sets in the categorial automata theory, CMUC 16 (1975), 21–35.
[18] V. TRNKOVÁ: Automata in categories, Lecture Notes in Comp. Sci. 32, 111–125.

1) CMUC = Commentationes Mathematicae Universitatis Carolinae.

Categorial realization theory. II Nerode equivalences

by Jiří Adámek, Prague

Contents

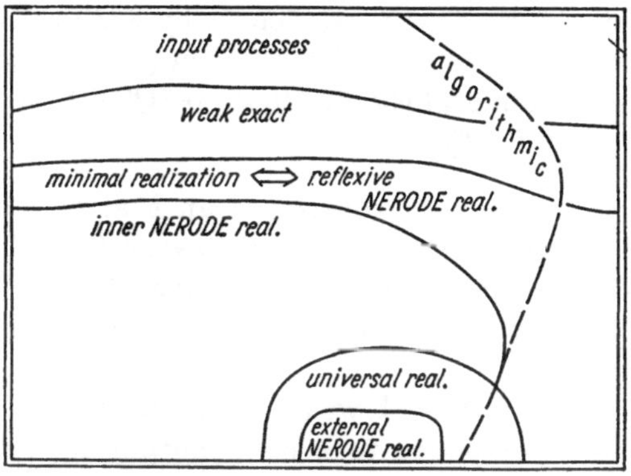

Introduction

This paper continues the presentation of results in the categorial theory of automata, started in the previous part [5]. The current part is completely self-contained; it is devoted to Nerode equivalences, studied by Arbib and Manes.

Generalizing the classical Nerode equivalence, Arbib and Manes have defined an interesting iterative construction of minimal realizations for input-process machines: *the* (*external*) Nerode *equivalence*. Later they found out that this construction fails even for tree machines. Thus, they defined a more elegant, but very abstract, "*reflexive* Nerode equivalence". But they did not present any general theorem on the existence of these concepts.

The aim of the present note is two-fold. Firstly, existence theorems for the above Nerode equivalences are proved. Roughly:

(a) External equivalences exist only in case that minimal realization is "universal", which is very rare.

(b) Reflexive equivalences always exist (if minimal realizations are guaranteed).

Secondly, we introduce a new concept, called *inner* NERODE *equivalence*, which has the pleasant algorithmic character of the external equivalence. Yet, under mild side conditions on the category we have:

(c) Inner equivalences always exist (if minimal realizations are guaranteed).

The obtained results are based on two papers: that, concerning universal realization, of V. TRNKOVÁ and the author [9], and the author's paper [2]. The results of the latter paper are generalized to triple-machines in [3, 4].

I. Preliminaries and external equivalence

I.1. Part I recapitulates some basic notions of [7] and the algorithmic input processes from [5]. Its purpose is to make this paper self-contained, and the reader may wish to skip it to Part II.

a) [Algorithmic] input processes

I.2. Given a category K and a functor $F: \mathsf{K} \to \mathsf{K}$, we denote by $\mathsf{K}(F)$ the category of F-algebras, i.e. pairs (Q, d) where $d: FQ \to Q$ is a K-morphism (ARBIB and MANES call them "dynamics"); by an F-homomorphism $f: (Q, d) \to (Q', d')$ is meant a K-morphism $f: Q \to Q'$ such that $f \cdot d = d' \cdot Ff$.

I.3. A *free algebra*, generated by a K-object X, consists of an F-algebra $(X^\#, \varphi^X)$ and a morphism $s^X: X \to X^\#$ which is universal as follows: for every F-algebra (Q, d) and every morphism $h: Q \to X$ there is a unique F-homomorphism $h^\#: (X^\#, \varphi^X) \to (Q, d)$ with $h^\# \cdot s^X = h$. If a free algebra exists for every X then F is called an *input process*.

Let F be an input process, then for each $f: X \to Y$ we have a natural $f^\#: X^\# \to Y^\#$ (more precisely: $s^Y \cdot f: X \to Y^\#$ can be extended to a homomorphism, denoted for simplicity just by $f^\#$).

I.4. Let K have finite sums and colimits of sequences.

Convention. For arbitrary $f: A \to X$, $g: B \to X$ we denote by $f \dot{\vee} g$ the natural morphism $f \dot{\vee} g: A \vee B \to X$.

Given an object X, consider the sequence

$$W_0 = X \xrightarrow{i_0} W_1 = X \vee FX \xrightarrow{i_1} W_2 = X \vee FW_1 \xrightarrow{i_2} \dots,$$

where i_0 is the canonical injection; $W_{n+1} = X \vee FW_n$ and $i_{n+1} = \mathrm{id}_X \vee Fi_n$. Let $X^\#$ (with $s_n: W_n \to X^\#$) denote the colimit of the above sequence. We say that F is an *algorithmic input process* if it preserves this colimit (for every object X). Then Fs_n: $FW_n \to FX^\#$ is a colimit, again, and we have a natural $\varphi^X: FX^\# \to X^\#$, defined by

$$\varphi^X \cdot Fs_n = s_{n+1} \cdot j_n \quad (j_n: FW_n \to W_{n+1} = X \vee FW_n \text{ is canonical}).$$

Then $(X^\#, \varphi^X)$ is a free algebra with respect to $s^X = s_0: X \to X^\#$:

Proposition. *For every F-algebra (Q, d) and every morphism $h\colon X' \to Q$ define $h^{\#}\colon X^{\#} \to Q$ by $h^{\#} \cdot s_n = h_{(n)}$ with*

$$h_{(0)} = h \quad \text{and} \quad h_{(n+1)} = h \dot{\vee} d \cdot Fh_{(n)}\,.$$

Then $h^{\#}$ is a unique homomorphism $(X^{\#}, \varphi^X) \to (Q, d)$ with $h^{\#} \cdot s_0 = h$.

Proof. Is easy; see [1].

b) Machines and minimal realization

I.5. Let $F\colon \mathsf{K} \to \mathsf{K}$ be an input process. Arbib and Manes define a machine as a tuple $M = (Q, d, Y, \beta, I, \tau)$ which forms a diagram in K:

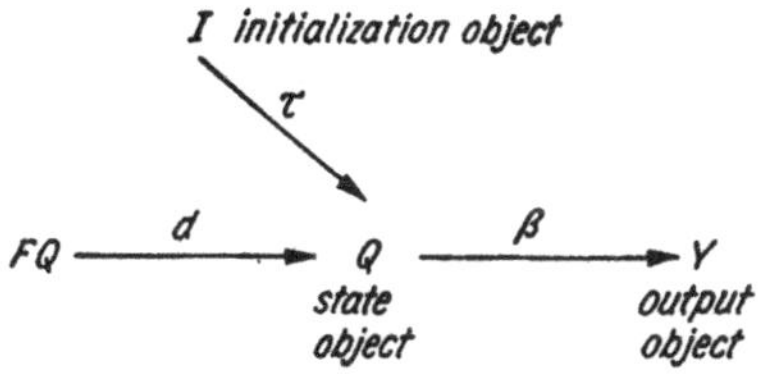

A *machine-morphism* is simply a transformation of diagrams, i.e. a triple $(\sigma, \sigma_Y, \sigma_I)\colon M \to M'$ such that the following diagram commutes

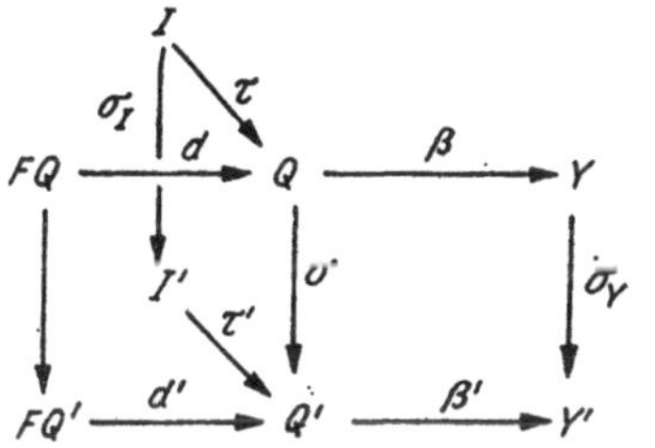

If $Y = Y'$, $I = I'$ and σ_Y, σ_I are identities, then we call σ a *simulation* and we write simply $\sigma\colon M \to M'$.

I.6. Given a machine M, we can extend $\tau\colon I \to Q$ to a homomorphism $r = \tau^{\#}\colon (I^{\#}, \varphi^I) \to (Q, d)$, called the *run map* of M. The morphism $f_M = \beta \cdot r\colon I^{\#} \to Y$ is the *behaviour* of M. Conversely, for a behaviour, i.e. a morphism of the form $f\colon I^{\#} \to Y$, M is a realization provided that $f_M = f$.

Let $\mathfrak{E}$ be an (arbitrary) class of epimorphisms of K. The machine M is *$\mathfrak{E}$-reachable*, if $r \in \mathfrak{E}$. In the present paper, we say simply "reachable" if $\mathfrak{E}$ is the class of all coequalizers.

Definition. Let $f\colon I^{\#} \to Y$ be a behaviour. Its *$\mathfrak{E}$-minimal realization* is a machine M which a) is its $\mathfrak{E}$-reachable realization and b) for every $\mathfrak{E}$-reachable realization M' there is a unique simulation $\sigma\colon M' \to M$.

An input process F is said to admit $\mathfrak{E}$-minimal realization if every behaviour has an $\mathfrak{E}$-minimal realization. We omit $\mathfrak{E}$ for the class of coequalizers. Thus *minimal = coequalizer-minimal.*

c) External NERODE *equivalence*

I.7. Example. $\mathsf{K} = \mathbf{Set}$, F is the functor, defined by $X \mapsto X \times \Sigma$ (Σ the "input" set), $f \mapsto f \times \mathrm{id}_\Sigma$. Then, for a singleton set $I = \{i\}$, machines are the usual MOORE sequential automata with initial state $q_0 = \tau(i)$. F is an input process and $I^\# = \Sigma^*$, where $\varphi^I\colon \Sigma^* \times \Sigma \to \Sigma$ is the concatenation-map and $s^I\colon \{i\} \to \Sigma^*$ maps i to Λ. Furthermore, F is well-known to admit minimal realizations; for a behaviour $f\colon \Sigma^* \to Y$ we define the NERODE equivalence E on Σ^*: uEv iff for arbitrary $x \in \Sigma^*$, $f(ux) = f(vx)$. Then the quotient set Σ^*/E is the state set for the minimal realization of f.

For categorial description, an equivalence E is a set (of pairs) together with the restricted projections $\alpha, \gamma\colon E \to \Sigma^*$ (thus uEv means that the pair $e = (u, v)$ is in E and $\alpha(e) = u$, $\gamma(e) = v$). The quotient set Σ^*/E is obtained as the coequalizer of α, γ:

$$E \underset{\gamma}{\overset{\alpha}{\rightrightarrows}} \Sigma^* \xrightarrow{\text{coeq}} \Sigma^*/E\,.$$

Also notice that NERODE equivalence is the biggest subset of $\Sigma^* \times \Sigma^*$ which fulfils: for each $(u, v) \in E$

0) $\quad f(u) = f(v)$,

1) $\quad f(u\sigma) = f(v\sigma)$ for every $\sigma \in \Sigma\ (= I \times \Sigma)$,

2) $\quad f(u\sigma_1\sigma_2) = f(v\sigma_1\sigma_2)$ for every $(\sigma_1, \sigma_2) \in \Sigma \times \Sigma\ (= I \times \Sigma \times \Sigma)$

etc.

I.8. Convention. Let $\mathfrak{A}$ be a collection of pairs of morphisms $\alpha, \gamma\colon E \to X$ in a category, with common X for each $(\alpha, \gamma) \in \mathfrak{A}$.

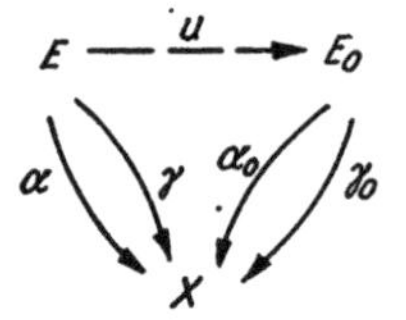

A pair $(\alpha_0, \gamma_0) \in \mathfrak{A}$ is said to be the *biggest pair* in $\mathfrak{A}$ if for every $(\alpha, \gamma) \in \mathfrak{A}$ there exists a unique morphism $u\colon E \to E_0$ with $\alpha = \alpha_0 \cdot u$ and $\gamma = \gamma_0 \cdot u$.

I.9. The above example of NERODE equivalence has led ARBIB and MANES in [7] to the following generalization. Let $f\colon I^\# \to Y$ be a behaviour. A pair of morphisms $\alpha, \gamma\colon E \to I^\#$ is said to be *externally f-equivalent* if

0) $\quad f \cdot \alpha = f \cdot \gamma\colon E \to Y$,

1) $\quad f \cdot \varphi^I \cdot F\alpha = f \cdot \varphi^I \cdot F\gamma\colon FE \to Y$,

$\quad\vdots$

n) $\quad f \cdot \alpha^{(n)} = f \cdot \gamma^{(n)}\colon F^nE \to Y$,

where $\alpha^{(n+1)} = \varphi \cdot F\alpha^{(n)}$ and $\alpha^{(0)} = \alpha$; analogously $\gamma^{(n+1)} = \varphi \cdot F\gamma^{(n)}$ and $\gamma^{(0)} = \gamma$.

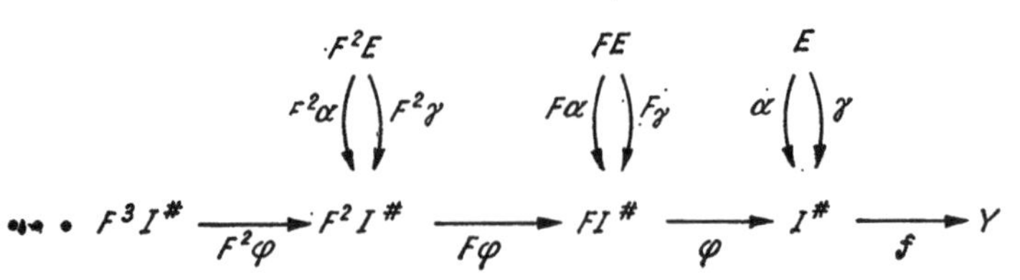

Definition [7]. The biggest externally f-equivalent pair is called the *external* NERODE *equivalence* of the behaviour f.

Theorem [7]. *Let F be an input process and $f\colon I^\# \to Y$ a behaviour, satisfying the following*

Postulate 1: f has the external NERODE equivalence $\alpha, \gamma\colon E \to I^\#$.

Postulate 2: α, γ have a coequalizer $r\colon I^\# \to Q$. [Note: since $f \cdot \alpha = f \cdot \gamma$, there is a $\beta\colon Q \to Y$ with $f = \beta \cdot r$.]

Postulate 3: there exists $d\colon FQ \to Q$, making r a homomorphism, i.e. with $d \cdot Fr = r \cdot \varphi^I$.

Postulate 4: Fe is epi for every coequalizer e.

Then f has a minimal realization, viz. $M = (Q, d, Y, \beta, I, r \cdot s^I)$.

We shall say that an input process F *admits external* NERODE *realization* if Postulates 1—4 hold for every behaviour.

Note. ARBIB and MANES use a weaker Postulate 4, viz.: Fe is epi for every run map e of any coequalizer-reachable machine. But we shall work with the above, more natural, formulation [this is necessary for Theorem III.2; but it really makes no difference, as Theorem III.4 shows].

II. Universality of minimal realization

II.1. External NERODE equivalence turns out to be closely related to universal realization, which is defined in [9] or [5], as a generalization of GOGUEN's concept: realization as a right adjoint to behaviour.

Let F be an input process, which admits $\mathfrak{E}$-minimal realization. Denote by **Mach** the category of $\mathfrak{E}$-reachable machines and machine morphisms; by **Beh** the category of behaviours (i.e. triples (Y, I, f) where $f\colon I^\# \to Y$) and behaviour-morphisms $(\sigma_Y, \sigma_I)\colon (Y, I, f) \to (Y', I', f')$, where $\sigma_Y\colon Y \to Y'$ and $\sigma_I\colon I \to I'$ are such morphisms that $f' \cdot \sigma_I^\# = \sigma_Y \cdot f$. Then "behaviour" is a functor $\boldsymbol{B}\colon$ **Mach** $\to$ **Beh**. But minimal realization presents, in general, no functor **Beh** $\to$ **Mach**.

Definition. We say that F admits $\mathfrak{E}$-*universal realization* if there exists a functor $\boldsymbol{R}\colon$ **Beh** $\to$ **Mach**, right adjoint to $\boldsymbol{B}$, such that $\boldsymbol{R}(Y, I, f)$ is the minimal realization of f (for an arbitrary behaviour (Y, I, f)).

II.2. It is proved in [9] that realization is seldom universal; e.g., it is universal for sequential machines (GOGUEN), but not for tree machines! How special universal realization is, follows from the theorem below: it implies preservation of unions, a very strong condition, indeed.

Convention. Let K have a factorization system $(\mathfrak{E}, \mathfrak{M})$. A functor F is said to preserve unions if for arbitrary morphisms $f_i\colon A_i \to X$, $i \in I$, we have:

$$\text{if} \quad m = \bigcup_{i \in I} \operatorname{im} f_i \quad \text{then} \quad \operatorname{im} Fm = \bigcup_{i \in I} \operatorname{im} Ff_i\,,$$

(where for $f = m, e$, $m \in \mathfrak{M}$ and $e \in \mathfrak{E}$, we denote $m = \operatorname{im} f$). This notion is, of course, relative: it depends on $(\mathfrak{E}, \mathfrak{M})$.

For the theorem below rather special conditions are put on the category K; we shall not go into details here, but we shall call these categories *suitable* and we exhibit a list of examples of suitable categories:

sets, tolerance spaces, topological spaces, compact T_2 spaces, posets, BANACH spaces, R-modules for any ring R, etc.

Among the conditions of "suitability" is: K is a concrete category with a factorization system $(\mathfrak{E}_s, \mathfrak{M})$ where $\mathfrak{E}_s$ are all surjective morphisms. Also, K is *connected*, i.e. hom $(A, B) \neq \emptyset$ for arbitrary non-initial objects A, B.

Theorem [9]. *Let* K *be a suitable category.*

Then every input process F, which preserves $\mathfrak{E}_s$ (i.e. $Fe \in \mathfrak{E}_s$ *if* $e \in \mathfrak{E}_s$) *and admits $\mathfrak{E}_s$-universal realization, preserves unions.*

II.3. Theorem. *External* NERODE *equivalence is universal.*

That means that every input process, which admits external NERODE realization, also admits (coequalizer-) universal realization.

Proof. Let us prove that for every behaviour-morphism $(\sigma_Y, \sigma_I): (Y_1, I_1, f_1) \to (Y_2, I_2, f_2)$, where $M_j = (Q_j, d_j, Y_j, \beta_j, I_j, \tau_j)$ is the minimal realization of f_j $(j = 1, 2)$, there exists a $\sigma: Q_1 \to Q_2$ such that $(\sigma, \sigma_Y, \sigma_I): M_1 \to M_2$ is a machine-morphism. Then we can define $\boldsymbol{R}$: Beh $\to$ Mach by $\boldsymbol{R}(Y_j, I_j, f_j) = M_j$ and $\boldsymbol{R}(\sigma_Y, \sigma_I) = (\sigma, \sigma_Y, \sigma_I)$; it is easy to verify that $\boldsymbol{R}$ is a correctly defined functor, which is right adjoint to $\boldsymbol{B}$.

Let $\alpha_1, \gamma_1: E_1 \to I_1^{\#}$ be the external NERODE equivalence of f_1 and let $r_1: I_1^{\#} \to Q_1$ be its coequalizer; analogously α_2, γ_2 and r_2. By hypothesis, r_j is the run-map of M_j $(j = 1, 2)$ and it actually suffices to prove that there exists a $\sigma: Q_1 \to Q_2$ with $r_1 \cdot \sigma = r_2 \cdot \sigma_I^{\#}$:

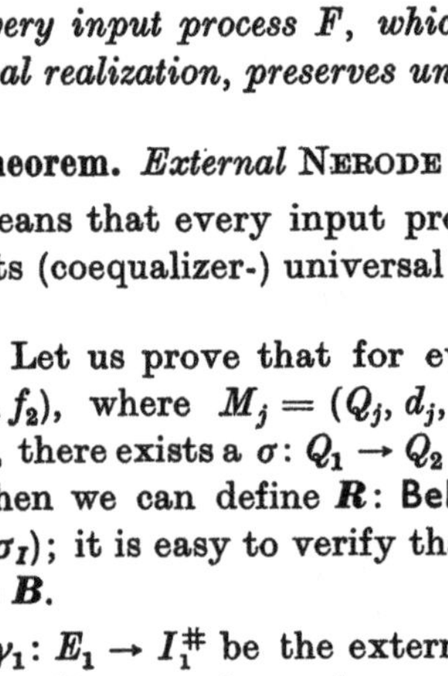

Since $r_1 = \operatorname{coeq}(\alpha_1, \gamma_1)$, we only have to show that $(r_2 \cdot \sigma_I^{\#})\, \alpha_1 = (r_2 \cdot \sigma_I^{\#}) \cdot \gamma_1$.

It follows from the f_1-equivalence of α_1, γ_1 that the pair $\sigma_I^{\#} \cdot \alpha_1$, $\sigma_I^{\#} \cdot \gamma_1: E_1 \to I_2^{\#}$ is f_2-equivalent. Indeed,

0) $\quad f_2 \cdot (\sigma_I^{\#} \cdot \alpha_1) = \sigma_Y \cdot f_1 \cdot \alpha_1 = \sigma_Y \cdot f_1 \cdot \gamma_1 = f_2 \cdot (\sigma_I^{\#} \cdot \alpha_2);$

1) $\quad f_2 \cdot \varphi^{I_2} \cdot F(\sigma_I^{\#} \cdot \alpha_1) = f_2 \cdot \sigma_I^{\#} \cdot \varphi^{I_1} \cdot F\alpha_1$

(because $\sigma_I^{\#}$ is a homomorphism)

$$= \sigma_Y \cdot (f_1 \cdot \varphi^{I_1} \cdot F\alpha_1) = \sigma_Y \cdot (f_1 \cdot \varphi^{I_2} \cdot F\gamma_1) = \ldots = f_2 \cdot \varphi^{I_2} \cdot F(\sigma_I^{\#} \cdot \gamma_1),$$

etc. The refore, there exists a $u : E_1 \to E_2$ such that $\sigma_I^{\#} \cdot \alpha_1 = \alpha_2 \cdot u$ and $\sigma_I^{\#} \cdot \gamma_1 = \gamma_2 \cdot u$. This finishes the proof:

$$r_2 \cdot (\sigma_I^{\#} \cdot \alpha_1) = r_2 \cdot \alpha_2 \cdot u = r_2 \cdot \gamma_2 \cdot u = r_2 \cdot (\sigma_I^{\#} \cdot \gamma_1) .$$

II.4. Theorem. *Let* K *have a factorization system* $(\mathfrak{E}, \mathfrak{M})$, *let it have unions and finite limits.*

Then every input process, which preserves unions, satisfies

Postulate 1: every behaviour has an external Nerode equivalence.

Proof. Let $\alpha_j, \gamma_j \colon E_j \to I^{\#}$ $(j \in J)$ be the collection of all externally f-equivalent pairs for a given $f \colon I^{\#} \to Y$. Denoting by $p', p'' \colon I^{\#} \times I^{\#} \to I^{\#}$ the projections, we get morphisms $b_j \colon E_j \to I^{\#} \times I^{\#}$ with $\alpha_j = p' \cdot b_j$ and $\gamma_j = p'' \cdot b_j$, $j \in J$. Put $b = \bigcup_{j \in J} \operatorname{im} b_j$, $b \colon E \to I^{\#} \times I^{\#}$. We shall prove that the pair $\alpha = p' \cdot b$, $\gamma = p'' \cdot b \colon E \to I^{\#}$ is also externally f-equivalent. Then it is the Nerode equivalence of f, of course.

0) $f \cdot \alpha = f \cdot \gamma$. Denote by $z_0 \colon Z_0 \to I^{\#} \times I^{\#}$ the equalizer of $f \cdot p'$, $f \cdot p''$. Since $z_0 \in \mathfrak{M}$ (it is easily seen that $\mathfrak{M}$ contains all equalizers) and since $f \cdot \alpha_j = f \cdot \gamma_j$, we have $\operatorname{im} b_j \subseteq Z_0$ for all j. Hence, $\operatorname{im} b \subseteq Z_0$, i.e. $f \cdot \alpha = f \cdot \gamma$.

1) $f \cdot \varphi^I \cdot F\alpha = f \cdot \varphi^I \cdot F\gamma$. Denote by $z_1 \colon Z_1 \to F(I^{\#} \times I^{\#})$ the equalizer of $f \cdot \varphi^I \cdot Fp'$ and $f \cdot \varphi^I \cdot Fp''$. Since $z_1 \in \mathfrak{M}$ and $f \cdot \varphi^I \cdot F\alpha_j = f \cdot \varphi^I \cdot F\gamma_j$, we have $\operatorname{im} Fb_j \subseteq Z_1$. By hypothesis, $\operatorname{im} Fb = \bigcup \operatorname{im} Fb_j$; hence $\operatorname{im} Fb \subseteq Z_1$. Therefore $f \cdot \varphi^I \cdot F\alpha = f \cdot \varphi^I \cdot F\gamma$. Etc.

Example. For sequential machines in closed categories of [8], i.e. for $F(-) = - \otimes \Sigma$, external Nerode equivalence always exists, since F preserves colimits. (But here a more natural construction of minimal realizations is exhibited in [8].)

Note. Now, we see that in a suitable category we have, roughly speaking:

$$\begin{array}{ccc} \text{surjective-universal} & & \text{coequalizer-universal} \\ \Downarrow & & \Uparrow \\ \text{preserves unions} & \Rightarrow & \text{external Nerode realization} . \end{array}$$

Unfortunately, only seldom the two classes, $\mathfrak{E}_s$ and coequalizers, are equal. But they are in *balanced* categories, i.e. such that every mono-epimorphism is an isomorphism. E.g., sets, R-modules and compact T_2-spaces form balanced, suitable categories.

III. Weak exact functors

III.1. Throughout this section K is a category with finite colimits, such that the sum-injection $A \to B \vee B$ is always a monomorphism.

Let $F \colon \mathsf{K} \to \mathsf{K}$ be a functor. A pair of morphisms $a, b \colon X \to X'$ in K is called a homomorphism-pair if there exist $d \colon FX \to X$ and $d' \colon FX' \to X'$ such that a and b are homomorphisms $(X, d) \to (X', d')$.

A pair $a, b \colon X \to X'$ is *reflexive* if there exists an $i \colon X' \to X$ such that $a \cdot i = b \cdot i = \mathrm{id}_{X'}$.

Definition. A functor F is *weak exact* if it preserves the coequalizer of any reflexive homomorphism-pair.

This property may seem strange on the first sight, but it is very useful in the NERODE-equivalence approach. On one hand, lots of functors are weak exact (see the next theorem), and on the other hand, Postulate 3 (I.9) follows from weak exactness.

III.2. It is a special property of reflexive pairs a, b that their pushout and coequalizer are "equal". This follows from the fact that if $p \cdot a = q \cdot b$ then $p = q$ (proof: $p = p \cdot a \cdot i = q \cdot b \cdot i = q$). Moreover, the morphisms a, b are coequalizers (even more: split epis). Hence, a functor is weak exact as soon as it preserves pushouts of homomorphisms, which are coequalizers. Now, every input process, admitting minimal realization, preserves the mentioned pushouts; this follows from the Reduction Theorem I in [2] (or Realization Theorem I in [5]) which states, roughly speaking, that a given input process F *admits minimal realization iff F preserves multiple pushouts of coequalizer-homomorphisms*. The assumptions in [2] contained: Fe is a coequalizer, provided that e is. But going through the proof, we see that Postulate 4 (I.9) is sufficient (Fe is epi if e is a coequalizer):

Theorem. *Every input process, which admits minimal realization and satisfies Postulate 4, is weak exact.*

Proof. An easy modification of Reduction Theorem I [2] where $\mathfrak{E}$ = all coequalizers.

III.3. Proposition. *Every weak exact input process fulfils Postulate* 3.

Proof. Let $\alpha, \gamma \colon E \to I^{\#}$ be the external NERODE equivalence of $f \colon I^{\#} \to Y$; let $r \colon I^{\#} \to Q$ be its coequalizer. To find $d \colon FQ \to Q$ with $d \cdot Fr = r \cdot \varphi^I$, it suffices to show that 1) $Fr = \operatorname{coeq}(F\alpha, F\gamma)$ and 2) $(r \cdot \varphi^I) \cdot F\alpha = (r \cdot \varphi^I) \cdot F\gamma$.

Ad 1): α, γ is a reflexive pair, because the pair $\mathrm{id}_{I^{\#}}, \mathrm{id}_{I^{\#}}$ is externally f-equivalent (hence, there is a $u \colon I^{\#} \to E$ with $\mathrm{id}_{I^{\#}} = \alpha \cdot u$, $\mathrm{id}_{I^{\#}} = \gamma \cdot u$). Moreover, the pair $\varphi^I \cdot F\alpha, \varphi^I \cdot F\gamma \colon FE \to I^{\#}$ is easily seen to be externally f-equivalent. Thus, there is a $v \colon FE \to E$ with $\varphi^I \cdot F\alpha = \alpha \cdot v$ and $\varphi^I \cdot F\gamma = \gamma \cdot v$. Therefore, α, γ is a reflexive homomorphism-pair $(E, v) \to (I^{\#}, \varphi^I)$ — by hypothesis, $F \operatorname{coeq}(\alpha, \gamma) = \operatorname{coeq}(F\alpha, F\gamma)$.

Ad 2) We have $r \cdot (\varphi^I \cdot F\alpha) = r \cdot \alpha \cdot v = r \cdot \gamma \cdot v = r \cdot (\varphi^I \cdot F\gamma)$.

Note. This proposition contradicts Theorem 3.14 of [6] which claims that for $\mathsf{K} = \mathbf{Set}$ and F = cartesian square ($FX = X \times X$) there exists a behaviour, satisfying Postulate 1 but not Postulate 3. (Nevertheless, F is easily seen to be weak exact.) The authors have made a mistake there. Going through their proof, the sets D_n are not what they are supposed to be, on which the error of the proof is based.

III.4. Finally, we shall show that Postulate 4 is very mild for connected categories (defined in II.2):

Theorem. *If* K *is connected, then every input process, admitting minimal realization, satisfies*

Postulate 4: Fe is epi for every coequalizer e.

Note. Let $\mathfrak{E}$ be a class of epis which (i) contains all split epis, (ii) $e \cdot s \in \mathfrak{E}$ whenever s is a split epi, $e \in \mathfrak{E}$ and (iii) $\mathfrak{E}$ is closed to finite sums. We shall actually show that, whenever F admits $\mathfrak{E}$-minimal realization, then Fe is epi for every $e \in \mathfrak{E}$.

Proof outline. Let $k: A \to B$ be a coequalizer. Given $g, h: FB \to C$ with $g \cdot Fk = h \cdot Fk$, we are to verify that $g = h$. We can assume that A is a non-initial object; then $\hom(B, A) \neq \emptyset$ and we choose any $t: B \to A$.

Put $\bar{A} = A \vee C$ with $v: C \to \bar{A}$ canonical; analogously $w: C \to \bar{B} = B \vee C$. Put $\bar{k} = k \vee \mathrm{id}_C: \bar{A} \to \bar{B}$. We shall define machines M, M_g and M_h with common output $\bar{B}$ and common initialization $\bar{A}$. First, define $d: F\bar{A} \to \bar{A}$ by

$$d = v \cdot g \cdot Fk \cdot F(\mathrm{id}_A \dot{\vee}\, t) \quad \text{(recall } \dot{\vee} \text{ from I.4)}$$

and then put $M = (\bar{A}, d, \bar{B}, \bar{k}, \bar{A}, \mathrm{id})$. Analogously, define $d_g, d_h: F\bar{B} \to \bar{B}$ by $d_g = w \cdot g \cdot F(\mathrm{id}_B \dot{\vee}\, k \cdot t)$ and $d_h = w \cdot h \cdot F(\mathrm{id} \dot{\vee}\, k \cdot t)$ and put $M_g = (\bar{B}, d_g, \bar{B}, \mathrm{id}, \bar{A}, \bar{k})$, $M_h = (\bar{B}, d_h, \bar{B}, \mathrm{id}, \bar{A}, \bar{k})$. Then $\bar{k}: M \to M$ and $\bar{k}: M \to M_h$ are simulations.

It is easy to check that M, M_g and M_h are reachable machines with a common behaviour $f: \bar{A}^{\#} \to \bar{B}$. Let M_0 be the minimal realization of f, say $M_0 = (Q_0, d_0, \bar{B}, \beta, \bar{A}, \tau)$; we have simulations $p_g: M_g \to M_0$ and $p_h: M_h \to M_0$. Looking at the outputs, we get $\beta \cdot p_g = \mathrm{id}_{\bar{B}}$; hence p_g is a split mono as well as epi — p_g is an isomorphism. It is easy to check that $p_g^{-1}: M_0 \to M_g$ is also a simulation, thus so is $p_h \cdot p_g^{-1}: M_g \to M_h$. Looking at the outputs, again, we get $p_h \cdot p_g^{-1} = \mathrm{id}_{\bar{B}}$, thus $p_h = p_g$ and so $M_g = M_h$. There follows easily $g = h$.

III.5. *Characterization theorem for external* NERODE *equivalence.*

Theorem. *If* K *is a connected category with finite colimits and monomorphic sum-injections, then the following holds for any input process F, admitting minimal realization:*

(i) *F satisfies Postulates* 3, 4;

(ii) *F satisfies Postulate* 1 *iff it admits external* NERODE *realization;*

(iii) *if so, F admits universal realization.*

Proof: (i) See III.3 and III.4; (ii) see III.2; (iii) see II.3.

Corollary. For suitable balanced categories K and for an input process F, preserving $\mathfrak{E}_s$, the following conditions are equivalent:

a) F admits external NERODE realization;

b) F admits universal realization;

c) F preserves unions.

Application: sets. The category **Set** of sets and mappings is suitable and balanced and every functor preserves $\mathfrak{E}_s$ (= all split epis!). A functor F preserves unions iff there are sets Σ_1, Σ_0 such that F is naturally equivalent to the functor

$$X \mapsto (X \times \Sigma_1) \vee \Sigma_0 \quad \text{and} \quad f \mapsto (f \times \mathrm{id}_{\Sigma_1}) \vee \mathrm{id}_{\Sigma_0}.$$

Then F-machines are precisely *sequential Σ-machines,* where $\Sigma = \Sigma_1 \cup \Sigma_0$ and the *Σ_0-inputs are constant.* Thus, these are the only machines in **Set** which have an external NERODE (or universal) realization.

Proof. See [9].

Application: vector spaces. Also the category **Vect** of vector spaces (over a fixed field) and linear maps is balanced, suitable and every functor preserves $\mathfrak{E}_s$. A functor F preserves unions iff there exist spaces Σ_0, Σ_1 such that F is naturally equivalent to the functor

$$X \to (X \otimes \Sigma_0) \times \Sigma_1; \qquad f \mapsto (f \otimes \mathrm{id}_{\Sigma_0}) \times \mathrm{id}_{\Sigma_1},$$

where $\otimes$ denotes the tensor product and $\times$ the direct product (biproduct). Then F-machines are *sequential Σ-machines, where $\Sigma = \Sigma_1 \times \Sigma_0$, with Σ_1-bilinear and Σ_0-linear inputs*. Thus, these are the only machines in **Vect**, which have external NERODE realization.

Proof. See [9].

IV. Inner NERODE equivalence

IV.1. A minor change in the definition of external f-equivalence leads to another iterative construction of minimal realizations, more general than the external NERODE equivalence. We call it, for contrast, inner NERODE equivalence.

Recall the definition of reflexive pairs (III.1) and of $\dot{\vee}$ (I.4). Throughout this part we assume that K has finite limits, finite colimits (with monomorphic sum-injections) and colimits of sequences.

IV.2. Let $f: I^\# \to Y$ be a behaviour. A pair $\alpha, \gamma: E \to I^\#$ is said to be inner f-equivalent if it is reflexive and fulfils

$$0)\ f \cdot \alpha = f \cdot \gamma : E \to Y,$$

$$1)\ f \cdot (\alpha \mathbin{\dot{\vee}} \varphi \cdot F\alpha) = f \cdot (\gamma \mathbin{\dot{\vee}} \varphi \cdot F\gamma) : E \vee FE \to Y,$$

$$\vdots$$

$$n)\ f \cdot \alpha_{(n)} = f \cdot \gamma_{(n)} = W_n \to Y,$$

where $W_0 = E$ and $\alpha_{(0)} = \alpha$, $\gamma_{(0)} = \gamma$; $W_{n+1} = E \vee FW_n$ and $\alpha_{(n+1)} = \alpha \mathbin{\dot{\vee}} \varphi^I \cdot F\alpha_{(n)}$, analogously $\gamma_{(n+1)} = \gamma \mathbin{\dot{\vee}} \varphi^I \cdot F\gamma_{(n)}$.

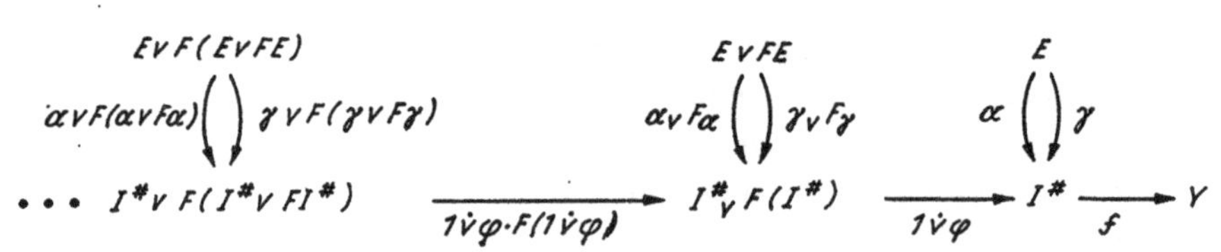

Note. The differences to the external equivalence are two: 1) only reflexive pairs are considered and 2) here $\alpha_{(n+1)} = \alpha \mathbin{\dot{\vee}} \varphi^I \cdot F\alpha_{(n)}$ while there $\alpha^{(n+1)} = \varphi^I \cdot F\alpha^{(n)}$. Concerning the sequential-machine motivation (I.7), the whole difference is that there the n-th condition ($f \cdot \alpha^{(n)} = f \cdot \gamma^{(n)}$) says: $f(ux) = f(vx)$ for every word x of length n. And here, the condition $f \cdot \alpha_{(n)} = f \cdot \gamma_{(n)}$ says: $f(ux) = f(vx)$ for every word x of length at most n.

Definition. The biggest inner f-equivalent pair is called the *inner* NERODE *equivalence* of f.

IV.3. Analogously as for external equivalence (see I.9), the four Postulates guarantee minimal realization. (Postulate 1: f has an inner NERODE equivalence α, γ; the others are unchanged.) We say, again, that F admits *inner* NERODE *realization* if it fulfils these four Postulates for every behaviour.

The difference in the existence of the two equivalences (i.e., in Postulates 1) is enormous:

Theorem. *Let F be a weak exact, algorithmic* (I.4) *input process. Then every behaviour with a minimal realization has an inner* NERODE *equivalence and F admits inner* NERODE *realization.*

Proof. Will be exhibited in the next part, see V.4.

IV.4. The condition "algorithmic input process" is a mild one. If K satisfies the conditions $c_1 - c_4$ below, then *every input process with minimal realizations is algorithmic*, by Realization Theorem in [2].

For the next corollary, K is assumed to fulfil:

c_1: K is cowell-powered and cocomplete;

c_2: K has finite limits; pullbacks commute with colimits of sequences;

c_3: K is connected (II.2);

c_4: sum-injections $A \to A \vee B$ are monomorphisms.

Corollary (*Characterization theorem for inner* NERODE *equivalence*).

Let K *fulfil* $c_1 - c_4$.

Then every input process, admitting minimal realization, admits inner NERODE *realization.*

IV.5. Inner NERODE equivalence strictly generalizes the external one. For any category K with finite sums we have:

Comparison Theorem. *Every external* NERODE *equivalence of a behaviour is its inner* NERODE *equivalence.*

Proof. Let $\alpha_0, \gamma_0 \colon E_0 \to I^\#$ be the external equivalence of $f\colon I^\# \to Y$. First, this pair is inner f-equivalent. Indeed, the reflexivity of α_0, γ_0 follows from the fact that the pair $\mathrm{id}_{I^\#}, \mathrm{id}_{I^\#}$ is externally f-equivalent (thus, it factorizes through α_0, γ_0). Moreover, there exists $v\colon FE_0 \to E_0$ such that $\alpha_0, \gamma_0\colon (E, v) \to (I^\#, \varphi^I)$ are homomorphisms (see the proof in III.3). Then the inner f-equivalence easily follows: 0) $f \cdot \alpha_0 = f \cdot \gamma_0$ is clear; 1) $f \cdot (\alpha_0 \dot{\vee} \varphi \cdot F\alpha_0) = f(\gamma_0 \dot{\vee} \varphi \cdot F\gamma_0)$ follows from 0) and the fact that $\varphi \cdot F\alpha_0 = \alpha_0 \cdot v$, hence $\alpha_0 \dot{\vee} \varphi \cdot F\alpha_0 = \alpha_0 \cdot (\mathrm{id}_E \dot{\vee} v)$, analogously for γ_0; etc.

Given another inner f-equivalent pair $\alpha, \gamma\colon E \to I$, it suffices to show that the pair is also externally f-equivalent. But, denoting by $j_n\colon FW_n \to W_{n+1} = E \vee FW_n$ the injections, it is easy to see that $\alpha_{(0)} = \alpha^{(0)}$, $\alpha_{(1)} = \alpha^{(1)} \cdot j_1$, $\alpha_{(2)} = \alpha^{(2)} \cdot j_2 \cdot Fj_1$ etc., analogously for $\gamma^{(n)}$, thus $f \cdot \alpha^{(n)} = f \cdot \gamma^{(n)}$ implies $f \cdot \alpha_{(n)} = f \cdot \gamma_{(n)}$.

V. Reflexive NERODE equivalence

V.1. Finally, we get the most general (but least transparent) version of NERODE equivalence, defined in [6].

We assume, again, that K has finite limits, finite colimits (with monomorphic sum-injections) and colimits of sequences. For every morphism $h: A \to B$ there exists its *kernel pair*, i.e. the biggest pair $p, q: E \to A$ with $h \cdot p = h \cdot q$. (Indeed, denote by $\pi', \pi'': A \times A \to A$ the projections, and let $e: E \to A \times A$ be the equalizer of $h \cdot \pi'$ and $h \cdot \pi''$; then the kernel pair is $p = \pi' \cdot e$, $q = \pi'' \cdot e$.)

V.2. Let $f: I^{\#} \to Y$ be a behaviour. A pair $\alpha, \gamma: E \to I^{\#}$ is said to be reflexively f-equivalent if it is reflexive and fulfils $f \cdot \alpha^{\#} = f \cdot \gamma^{\#}$ (where $\alpha^{\#}: (E^{\#}, \gamma^{E}) \to (I^{\#}, \varphi^{I})$ is the free extension of α; analogously $\gamma^{\#}$).

Definition. The biggest reflexively f-equivalent pair is called the *reflexive* NERODE *equivalence* of f.

Note. In [6] a pair α, γ is called abstractly f-equivalent if $f \cdot \alpha^{\#} = f \cdot \gamma^{\#}$. (Then the reflexive equivalence is the biggest reflexive, abstractly f-equivalent pair.) The biggest of *all* abstractly f-equivalent pairs is there called the abstract NERODE equivalence of f.

V.3. The same remark for reflexive equivalence, as IV.3 for inner equivalence, is valid. The four Postulates guarantee minimal realization. *F admits reflexive* NERODE *realization* iff it satisfies these four Postulates. Postulates 2—4 cause no difficulty; Postulate 3 follows from Postulate 1 (the existence of reflexive equivalence) as in III.3 with an analogous proof. For any category K with kernel pairs and coequalizers we have:

Theorem. *Let F be a weak exact input process.*

Then every behaviour with a minimal realization has a reflexive NERODE *equivalence.*

Proof. Let $f: I^{\#} \to Y$ be a behaviour with a minimal realization $M = (Q, d, Y, \beta, I, \tau)$. Let $r = \tau^{\#}: I^{\#} \to Q$ be the run map and let $\alpha, \gamma: E \to I^{\#}$ be its kernel pair. We shall show that α, γ is the reflexive NERODE equivalence of f. Firstly, α, γ is a reflexive pair (because the pair $\mathrm{id}_{I^{\#}}, \mathrm{id}_{I^{\#}}$ satisfies $r \cdot \mathrm{id}_{I^{\#}} = r \cdot \mathrm{id}_{I^{\#}}$); secondly, the run map is a homomorphism $r: (I^{\#}, \varphi^{I}) \to (Q, d)$, therefore, clearly $(r \cdot \alpha)^{\#} = r \cdot \alpha^{\#}: (E^{\#}, \varphi^{E}) \to (I^{\#}, \varphi^{I}) \to (Q, d)$; also $(r \cdot \gamma)^{\#} = r \cdot \gamma^{\#}$. This shows that $f \cdot \alpha^{\#} = (\beta \cdot r) \cdot \alpha^{\#} = \beta \cdot (r \cdot \alpha)^{\#} = \beta \cdot (r \cdot \gamma^{\#}) = f \cdot \gamma^{\#}$.

Given another reflexively f-equivalent pair $\alpha_1, \gamma_1: E_1 \to I^{\#}$, we are to find $u: E_1 \to E$ with $\alpha_1 = \alpha \cdot u$ and $\gamma_1 = \gamma \cdot u$. Let $r_1: I^{\#} \to Q$ be the coequalizer of the homomorphism-pair $\alpha_1^{\#}, \gamma_1^{\#}$. Since α_1, γ_1 is reflexive, so is $\alpha_1^{\#}, \gamma_1^{\#}$. By weak exactness, F preserves the coequalizer of $\alpha_1^{\#}, \gamma_1^{\#}$. Since $(r_1 \cdot \varphi^{I}) \cdot F\alpha_1^{\#} = r_1 \cdot \alpha_1^{\#} \cdot \varphi^{E_1} = r_1 \cdot \gamma_1^{\#} \cdot \varphi^{E_1} = (r_1 \cdot \varphi^{I}) \cdot F\gamma_1^{\#}$ and since $Fr_1 = \mathrm{coeq}\,(F\alpha_1^{\#}, F\gamma_1^{\#})$, we get $d_1: FQ_1 \to Q_1$ with $r_1 \cdot \varphi^{I} = d \cdot Fr_1$, i.e. $r_1: (I^{\#}, \varphi^{I}) \to (Q_1, d_1)$. Since, further, $f \cdot \alpha_1^{\#} = f \cdot \gamma_1^{\#}$, we get $\beta_1: Q_1 \to Y$ with $f = \beta_1 \cdot r_1$. Now,

$$M_1 = (Q_1, d_1, Y, \beta_1, I, r_1 \cdot s^{I})$$

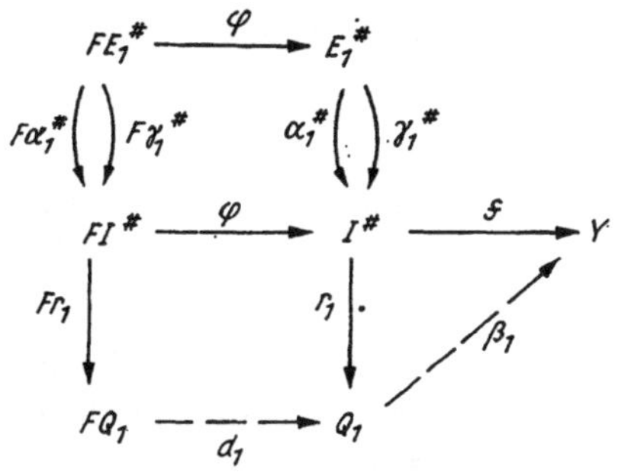

is a reachable realization of f (since the run map of M_1 is clearly r_1) and so we have a simulation $\sigma\colon M_1 \to M$. In particular, $\sigma \cdot r_1 = r$. Therefore, $r \cdot \alpha_1 = r \cdot \gamma_1$ and, since α, γ is a kernel pair of r, we get $u\colon E_1 \to E$ with $\alpha_1 = \alpha \cdot u$; $\gamma_1 = \gamma \cdot u$, as required.

Corollary (*Characterization Theorem for reflexive* NERODE *equivalence*).

Let $F\colon \mathsf{K} \to \mathsf{K}$ be an input process; let either K be connected or Fe be epi for any coequalizer e.

If F admits minimal realization, it admits reflexive NERODE *realization.*

Proof. By III.4 we can use III.2: F is weak exact. Apply III.3 and the above theorem.

Thus, we see that the reflexive equivalence is very general: we only need the initial hypothesis on K (V.1) and less than the preservation of epis!

V.4. Proof of Theorem IV.3. Since F is an algorithmic input process, we have $E^{\#} = \operatorname{colim} W_n$ and $\alpha^{\#} = \operatorname{colim} \alpha_{(n)}$, $\gamma^{\#} = \operatorname{colim} \gamma_{(n)}$ by Proposition I.4. Thus, a pair is reflexively f-equivalent iff it is inner f-equivalent. This shows that the reflexive NERODE equivalence (which exists by the above theorem) is the inner equivalence, as well.

Thus, we see that for an algorithmic input process the reflexive and inner equivalences coincide. For an arbitrary input process and for K, having finite sums, the following can be proved analogously as IV.5.

Comparison Theorem. *Every inner (as well as every external)* NERODE *equivalence of a behaviour is its reflexive* NERODE *equivalence.*

VI. Concluding remarks

VI.1. We have studied three concepts of NERODE equivalence. These are, when ordered by their generality (see IV.5 and V.4):

$$\text{external} \subsetneqq \text{inner} \subsetneqq \text{reflexive}\ .$$

It turns out that the existence of a NERODE equivalence (Postulate 1) guarantees all the rest: minimal realization is a coequalizer of the NERODE equivalence. (See III.5 for the external case; analogously for the remaining two.) The external equivalence exists rarely, because it implies universality of minimal realization (II.3).

The other way round: the external equivalence is a good tool of description of universal minimal realization (see II.4).

VI.2. The four Postulates, formulated by ARBIB and MANES (I.9) turn out to be very well chosen. While they are sufficient for the existence of minimal realization, as proved in [7], we have shown their necessity, too. See IV.3 for V.3 for Postulate 1, and III.2—III.4 for Postulates 3 and 4; the Postulate 2 is a mild one.

VI.3. In the more general approach of monadic machines, studied e.g. in [3, 4], the reflexive NERODE equivalence can also be defined. Results, analogous to the Characterization theorem V.3, hold also in this generality.

References

[1] J. ADÁMEK: Free algebras and automata realizations in the language of categories, Comm. Math. Univ. Carolinae **15** (1974), 589—602.
[2] J. ADÁMEK: Realization theory for automata in categories, to appear in J. Pure Appl. Algebra (revised version).
[3] J. ADÁMEK: On the cogeneration of algebras, to appear.
[4] J. ADÁMEK: Cogeneration of algebras in regular categories, Bull. Austral. Math. Soc. **15** (1976), 355—370.
[5] J. ADÁMEK: Categorial realization theory. I, this. vol. 111—122.
[6] D. O. ANDERSON; M. A. ARBIB; E. G. MANES: Foundations of system theory: finitary and infinitary conditions, Tech. Report.
[7] M. A. ARBIB; E. G. MANES: Machines in a category, an expository introduction, SIAM Review **16** (1974), 163—192.
[8] H. EHRIG; K. D. KIERMEIER; H.-J. KREOWSKI, W. KÜHNEL: Universal theory of automata, Teubner, Stuttgart 1974.
[9] V. TRNKOVÁ; J. ADÁMEK: Realization is not universal, to appear.

The skeleton of minimal realization

by HARTMUT EHRIG and HANS-JÖRG KREOWSKI, Berlin

Abstract

The problem of minimal realization in automata and systems theory, including reduction, minimization, reachability and realization, is considered. In most cases this problem can be reduced to the construction of generalized image-factorizations $I \to UQ' \to UQ$ and $UQ \to U\overline{Q} \to Y$ of morphisms $I \to UQ$ (initial states) and $UQ \to Y$ (output) where $U\colon \mathsf{D} \to \mathsf{K}$ is a forgetful functor from a category D of dynamics to a base category K. In the axiomatic minimal realization theories, known up to now, U is assumed to have a left and a right adjoint such that generalized image-factorizations can be reduced to usual image-factorizations in D. For several interesting examples, however, these assumptions are too strong. In this paper now, weaker conditions for the type of the morphisms and the construction of generalized image-factorizations are given and applied to the problems mentioned above.

Introduction

What is the basic concept behind all the approaches to minimal realization in the literature? GOGUEN's "Minimal-Realization-Principle" in [13] states that minimal realization is a right adjoint and right inverse of the behaviour functor. This is only true for reachable systems. But reachability, as well as observability and in several cases minimal realization can be obtained by $\mathfrak{E}$-$\mathfrak{M}$-factorization of the reachability and observability morphisms. These morphisms can be constructed using the left and the right adjoint of the forgetful functor $U\colon \mathsf{D} \to \mathsf{K}$ where D is a category of dynamics and K a base category. This is the basic idea of the "Axiomatic Minimal-Realization-Theorem" in [5] which is implicit already in [6] for machines in a hyperdoctrine, in [14], [10], and [9] for automata in closed categories, and in [4] for adjoint machines. But what about the following examples which do not fit in this framework?

1. Algebra-Automata in [3]
2. Automata in Pseudoclosed Categories in [10], [9], and [11] including Nondeterministic and Stochastic Automata,
3. Nonlinear Smooth Dynamical Systems in [20],
4. Mealy-Automata with initial states in [9],
5. Monoid-Automata (Minimization of the Transition-Monoid) in [9],
6. Multidecomposable Systems in [2],
7. The Canonical Realization Theory for Feedback Systems in [7].

In all these cases there are also unique constructions for reachability, reduction, minimization and minimal realization, or at least for a part of these problems. For nondeterministic and stochastic automata, for example, there is a unique reachability and reduction construction but no unique minimization or minimal realization.

In section 1 of this paper we extend the "Axiomatic Minimal-Realization-Approach" in [5] to include also examples 1—3 above. For this purpose we construct "image-factorizations" $I \to UQ$ and $UQ \to U\overline{Q} \to Y$ of the initial state morphism $\tau: I \to UQ$ and of the output morphism $\beta: UQ \to Y$ respectively without assuming left or right adjoints of U. Similar factorizations were studied in [16] before.

Unfortunately, this concept is still not general enough to cover examples 4—7 above. But they can be treated generalizing morphisms $I \to UQ$ and $UQ \to Y$ to "F-morphisms" in section 2. Given a profunctor $F: \mathsf{G}^0 \times \mathsf{K} \to \mathsf{Set}$ an F-morphism f from G to K, written $F: \mathsf{G} \to \mathsf{K}$, is an element $f \in F(\mathsf{G}, \mathsf{K})$. (In the special case $\mathsf{G} = \mathsf{K}$ and $F = \text{Mor}\ \mathsf{K}$ F-morphisms are ordinary K-morphisms). F-morphisms, especially their factorization properties, are studied in [8]. An introduction to F-morphisms and the reformulation of section 1 in this new terminology is given in section 2.

Finally, in section 3 we give a table of several interesting examples (more or less all the examples studied in the literature mentioned in our references) classified with respect to their factorization properties and hence with respect to their properties concerning reachability, reduction, minimization and minimal realization. (But we don't consider the non-unique minimization constructions in the nondeterministic case studied in [11] and [5].) Last but not least we give a positive solution of the open problem stated in [2] concerning the uniqueness of minimal realization for multi-decomposable systems. Again, the solution is given by an image-factorization. Hence:

The skeleton of minimal realization is nothing else but image factorization.

1. The Arbib-Manes-Bainbridge-skeleton

The Arbib-Manes-Bainbridge-skeleton of minimal realization, explicitly in [5], is the idea to start with a category D of dynamics together with a forgetful functor $U: \mathsf{D} \to \mathsf{K}$ from D to a base category K which has a left and a right adjoint, called free and cofree dynamics, respectively. Moreover an $\mathfrak{E}$-$\mathfrak{M}$-factorization is given in K and it is assumed that it can be lifted by U to the category D of dynamics.

In this section we will start with the same situation, without assuming left and right adjoints. Nevertheless, we will be able to formulate and solve most of the problems concerning reduction, minimization and minimal realization using weaker assumptions. The terminology is similar to Arbib-Manes (cf. [4, 3]) as far as possible: However, we have not the sign @ and use § instead, and we keep writing $g \cdot f$ for first f and then g, UQ for functor U applied to Q, and K^0 for the dual category.

1.1. General assumptions. Let D be a category, called category of *dynamics*, with $\mathfrak{E}$-$\mathfrak{M}$-factorization (in some cases we only need classes $\mathfrak{E}$ or $\mathfrak{M}$ in D), K another category, called base category, and $U: \mathsf{D} \to \mathsf{K}$ a functor, called forgetful functor. A system or machine in this context is a construct $S = (Q, I, Y, \tau, \beta)$ where

— Q is an object in D (state object),

— I (initial state object) and Y (output object) are objects in K,

— τ (initial state morphism) and β (output morphism) are K-morphism of the form

$$I \xrightarrow{\tau} UQ \xrightarrow{\beta} Y .$$

We will say that τ is an IN-U-morphism and β an OUT-U-morphism because the target of τ and the source of β is of the form UQ with Q in D.

We are able to transform the classes $\mathfrak{E}$ and $\mathfrak{M}$ to IN-U and OUT-U-morphisms in the following way:

$\mathfrak{E}_{\mathrm{IN}}$ is the class of all IN-U-morphisms $\tau: I \to UQ$ such that for each "IN-U-factorization" $\tau = Uf \circ \tau'$ with f in D we have f in $\mathfrak{E}$.

Dually $\mathfrak{M}_{\mathrm{OUT}}$ is the class of all OUT-U-morphisms $\beta: UQ \to Y$ such that for each "OUT-U-factorization" $\beta = \beta' \circ Uf$ with f in D we have f in $\mathfrak{M}$.

Consequently we say that we have an $\mathfrak{E}_{\mathrm{IN}}$-$\mathfrak{M}$-factorization if each IN-U-morphism $\tau: I \to UQ$ has a unique (up to D-isomorphism) IN-U-factorization $\tau = Uf \circ \tau'$ with τ' in $\mathfrak{E}_{\mathrm{IN}}$ and f in $\mathfrak{M}$.

Dually we have an $\mathfrak{E}$-$\mathfrak{M}_{\mathrm{OUT}}$-factorization if each OUT-U-morphism $\beta: UQ \to Y$ has a unique (up to D-isomorphism) OUT-U-factorization $\beta = \beta' \circ Uf$ with f in $\mathfrak{E}$ and β' in $\mathfrak{M}_{\mathrm{OUT}}$.

Remark: IN-U-factorizations and dually OUT-U-factorizations are due to [16] where however, $\mathfrak{E}_{\mathrm{IN}}$ corresponds to a class of "generating morphisms" and $\mathfrak{M}$ to the class of (extremal) monomorphisms.

Note, that $\mathfrak{E}_{\mathrm{IN}}$ is not closed under composition with $\mathfrak{E}$ from right in general, even if $\mathfrak{E}$ is closed under composition. Hence we don't have a diagonal-morphism in general. In the special case $U = \mathrm{id}_{\mathsf{D}}$, where only a class $\mathfrak{E}$ is given in D, we have a terminal factorization situation in the sense of [15].

In the following the objects I and Y are fixed and a morphism $f: S \to S'$ of systems is a D-morphism $f: Q \to Q'$ such that $Uf \circ \tau = \tau'$ and $\beta' \circ Uf = \beta$.

1.2. Reduced and simple systems. Let us call a system *reduced* if β is in $\mathfrak{M}_{\mathrm{OUT}}$ and *simple* if τ is in $\mathfrak{E}_{\mathrm{IN}}$. It is easy to see that S is reduced iff for each morphism $f: S \to S'$ of systems with f in $\mathfrak{E}$, f is already an isomorphism. Dually S is simple iff S has no proper $\mathfrak{M}$-subobject. Note, that we have not used the notation "observable" and "reachable", which will be introduced in 1.5 and 1.5*, because in general "reduced" is weaker than "observable" and "simple" weaker than "reachable". Nevertheless, we have most of the results which are expected for reduced systems provided that we have an $\mathfrak{E}$-$\mathfrak{M}_{\mathrm{OUT}}$-factorization of OUT-$U$-morphisms. Sufficient conditions for the existence of such a factorization will be given below.

1.3. Theorem (reduced systems). *Given an $\mathfrak{E}$-$\mathfrak{M}_{\mathrm{OUT}}$-factorization of* OUT-$U$-*morphisms we say that a reduction is a morphism $f: S \to S'$ of systems with $f: Q \to Q'$ in $\mathfrak{E}$.*

Then we have:

1. *Existence and uniqueness: For each system S there is up to isomorphism a unique reduced system RS together with a reduction $f: S \to RS$.*

2. *Universal property: The reduction $f: S \to RS$ satisfies the following universal property: Each reduction $f': S \to S'$ can be extended uniquely by a reduction $\overline{f}: S' \to RS$ to the "universal" reduction f such that $f = \overline{f} \circ f'$.*

3. *Minimality (in the sense of* BAINBRIDGE's *Minimality Theorem in* [7]): *Given a system S' and a morphism $f': S \to S'$ (not necessarily a reduction) then RS is a "subquotient" of S'. More precisely, there is a system S'' and morphisms $e'': S'' \to RS$ with $e'' \in \mathfrak{E}$ and $m'': S'' \to S'$ with $m'' \in \mathfrak{M}$.*

Remarks: Note, that the universal property implies that the $\mathfrak{E}$-$\mathfrak{M}_{OUT}$-factorization of $\beta: UQ \to Y$ is a terminal factorization with respect to the class $\mathfrak{E}$ in the sense of [15]. It corresponds also to the factorization in BAINBRIDGE's reduction theorem in [7]. On the other hand, strictly speaking, the universal problem is not a universal but a "special problem" in the sense of [8]. The corresponding universal problem, the reflexivity of the subcategory of reduced systems, is solvable if the class $\mathfrak{M}_{OUT}$ is closed under composition with $\mathfrak{M}$ from left.

Proof: 1. Given a system $S = (Q, I, Y, \tau, \beta)$ and an $\mathfrak{E}$-$\mathfrak{M}_{OUT}$-factorization of $\beta = \overline{\beta} \circ Uf$ the reduced system RS of S is defined by $RS = (\overline{Q}, I, Y, Uf \circ \tau, \overline{\beta})$ such that $f: S \to RS$ becomes a reduction ($f \in \mathfrak{E}$) and RS is reduced ($\overline{\beta} \in \mathfrak{M}_{OUT}$). The uniqueness is a corollary of the universal property.

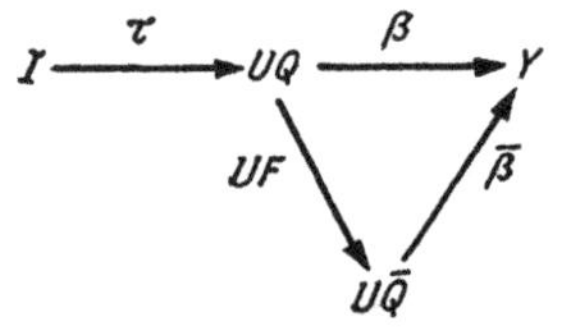

2. Given a reduction $f': S \to S'$ and an $\mathfrak{E}$-$\mathfrak{M}_{OUT}$-factorization $\beta' = \beta'' \circ Uf''$ we obtain another $\mathfrak{E}$-$\mathfrak{M}_{OUT}$-factorization $\beta = \beta'' \cdot U(f'' \cdot f')$ of β. Hence, by uniqueness there is a D-isomorphism $\varphi: Q'' \xrightarrow{\sim} \overline{Q}$ such that the following diagram commutes.

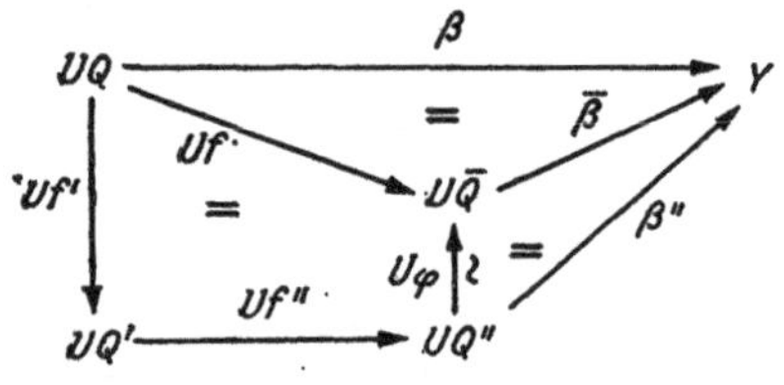

Hence, $\overline{f} = \varphi \circ f'': S' \to R(S)$ is the required unique reduction.

3. Let $f = m'' \cdot e$ be an $\mathfrak{E}$-$\mathfrak{M}$-factorization in D then there is a unique reduction $e'': S'' \to R(S)$ by 1. leading to the following commutative diagram:

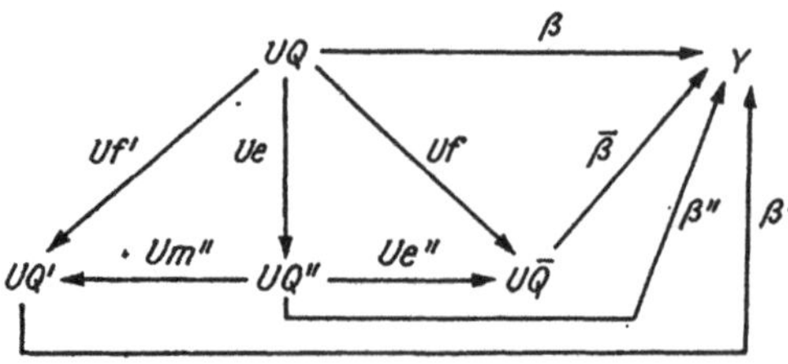

The dual theorem of 1.3 will be only sketched.

1.3*. Theorem (simple systems). *Given an $\mathfrak{E}_{IN}$-$\mathfrak{M}$-factorization of* IN-U*-morphisms we have:*

1. *The "simplification-problem" is solvable.*
2. *"Simplified" systems are unique up to isomorphism.*
3. *For each system S' with morphism $f: S' \to S$ the simplified systems $C(S)$ of S is an $\mathfrak{M}$-subobject of an $\mathfrak{E}$-quotient of S'.*

1.4. Theorem ($\mathfrak{E}$-$\mathfrak{M}_{OUT}$-factorization). *In each of the following cases we have an $\mathfrak{E}$-$\mathfrak{M}_{OUT}$-factorization of* OUT-U*-morphisms:*

1. D is cowell-powered and has cointersections of $\mathfrak{E}$-morphisms which are preserved by $U: \mathsf{D} \to \mathsf{K}$.

If, in addition, D has coequalizers, which are preserved by U, and $\mathfrak{M}$ is the class of all monomorphisms in D then we have $\mathfrak{M}_{OUT}$ is closed under composition with $\mathfrak{M}$ from left which implies the existence of diagonal morphisms.

2. D has kernel-pairs relative to OUT-U-morphisms and coequalizers of these kernel-pairs which are preserved by $U: \mathsf{D} \to \mathsf{K}$, provided that $\mathfrak{E}$ is the class of all coequalizers in D.

3. $U: \mathsf{D} \to \mathsf{K}$ has a right adjoint $\Box_S: \mathsf{K} \to \mathsf{D}$. In this case an OUT-$U$-morphism $\beta: UQ \to Y$ is in $\mathfrak{M}_{OUT}$ iff the adjoint morphism $\sigma: Q \to Y$ is in $\mathfrak{M}$.

Remarks: Part 1 of the theorem is a modified dual version of theorem 2.2 in [16] (or theorem 35.12 in [17]) used in proposition 7.3 [9] for the construction of reduced automata in pseudoclosed categories. Part 2 is closely related to the NERODE *Realization theorem* in [3] which is used for the construction of reduced and minimal algebra automata. Part 3 shows that the well-known construction of $\mathfrak{E}$-$\mathfrak{M}$-factorizations in D is only one special case for the construction of $\mathfrak{E}$-$\mathfrak{M}_{OUT}$-factorizations. This case will be discussed below in more detail.

Proof: 1. Given an OUT-U-morphism $\beta: UQ \to Y$ let $f_i: Q \to Q_i$ $(i \in I)$ a representative set of all D-morphisms $f': Q \to Q'$ in $\mathfrak{E}$ such that there is an OUT-U-morphism $\beta': UQ' \to Y$ satisfying $\beta = \beta' \circ Uf'$. Now, let $f: Q \to \overline{Q}$ be the cointersection of the family f_i $(i \in I)$. Using the universal properties of the cointersection and the $\mathfrak{E}$-$\mathfrak{M}$-factorization in D it is easy to see that also f belongs to $\mathfrak{E}$ (cf. proof of prop. 7.3 in [9].) Since U preserves cointersections there is also a unique $\overline{\beta}: U\overline{Q} \to Y$ satisfying $\beta = \overline{\beta} \circ Uf$. In order to show $\overline{\beta} \in \mathfrak{M}_{OUT}$ let us consider an arbitrary OUT-U-factorization $\overline{\beta} = \beta' \circ Uf'$ of $\overline{\beta}$. We have to show: $f' \in \mathfrak{M}$. Now, let $f' = m \circ e$ be an $\mathfrak{E}$-$\mathfrak{M}$-factorization of f'. Hence we have $\beta = \beta' \circ U(m) \circ U(e \circ f)$ with $e \circ f \in \mathfrak{E}$ such that $e \circ f = f_{i_0}$ for some $i_0 \in I$ without loss of generality. But then using the construction of f we have that e is an isomorphism and hence $f' = m \circ e \in \mathfrak{M}$. Given another $\mathfrak{E}$-$\mathfrak{M}_{OUT}$-factorization $\beta = \beta_0 \circ Uf_0$ there is a unique $\varphi: Q_0 \to \overline{Q} \in \mathfrak{E}$ with $\varphi \circ f_0 = f$ and $\overline{\beta} \circ U\varphi = \beta_0$ by construction of the cointersection. But $\beta_0 \in \mathfrak{M}_{OUT}$ implies also $\varphi \in \mathfrak{M}$ such that φ becomes a D-isomorphism. In order to show that $\mathfrak{M}_{OUT}$ is closed under composition with $\mathfrak{M}$ from left let $\beta_2 \circ Uf_2 = \beta \circ Uf_1$, with $\beta \in \mathfrak{M}_{OUT}$ and $f_1 \in \mathfrak{M}$. We have to show that f_2 is a monomorphism. Hence, let $f_3, f_4 : Q_3 \to Q_1$ be arbitrary D-morphisms with $f_2 \circ f_3 = f_2 \circ f_4$ and $d: Q \to Q_4$ the coequalizer of $f_1 \circ f_3$ and $f_1 \circ f_4$. Since U preserves coequalizers there is a unique $\beta_4: UQ_4 \to Y$ satisfying $\beta = \beta_4 \circ Ud$. Now, $\beta \in \mathfrak{M}_{OUT}$ implies $d \in \mathfrak{M}$. Using also $f_1 \in \mathfrak{M}$ and $d \circ f_1 \circ f_3 = d \circ f_1 \circ f_4$ we obtain $f_3 = f_4$.

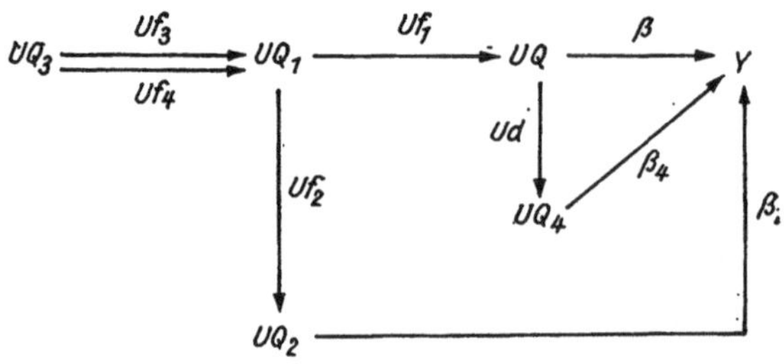

2. Let $(f_1, f_2): Q_1 \rightrightarrows Q$ be the kernel pair relative $\beta: UQ \to Y$ and $d: Q \to \overline{Q}$ the coequalizer of (f_1, f_2). Since U preserves this coequalizer there is a unique $\overline{\beta}: U\overline{Q} \to Y$ satisfying $\beta = \overline{\beta} \circ Ud$. In order to show $\overline{\beta} \in \mathfrak{M}_{\mathrm{OUT}}$ let $\overline{\beta} = \beta' \circ Uf'$ and $f' = m \circ e$ an $\mathfrak{E}$-$\mathfrak{M}$-factorization of f'. It remains to show that e is an isomorphism. By assumption $e \circ d \in \mathfrak{E}$ is again a coequalizer, say of $(f_3, f_4): Q_3 \rightrightarrows Q_4$. Hence, by construction of (f_1, f_2) there is a unique $f_5: Q_3 \to Q_1$ satisfying $f_1 \circ f_5 = f_3$ and $f_2 \circ f_5 = f_4$. Thus $d \circ f_3 = d \circ f_4$ such that there is a unique $f'': Q'' \to \overline{Q}$ with $f'' \circ e \circ d = d$. Now $d \in \mathfrak{E}$ implies that e is an isomorphism showing that $\overline{\beta}$ is in $\mathfrak{M}_{\mathrm{OUT}}$. Using a similar proof we can show that the $\mathfrak{E}$-$\mathfrak{M}_{\mathrm{OUT}}$-factorization $\beta = \overline{\beta} \circ Ud$ is unique up to isomorphism.

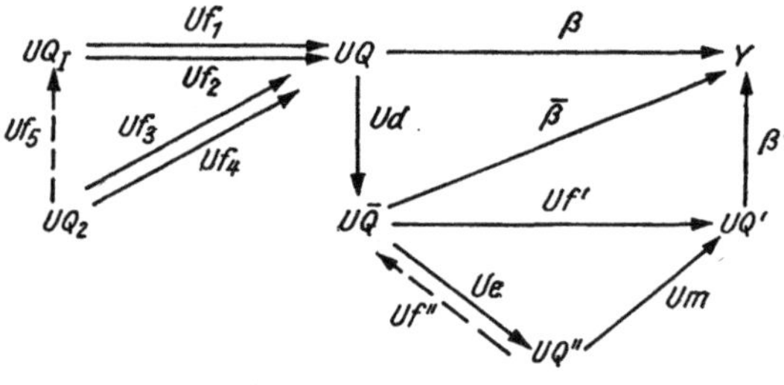

3. Using the couniversal properties of the right adjoint we obtain a unique adjoint morphism $\sigma: Q \to Y_{\S}$ in D satisfying $v_Y \circ U\sigma = \beta$. Let $\sigma = m \circ e$ be an $\mathfrak{E}$-$\mathfrak{M}$-factorization then it is easy to see that $\beta = (v_Y \circ Um) \circ Ue$ is an $\mathfrak{E}$-$\mathfrak{M}_{\mathrm{OUT}}$-factorization of β and that it is uniquely determined up to isomorphism.

If $\beta \in \mathfrak{M}_{\mathrm{OUT}}$ then $e \in \mathfrak{E} \cap \mathfrak{M}$ is an isomorphism and hence $\sigma \in \mathfrak{M}$. Vice versa $\sigma \in \mathfrak{M}$ implies $\beta \in \mathfrak{M}_{\mathrm{OUT}}$ because $g \circ f \in \mathfrak{M}$ implies $f \in \mathfrak{M}$.

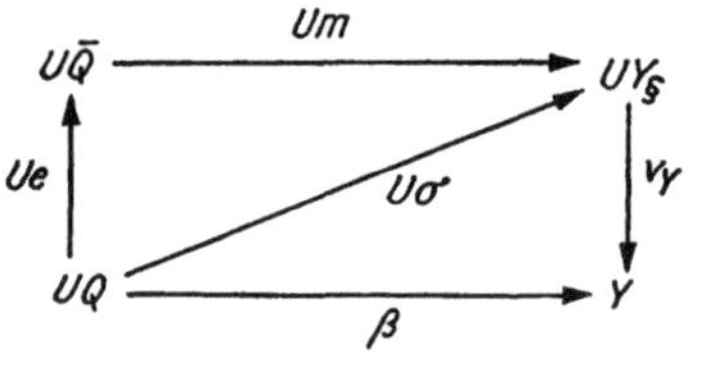

Dually we have

1.4*. Theorem ($\mathfrak{E}_{\mathrm{IN}}$-$\mathfrak{M}$-factorization). *In each of the following cases we have an $\mathfrak{E}_{\mathrm{IN}}$-$\mathfrak{M}$-factorization of IN-$U$-morphisms:*

1. *D is well-powered and has intersections of $\mathfrak{M}$-morphisms which are preserved by U: D $\to$ K.*

2. D *has cokernel pairs relative to* IN-U*-morphisms and equalizers of these cokernel pairs which are preserved by* U: D → K, *provided that* $\mathfrak{M}$ *is the class of all equalizers in* D.

3. U: D → K *has a left adjoint* $\square^{\S}$: K → D. *In this case an* IN-U*-morphism* τ: $I \to UQ$ *is in* $\mathfrak{E}_{\mathrm{IN}}$ *iff the adjoint morphism* $r: I^{\S} \to Q$ *is in* $\mathfrak{E}$.

Remark: Part 1 of the theorem can be applied to initial automata in pseudoclosed categories (cf. [9]) where D is the category K-K′-Medv (cf. 3.8). U: K-K′-Medv → K has no left adjoint because the unique morphism from the free automaton to the given one is not in the subcategory K′ in general. But if K′ has intersections which are preserved by I: K′ → K, and this is true for relational (K-K′ = Rel-PD), nondeterministic (ND-Set), and stochastic (Stoch-Set), then the assumptions of theorem 1.4*.1 are satisfied such that theorem 1.3* can be applied leading to a "reachability construction" for initial automata in pseudoclosed categories which coincides with that in [9] theorem 9.7. Note, that U: K-K′-Medv → K also fails to have a right adjoint but the assumptions of theorem 1.4.1 are satisfied in the examples mentioned above. Hence, initial automata in pseudoclosed categories provides a nice example which doesn't fit in the theory with adjoint functors but in our framework. As mentioned above the same is true for algebra automata.

In the following we will study the case that U: D → K has a right adjoint separately from that of a left adjoint. In both cases we are able to define a meaningful behaviour for systems and a complete Minimal-Realization-Theory. But both behaviour notations are equivalent if we have both adjoints.

1.5. Observable systems. For this subsection we assume that U: D → K has a right adjoint $\square_{\S}$: K → D ($\square$ stands for a blank). In this case the unique extension $\sigma: Q \to Y_{\S}$ of $\beta: UQ \to Y$ is called *observability map* of S satisfying

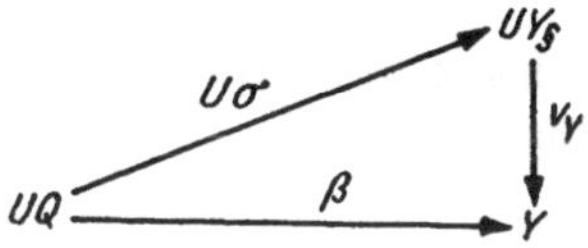

where $v_Y: UY_{\S} \to Y$ is the couniversal morphism of the adjunction.

A system is called *observable* if σ belongs to $\mathfrak{M}$. Hence, a system is reduced iff it is observable (cf. theorem 1.4.3). But note, that this is only true in the case of the existence of a right adjoint (which is not true for automata in pseudoclosed categories, where, however, observability can be defined in another way).

Now, the *behaviour* of a system is the morphism

$$I \xrightarrow{\tau} UQ \xrightarrow{U\sigma} UY_{\S}\,.$$

Before we give the Minimal-Realization-Theory in this case let us first consider systems without initial states. In that case the behavior is the K-subobject $U\overline{Q}$ of $UY_{\S}$ where $Q \xrightarrow{e} \overline{Q} \xrightarrow{m} Y_{\S}$ is an $\mathfrak{E}$-$\mathfrak{M}$-factorization of the observability map $\sigma: Q \to Y_{\S}$. In fact, all the constructions concerning reduction, minimization and realization done for automata in closed categories in chapters 4 and 5 of [9] can be extended to our case immediately.

Especially we have

1.6. Theorem (observable systems). *For each system (without initial states) there is an equivalent observable one which is unique up to isomorphism. Moreover, there is a canonical unique minimal realization (Generalization of Theorem* 5.3 *in* [9]).

Now let us turn back to systems with initial states.

1.7. Theorem (simple and observable systems). *Assume that* $U: \mathsf{D} \to \mathsf{K}$ *has a right adjoint and that there is an* $\mathfrak{E}_{\mathrm{IN}}$-$\mathfrak{M}$*-factorization of* IN-$U$*-morphisms.*

Then we have:

1. *For each system* S *there is an equivalent simple and observable system* S' *which is a subobject of a quotient of* S.

2. *Equivalent simple and observable systems are isomorphic.*

3. *For each behaviour* $I \to UY_{\S}$ *the* $\mathfrak{E}_{\mathrm{IN}}$-$\mathfrak{M}$*-factorization leads (up to isomorphism) to a unique minimal (i.e. simple and observable) realization.*

4. *A system is simple and observable iff it is an initial object in the subcategory of all equivalent observable systems.*

5. *A system is reduced iff it is observable.*

Proof: 1. Take first an $\mathfrak{E}$-$\mathfrak{M}$-factorization $m \circ e = \sigma$ of σ and then an $\mathfrak{E}_{\mathrm{IN}}$-$\mathfrak{M}$-factorization of $Ue \circ \tau$.

2.—4. are easy consequences of the existence and uniqueness of $\mathfrak{E}_{\mathrm{IN}}$-$\mathfrak{M}$-factorizations of the behaviour or the initial state morphism.

5. cf. theorem 1.4.3.

Dually to 1.5—1.7 we have:

1.5*. Reachable systems. Assuming that $U: \mathsf{D} \to \mathsf{K}$ has a left adjoint $\Box^{\S}: \mathsf{K} \to \mathsf{D}$ we obtain a unique extension $r: I^{\S} \to Q$, called *reachability map*, of $\tau: I \to UQ$ satisfying

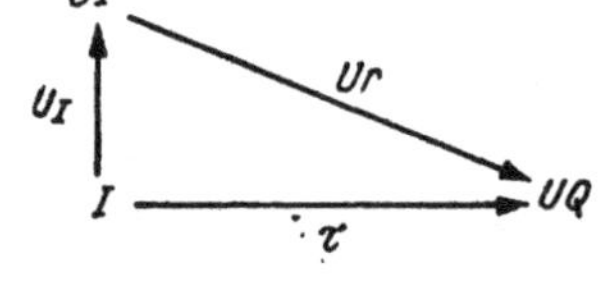

where $u_I: I \to UI^{\S}$ is the universal morphism of the adjunction.

A system is called *reachable* if r belongs to $\mathfrak{E}$. Hence, a system is simple iff it is reachable. The *behaviour* of a system is the morphism

$$UI^{\S} \xrightarrow{Ur} UQ \xrightarrow{\beta} Y .$$

For systems without output we define the behaviour to be K-quotient object UQ' of $UI^{\S}$ where $I^{\S} \xrightarrow{e} Q' \xrightarrow{m} Q$ is an $\mathfrak{E}$-$\mathfrak{M}$-factorization of r. Then we get the following dual theorem of 1.6.:

1.6*. Theorem (reachable systems). *For each system (without output) there is an equivalent reachable one which is unique up to isomorphism.*

For systems with output we obtain:

1.7*. Theorem (reachable and reduced systems). *If $U\colon \mathsf{D} \to \mathsf{K}$ has a left adjoint and there is an $\mathfrak{E}$-$\mathfrak{M}_{\mathrm{OUT}}$-factorization of* OUT-$U$-*morphisms then we have:*

1. *For each system S there is an equivalent reachable and reduced system S' which is a quotient of a subobject of S.*
2. *Equivalent reachable and reduced systems are isomorphic.*
3. *Existence of a unique minimal (i.e. reachable and reduced) realization.*
4. *A system is reachable and reduced iff it is "minimal" in the sense that it is a terminal object in the subcategory of all equivalent reachable systems.*
5. *A system is simple iff it is reachable.*

Remark: 1.7*.4 is exactly the "minimality" shown for algebra automata in [3].

Now, let us assume that we have both adjoints of $U\colon \mathsf{D} \to \mathsf{K}$:

1.8. Reachable and observable systems. Using the notation of 1.5 and 1.5* it remains to show that both behaviour notations are equivalent. All the other results are already included in 1.7 and 1.7*. The only new aspect is that Minimal-Realization can be obtained by $\mathfrak{E}$-$\mathfrak{M}$-factorization of the *extended behaviour morphism* $I^{\S} \to Y_{\S}$ which is the extension as well of a behaviour morphism $I \to UY_{\S}$ in 1.5 with respect to left adjoint as of a behaviour morphism $UI^{\S} \to Y$ in 1.5* with respect to the right adjoint.

1.9. Corollary. *There is a natural bijective correspondence between behaviour morphisms $I \to UY_{\S}$ in* 1.5, *$UI^{\S} \to Y$ in* 1.5* *and extended behaviour morphisms $I^{\S} \to Y_{\S}$ due to the following natural equivalences associated with the adjoint situations $\Box^{\S} \dashv U \dashv \Box_{\S}$:*

$$\mathrm{Mor}\,\mathsf{K}(I, UY_{\S}) \cong \mathrm{Mor}\,\mathsf{D}(I^{\S}, Y_{\S}) \cong \mathrm{Mor}\,\mathsf{K}(UI^{\S}, Y)\,.$$

For the results concerning reduction, minimization, reachability and minimal realization confer 1.7 and 1.7*.

2. The skeleton of minimal realization in F-morphism terminology

As mentioned in the introduction there are some interesting examples (4—7 in the introduction) which don't fit in the framework of section 1: To include Mealy-automata it would be sufficient to replace the functor $U\colon \mathsf{D} \to \mathsf{K}$ by two different functors $U_1\colon \mathsf{D} \to \mathsf{K}$ (the usual forgetful functor $U\colon \mathsf{K}\text{-Medv} \to \mathsf{K}$) and $U_2\colon \mathsf{D} \to \mathsf{K}$ (the functor $U\text{-}\otimes I'\colon \mathsf{K}\text{-Medv} \to \mathsf{K}$, where I' is the input object, such that an OUT-U_2-morphism becomes a K-morphism $\beta\colon Q \otimes I' \to Y$) and to assume that U_1 has a left and U_2 a right adjoint. In fact, section 1 could be extended to this case without any difficulties. Moreover, IN-U-morphisms and OUT-U-morphisms are special cases of F-morphisms in the sense of [8] and section 1 remains valid even for F-morphisms. In fact, F-morphisms seem to be an appropriate setting because all the examples 4—7 in the introduction and also the special case of section 1 can be formulated in this framework nicely. The necessary categorical theory of F-morphisms, concerning factorization and solution of universal and couniversal problems, is already solved in [8]. Especially F-morphism terminology allows to forget the forgetful functor U in the notation of section 1.

2.1. Review of F-morphisms. The basic feature of F-morphisms, like bilinear mappings, is the fact that there is in general no composition between F-morphisms, but only with morphisms from one category from the left and with those from another from the right.

Given a functor $F: \mathsf{G}^0 \times \mathsf{K} \to \mathsf{Set}$, also called *profunctor*, each element $f \in F(G, K)$ is called *F-morphism* from G to K, written $f: G \to K$.

For morphisms $g: G' \to G$ in G and $k: K \to K'$ in K a formal composition $k \circ f \circ g$, written as usual

$$G' \xrightarrow{g} G \xrightarrow{f} K \xrightarrow{k} K'$$

is defined by $k \circ f \circ g := F(g, k)(f) \in F(G', K')$.

In [8] it is shown that F-morphisms are also "real" morphisms in a suitable connection category V_F but for our purpose it is better to regard them as special functions, or families of functions between objects G and K of different categories. For example an IN-U-morphism $\tau: I \to UQ$ is an IN-morphism $\tau: I \to Q$ with respect to the functor $\mathrm{IN} = \mathrm{Mor}\, \mathsf{K}(-, U-): \mathsf{K}^0 \times \mathsf{D} \to \mathsf{Set}$ and an OUT-U-morphism $\beta: UQ \to Y$ is an OUT-morphism $\beta: Q \to Y$ with respect to the functor $\mathrm{OUT} = \mathrm{Mor}\, \mathsf{K}(U-, -): \mathsf{D}^0 \times \mathsf{K} \to \mathsf{Set}$.

In analogy to the existence of a left adjoint functor for U we will assume that the "universal problem" of the functor IN, is solvable and a right adjoint of U corresponds to the fact that the "couniversal problem" of OUT, is solvable: Reviewing [8] we say that the *universal problem of a functor* $F: \mathsf{G}^0 \times \mathsf{K} \to \mathsf{Set}$, short UP($F$), is solvable if for all G in G there is a $G^\S$ in K and a universal F-morphism $u_G \in F(G, G^\S)$ such that for all K in K and all $f \in F(G, K)$ there is a unique $k: G^\S \to K$ in K satisfying $k \circ u_G = f$.

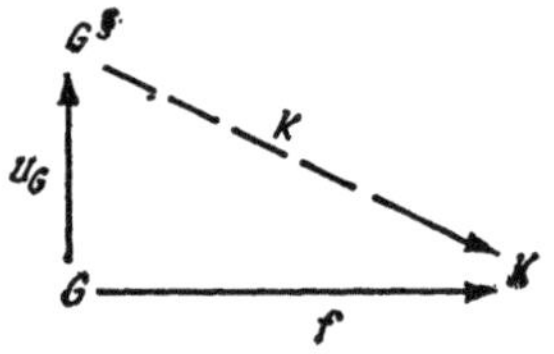

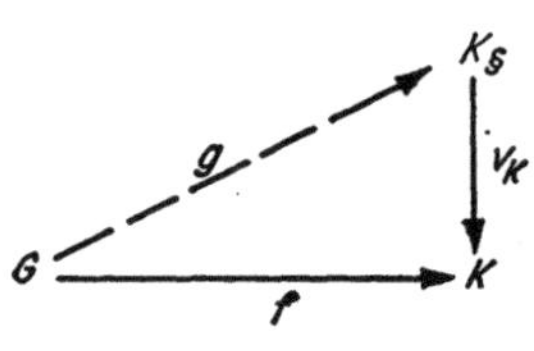

Dually, the *couniversal problem* of F, short Co UP(F), is solvable if for all K in K there is a $K_\S$ in G and a couniversal F-morphism $v_K \in F(K_\S, K)$ such that for all G in G and all $f \in F(G, K)$ there is a unique $g: G \to K_\S$ in G satisfying $v_K \circ g = f$.

A detailed discussion for the solvability of UP(F) and Co UP(F) is given in [8].

Of course, for $\mathrm{IN} = \mathrm{Mor}\, \mathsf{K}(-, U-)$ and $\mathrm{OUT} = \mathrm{Mor}\, \mathsf{K}(U-, -)$ UP(IN) and Co UP(OUT) are solvable iff U has a left and a right adjoint, respectively.

Now, let us see how Bainbridge's canonical realization theory of feedback systems can be formulated using F-morphisms:

2.2. Bainbridge's canonical realization theory. In [7] natural transformations $\Phi: \Phi_B^A \otimes_B \Psi_C^B \to X_C^A$ are considered and factorized in the form

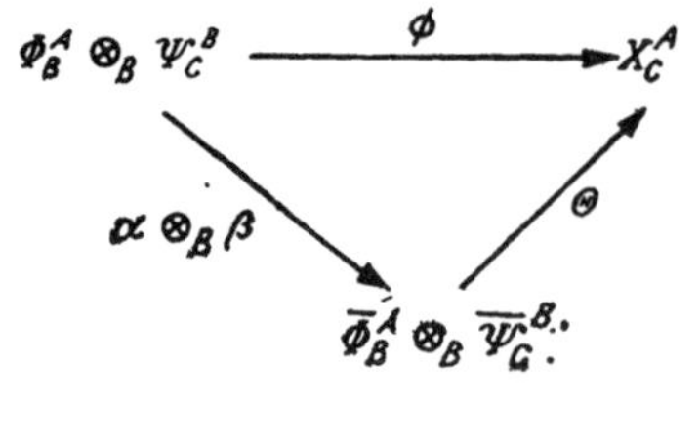

where Φ_B^A as object and α as morphism belongs to the category Funct_B^A of functors $\Phi\colon \mathsf{A}^0 \times \mathsf{B} \to \mathsf{Set}$ and natural transformations and $\otimes_B$ is a functor $\otimes_B\colon \mathsf{Funct}_B^A \times \times \mathsf{Funct}_C^B \to \mathsf{Funct}_C^A$ defined by the coend-construction $\Phi_B^A \otimes_B \Psi_C^B = \int^B \Phi_B^A \times \Psi_C^B$.

Such morphisms and factorizations are nicely to express in the F-morphism terminology taking

$$\mathsf{G} = \mathsf{Funct}_B^A \times \mathsf{Funct}_C^B\,, \qquad \mathsf{K} = \mathsf{Funct}_C^A$$

and $F\big((\Phi_B^A, \Psi_C^B), X_C^A\big)$ the set of all natural transformations $\theta\colon \Phi_B^A \otimes_B \Psi_C^B \to X_C^A$. The factorization above is a factorization of an F-morphism Φ in a G-morphism (α, β) followed by the F-morphism θ.

In this way BAINBRIDGE's Reduction and Minimality Theorems become special cases of 2.6.4 and 1.3.3 which remains true in our new context.

The formulation of monoid automata, multidecomposable systems in F-morphism-terminology, and the solution of the open problem in [2] will be given in section 3.

Now, we will reformulate section 1.

2.3. General assumptions. Let D, K_1 and K_2 be categories and $\mathrm{IN}\colon \mathsf{K}_1^0 \times \mathsf{D} \to \mathsf{Set}$ and $\mathrm{OUT}\colon \mathsf{D}^0 \times \mathsf{K}_2 \to \mathsf{Set}$ arbitrary functors then a system in this context is a construct $S = (Q, I, Y, \tau, \beta)$ where Q, I and Y are objects in D, K_1 and K_2, respectively, τ an IN-morphism from I to Q, and β an OUT-morphism from Q to Y

$$I \xrightarrow{\tau} Q \xrightarrow{\beta} Y\,.$$

Given an $\mathfrak{E}$-$\mathfrak{M}$-factorization in D let $\mathfrak{E}_{\mathrm{IN}}$ be the class of all IN-morphisms $\tau\colon I \to Q$ such that for each factorization $\tau = f \circ \tau'$ with f in D we have f in $\mathfrak{E}$.

We will say that we have an $\mathfrak{E}_{\mathrm{IN}}$-$\mathfrak{M}$-factorization of IN-morphisms if each $\tau \in \in \mathrm{IN}\,(I, Q)$ has a unique (up to D-isomorphism) factorization $\tau = f \circ \tau'$ with τ' in $\mathfrak{E}_{\mathrm{IN}}$ and f in $\mathfrak{M}$.

Dually $\mathfrak{M}_{\mathrm{OUT}}$ and $\mathfrak{E}$-$\mathfrak{M}_{\mathrm{OUT}}$-factorizations are defined.

Again I in K_1 and Y in K_2 will be fixed and a morphism $f\colon S \to S'$ of systems is a D-morphism $f\colon Q \to Q'$ satisfying $f \circ \tau = \tau'$ and $\beta' \circ f = \beta$. Note, that the explicit meaning of $f \circ \tau = \tau'$ is $\mathrm{IN}\,(I, f)\,(\tau) = \tau'$ and that of $\beta' \circ f = \beta$ is $\mathrm{OUT}\,(f, Y)\,(\beta') = \beta$. In the following we will not use different symbols for the different sorts of composition of IN, OUT and D-morphisms. Moreover, the usual composition rules, like associativity, remain valid if the composition is well-defined (cf. [8]). Finally, let us point out that we don't need a formal composition of IN-morphisms with OUT-morphisms. For example, we don't want to write $\beta \circ \tau\colon I \to Y$ instead of $I \xrightarrow{\tau} Q \xrightarrow{\beta} Y$ because the structure of dynamics object Q in D is essential.

Now, we are able to translate section 1 into the F-morphism terminology due to the following:

2.4. Translation rules.

Replace

a) "OUT-U-morphisms"	by	"OUT-morphisms"
b) "IN-U-morphisms"	by	"IN-morphisms"
c) "U has a right adjoint"	by	"CoUP(OUT) is solvable"
d) "U has a left adjoint"	by	"UP(IN) is solvable"

e) Delete the symbol U in all formulas and diagrams.

2.5. Reduced and simple systems. Read subsections 1.2, 1.3, and 1.3* using the translation rules 2.4.

In analogy to 1.4 and 1.4* we have:

2.6. Theorem ($\mathfrak{E}$-$\mathfrak{M}_{OUT}$-factorization). *In each of the following cases we have an $\mathfrak{E}$-$\mathfrak{M}_{OUT}$-factorization of* OUT-*morphisms*:

1. D *is cowell-powered and has cointersections, i.e. intersections in* D^0, *which are preserved by* OUT(—, Y): $D^0 \to$ Set.

2. D *has kernel pairs relative to* OUT-*morphisms and coequalizers of these kernel pairs, which are preserved as equalizers in* D^0 *by* OUT(—, Y): $D^0 \to$ Set, *provided that* $\mathfrak{E}$ *is the class of all coequalizers in* D.

3. CoUP(OUT) *is solvable. In this case an* OUT-*morphism* $\beta: Q \to Y$ *is in* $\mathfrak{M}_{OUT}$ *iff the adjoint* D-*morphism* $\sigma: Q \to Y_\S$ *is in* $\mathfrak{M}$.

4. *Let* $G = G_1 \times G_2$, $K = G_3$, *and* $F = \text{Mor}\, G_3(S(-, -), -)$, $F = \text{Mor}\, G_3(S(-, -)-)$: $(G_1 \times G_2)^0 \times G_3 \to$ Set, *where* $S: G_1 \times G_2 \to G_3$ *is a functor which is left adjoint to* $T_1: G_1 \times G_3 \to G_2$ *(for fixed* G_1 *in* G_1*) and also left adjoint to* $T_2: G_2 \times G_3 \to G_1$ *(for fixed* G_2 *in* G_2*)*. (Note, that this is exactly the situation of the reduction theorem in [7] with $S = \otimes_B$, $T_1 = \text{Hom}^A$ and $T_2 = \text{Hom}_C$.)

In order to show 2.4.4 we can extend the proof of the reduction theorem in [7] according to our assignments in 2.2 and $S = \otimes_B$, $T_1 = \text{Hom}^A$ and $T_2 = \text{Hom}_C$. Note, that the terminality of the factorization in [7] is equivalent to a (unique) $\mathfrak{E}$-$\mathfrak{M}_{OUT}$-factorization, where $\mathfrak{E} = \mathfrak{E}_1 \times \mathfrak{E}_2$ and $\mathfrak{M} = \mathfrak{M}_1 \times \mathfrak{M}_2$ for given $\mathfrak{E}_i$-$\mathfrak{M}_i$-factorizations in G_i ($i = 1, 2$).

2.6*. Theorem ($\mathfrak{E}_{IN}$-$\mathfrak{M}$-factorization). *In each of the following cases we have an* $\mathfrak{E}_{IN}$-$\mathfrak{M}$-*factorization of* IN-*morphisms*:

1. D is well-powered and has intersections which are preserved by IN (I, —): D $\to$ Set.

2. D has cokernel pairs relative to IN-morphisms and equalizers of these cokernel pairs which are preserved by IN (I, —): D $\to$ Set, provided that $\mathfrak{M}$ is the class of all equalizers in D.

3. UP(IN) is solvable. In this case an IN-morphism $\tau: I \to Q$ is in $\mathfrak{E}_{IN}$ iff the adjoint D-morphism $r: I^\S \to Q$ is in $\mathfrak{E}$.

4. Dual to 2.6.4.

2.7. Reachable and observable systems. Read subsections 1.5—1.7, 1.5*—1.7* 1.8, and 1.9 using the translation rules 2.4.

Since the adjoint situations $\square^\S \dashv U \dashv \square_\S$ are replaced by the statements UP(IN) and CoUP(OUT) solvable the translation of the bijective correspondence in 1.9 is the following:

$$\text{IN}(I, Y_\S) \cong \text{Mor}\, D(I^\S, Y_\S) \cong \text{OUT}\,(I^\S, Y)\,.$$

The correctness of this translation is shown in [8], Satz 4.2.

IN	OUT		
	OBSERVABILITY $U \dashv \square_{\S}$ 1.5—1.9 CoUP (OUT) 2.7	REDUCTION $\mathfrak{E}$-$\mathfrak{M}_{OUT}$ 1.2—1.4, 2.5—2.6	NO OUTPUT
REACHABILITY $\square^{\S} \dashv U$ 1.5*—1.7* 1.8—1.9 UP(IN) 2.7	— Initial MEALY- and MOORE-Automata in Closed Categories (3.1) — Adjoint Machines (3.2) — Machines in a Hyperdoctrine (3.3) — Initial Monoid Automata (3.4) — Continuous-Time-Systems (3.5) — Implicit λ-machines (3.6)	— Postulate 1—4 Machines (3.7) — Algebra Automata (3.7)	— MEDVEDEV-Automata in Closed Categories (3.1)
SIMPLIFICATION $\mathfrak{E}_{IN}$ — $\mathfrak{M}$ 1.2, 1.3*—1.4* 2.5, 2.6*	— Multidecomposable Systems ? (3.10) — Semilinear bicontinuous systems (3.5)	— Initial Automata in Pseudoclosed Categories (3.8) — Nonlinear Smooth Systems (3.5)	— MEDVEDEV-Automata in Pseudoclosed Categories (3.8)
NO INITIAL STATES	— Automata in Closed Categories (3.1) — Monoid Automata (3.4)	— Automata in Pseudoclosed Categories (3.8) — Feedback Systems (3.9)	—

3.0. Classification of automata, machines and systems with respect to properties of OBSERVABILITY, REDUCTION, REACHABILITY, SIMPLIFICATION, and MINIMAL REALIZATION (framed in bold line)

Remark: In the special case of section 1 we have IN = Mor K(—, U—) and OUT = Mor K(U—, —). Thus, we have trivially that CoUP(IN) and UP(OUT) are solvable. But these assumptions are absolutely not necessary for our theory although, as far as we know, they are satisfied for all interesting examples in automata and systems theory, except for the minimization of the transition monoid (cf. 3.4 and [9] 5.7).

3. Applications

In table 3.0 on p. 149 are putted together of several interesting examples in automata and systems theory showing how they fit in the general frame of sections 1 and 2. For each type the corresponding properties (existence of adjoints, $\mathfrak{E}_{IN}$-$\mathfrak{M}$-factorizations, etc.) are sketched below. In most cases the properties are already shown in the literature so that we only give the references. The classification of multidecomposable systems is not completely sure. But in 3.10 we give a modified version that fits in and in 3.11 a solution for the open problem in [2] using an $\mathfrak{E}$-$\mathfrak{M}_{OUT}$-factorization to obtain "reduced connections". Numbers in table 3.0 are references to this paper.

3.1. Automata in closed categories. Initial Moore-automata in closed categories, including deterministic, partial, bilinear and topological automata, were first studied in [14]. In this case D is the category K-Medv of Medvedev-automata in K. The existence of left and right adjoints for U: K-Medv → K is well-known, implicitly in [9]. In the case of Mealy-automata we have to take $U_1 = U\text{-}\otimes I$: K-Medv → K for the output part, the right adjoint is given in [9], Lemma 4.4, and for initial Mealy-automata we have to take again U (different from U_1, so that only section 2 is applicable!). The left adjoint is given more or less in [9], 9.3. For noninitial automata confer also 1.5 and 1.6.

3.2. Adjoint machines. For adjoint machines in [4] we take D = Dyn(X) and U: Dyn(X) → K the forgetful functor which by definition has left and a right adjoint. In addition to initial Moore-automata in closed categories (3.1) an interesting example of this case is given by linear automata.

3.3. Machines in a hyperdoctrine. Following [6] we have D = $P(X)$, K = $P(E)$ U = sub u where $u: E \to X$ is a "behaviour scheme" which implies that there are left and right adjoints $\square^{\S} = \sum u$ and $\square_{\S} = \Pi u$ of U = sub u.

3.4. Monoid automata. Replacing the free monoid I^* in the theory of (initial) Moore-automata in closed categories by an arbitrary monoid M, such that $d: S \otimes \otimes M \to S$ becomes a right action, we obtain (initial) monoid automata. These are important models for continuous-time systems as shown in [12] where also left and right adjoints for the forgetful functor U: K-Ract$_M$ → K are given (cf. also [9], 5.7).

Considering monoid-automata with respect to their transition monoid we obtain another interesting example (cf. [9], 5.6). Let us take D = K-Mon, the category of monoids in K, and OUT: K-Mon^0 × K → Set where for each monoid M and K-object S OUT (M, S) is the set of all right action morphisms $d: S \otimes M \to S$. (Note, that OUT is not of the form Mor $\mathsf{K}(U-, -)$. Following [9], 5.6 we have

$$\text{OUT}(M, S) \cong \text{Mor } \mathsf{K\text{-}Mon}(M, \langle S, S\rangle)$$

where $\langle S, S\rangle$ is the endomorphism monoid, such that CoUP(OUT) is solvable. Thus we obtain the transition morphism $t\colon M \to \langle S, S\rangle$ as adjoint morphism of d. The behaviour in this case, the image of t, is the transition monoid. Thus, we obtain the minimization of the transition monoid in [9] 5.7.

3.5. Continuous-time-systems. In [12] it is shown that the state-transition function of constant continuous-time-systems in the sense of [18] is a right action of a monoid on the state space. The monoid is built up from the time set and the input functions. Hence, these systems can be regarded as monoid automata in the sense of the first part of 3.4, provided that only a discrete or topological structure on the state space X is considered. Assuming X to be an R-module and the state-transition function φ to be linear in this variable we obtain "semilinear" continuous-time-systems, a generalization of the bilinear systems in [19]. For semilinear systems a right and left adjoint of the corresponding forgetful functor $U\colon \mathsf{Mod}_R\text{-}\mathsf{Ract}_M \to \mathsf{Mod}_R$ is given in [12].

If, in addition, X is a topological space and φ continuous in this variable, we obtain semilinear "bicontinuous" systems. Again we have a right adjoint but difficulties with the left adjoint. In any case, however, there is a natural $\mathfrak{E}_{\text{IN}}$-$\mathfrak{M}$-factorization.

Finally, considering nonlinear (smooth) systems in the sense of [20], where X is a C^∞-manifold we presumably neither have a right nor a left adjoint. But using only some of the constructions in [20] suitable $\mathfrak{E}_{\text{IN}}$-$\mathfrak{M}$- and $\mathfrak{E}$-$\mathfrak{M}_{\text{OUT}}$-factorizations are given in [12] leading to the unique constructions for reachability and observability (simplification and reduction in our terminology) in [20].

3.6. Implicit λ-machines. Implicit λ-machines together with suitable left and right adjoints are defined in [5]. Note, however, that two different forgetful functors are used for implicit λ-machines such that also this example — intended to motivate the "Axiomatic Minimal-Realization-Theorem" — actually is not in the "Arbib-Manes-Bainbridge-Skeleton" of section 1.

3.7. Algebra-automata. Algebra automata are machines in the sense of [3] where $\mathsf{D} = \mathsf{Dyn}(X)$ is the category Ω-Alg of Ω-algebras. There is a well-known left adjoint of the forgetful functor $U\colon \Omega\text{-}\mathsf{Alg} \to \mathsf{Set}$ but no right adjoint. For machines satisfying postulates 1—4 an $\mathfrak{E}$-$\mathfrak{M}_{\text{OUT}}$-factorization of the behaviour morphism $UI^{\$} \to Y$ is constructed in [3]. The construction is more or less that of 1.4.2. For algebra-automata these postulates are verified in [3]. But also the construction of $\mathfrak{E}$-$\mathfrak{M}_{\text{OUT}}$-factorizations in 1.4.1. can be applied because $U\colon \Omega\text{-}\mathsf{Alg} \to \mathsf{Set}$ preserves cointersections (but not coproducts in general).

3.8. Automata in pseudoclosed categories. Automata in pseudoclosed categories, including nondeterministic, relational, stochastic and relational topological automata, are defined in [10] and [9]. The category D in this case is the category $\mathsf{K\text{-}K'\text{-}Medv}$ of Medvedev-automata in K where the morphisms are in the subcategory K' componentwise. $U\colon \mathsf{K\text{-}K'\text{-}Medv} \to \mathsf{K}$ in general has no right adjoint nor a left adjoint. (Although $U\colon \mathsf{K\text{-}Medv} \to \mathsf{K}$ has a left adjoint we prefer to use $\mathsf{K\text{-}K'\text{-}Medv}$ in order to have cointersections.)

Due to the remark in 1.4* we have an $\mathfrak{E}_{IN}$-$\mathfrak{M}$-factorization of IN-U-morphisms by 1.4*.1, provided that K' has and the inclusion functor I: $\mathsf{K}' \to \mathsf{K}$ preserves intersections, and an $\mathfrak{E}$-$\mathfrak{M}_{OUT}$-factorization of OUT-U-morphisms (or OUT-(U-$\otimes$ I)-morphisms in the case of Mealy-automata) by 1.4.1, provided that K' has cointersections and is cowell-powered. In fact, the reduction theorem 7.3 in [9] is a special case of 1.4.1.

3.9. Feedback systems. Following [7] the reduction theorem in [7], which is a speciel case of our construction 2.6.4 for $\mathfrak{E}$-$\mathfrak{M}_{OUT}$-factorizations, can be applied to feedback systems defined in [7]. In 2.2 it is shown that the corresponding canonical realization theory fits in our concept of F-morphisms.

3.10. Multidecomposable systems. Let us consider the following modified version of k-line-systems in [2]. $(\mathsf{K}, \otimes)$ is an arbitrary closed category and $\mathsf{D} = \mathsf{Dyn}(\mathrm{Id}_{\mathsf{K}})$. The initial state object I is replaced by the family of objects I^{α} defined in [2] where α ranges over all proper subsets of $K = \{1, 2, \ldots, k\}$. The initial state morphism is replaced by the family G_{α}: $I^{\alpha} \to Q$. Our modification is that we start with the morphisms G_{α} whereas in [2] G_{α} is defined by $G_{\alpha} = J_{\alpha} \circ (f_{\alpha} \otimes \mathrm{id})$ where f_{α}: $I^{[\alpha]} \to Y_{\alpha}$ is the response function of the $[\alpha]$-line subsystem M_{α} and J_{α}: $Y_{\alpha} \otimes I^{\bar{\alpha}} \to Q$ a given "connection"-morphism.

Thus, in the terminology of section 2 we have:

$$\mathrm{IN}\big((I^{\alpha})_{\alpha \subset K}, Q\big) = \text{family of all } G_{\alpha}: I^{\alpha} \to Q\,.$$

The "Principle of Simple k-Reduction" in [2] shows exactly that UP(IN) is solvable leading to the reachability morphism r: $I^{K} \to Q$. On the other hand the forgetful functor U: $\mathsf{Dyn}(\mathrm{Id}_{\mathsf{K}}) \to \mathsf{K}$ has a right adjoint such that we have CoUP(OUT) solvable for OUT-Mor $\mathsf{K}(U\text{-}, -)$ and all the results are applicable.

It is still open whether the original hierarchical definition of k-line systems fits in our concept. But we can give a solution to the open problem of uniqueness of the canonical realization.

3.11. Solution of the open problem in [2]. The definition 3.3 of *canonical* k-line systems in [2] is too weak. We have to add the following restriction for the output objects Y_{α} of M_{α} and for the connections J_{α}: $Y_{\alpha} \otimes I^{\bar{\alpha}} \to Q$:

We say that the k-line system has *reduced connections* if we have $f_{\alpha} \in \mathfrak{E}$ for the response morphisms f_{α} of M_{α} and $\bar{J}_{\alpha} \in \mathfrak{M}$ for the adjoint morphism $\bar{J}_{\alpha}$: $Y_{\alpha} \to Q^{I^{\bar{\alpha}}}$ of J_{α} for all proper subsets α of K. Now, the construction of the canonical realization in [2] must be modified in the following way:

We don't take the universal factorization $G_{\alpha} = \varepsilon_{Q} \circ (f_{\alpha} \otimes \mathrm{id})$ of G_{α} but the unique $\mathfrak{E}$-$\mathfrak{M}_{OUT}$-factorization with respect to the functor OUT $=$ Mor $\mathsf{K}(-\otimes I^{\bar{\alpha}}, -)$, say $G_{\alpha} = J_{\alpha} \circ (\bar{f}_{\alpha} \otimes \mathrm{id})$. Since CoUP(OUT) is solvable we can get this by an $\mathfrak{E}$-$\mathfrak{M}$-factorization $I^{[\alpha]} \xrightarrow{\bar{f}_{\alpha}} Y_{\alpha} \xrightarrow{\bar{J}_{\alpha}} Q^{I^{\bar{\alpha}}}$ of $f_{\alpha}^{\urcorner}$. Thus we have a reduced connection.

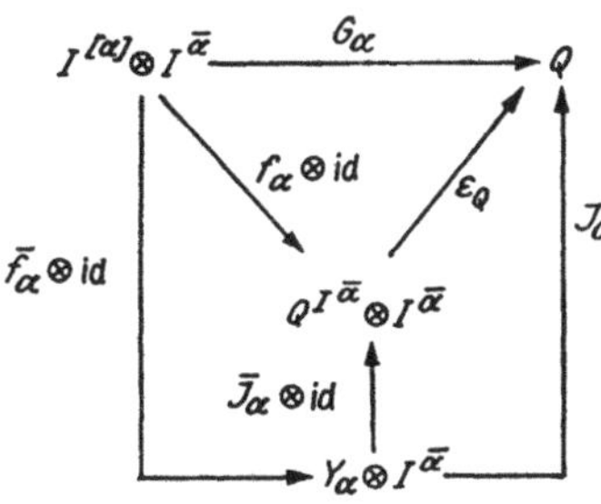

The uniqueness of canonical realization (with reduced connections) is straight forward using the uniqueness of $\mathfrak{E}$-$\mathfrak{M}_{\mathrm{OUT}}$-factorizations.

Acknowledgements. 1. The material of this paper was given as a contribution to the "Workshop on Algebraic Systems Theory" organized in June 1976 by M. Arbib and E. G. Manes at the University of Massachusetts in Amherst and we are grateful to them for several useful suggestions.

2. We are sure that several of the results given by the group Adámek-Koubek-Reiterman-Trnková for Arbib-Manes-machines remain true in the more general context of this paper. Especially they have shown in [21] and [1] that several sufficient conditions for reduction, minimization and minimal realization are also necessary. Due to a private communication of J. Adámek, for example, the condition that U preserves cointersections in 1.4 is also necessary for reduction in theorem 1.3.

References

[1] Adámek, J.: Automata and Categories: Finiteness contra Minimality, Lecture Notes in Comp. Sci. 32 (1975), 160—166.

[2] Anderson, B. D. O.; Arbib, M. A.; Manes, E. G.: Foundations of System Theory: Multidecomposable Systems Research Report (Univ. of Mass.) 1975.

[3] Arbib, M. A.; Manes, E. G.: Machines in a Category: An expository introduction, COINS Technical Report 71C-6, SIAM Review 16 (1974), 163—192.

[4] — —: Adjoint Machines, State-behaviour Machines and Duality, COINS Technical Report 73B-1, J. Pure Appl. Alg. 6 (1975), 313—344.

[5] — —: Fuzzy Machines in a Category, COINS Technical Report 75B-1.

[6] Bainbridge, E. S.: A Unified Minimal Realization Theory with Duality, Technical Report No. 140 (1972), Univ. of Michigan.

[7] —: Feedback and Generalized Logic, Research Report (Univ. of Ottawa) 1975, to appear in Information and Control.

[8] Ehrig, H.: F-Morphismen, Math. Nachr. 59 (1974) 75—93.

[9] Ehrig, H.; Kiermeier, K. D.; Kreowski, H.-J.; Kühnel, W.: Universal Theory of Automata: A Categorical Approach, Teubner-Verlag, Stuttgart 1974.

[10] Ehrig, H.; Kreowski, H.-J.: Systematic Approach to Reduction and Minimization in Automata and System Theory, Technical Report 73-16, Technical University Berlin, Journ. Comp. Syst. Sci. 12 (1976), 269—304.

[11] —: Minimization Concepts of Automata in Pseudoclosed Categories, Proc. Int. Symp. Algeb. Syst. Theory, Udine 1975, Lecture Notes in Econ. Math. Syst. 131 (1976), 359—374.

[12] Ehrig, H.; Kühnel, W.: Categorical Approach to Nonlinear Constant Continuous Time Systems, Technical Report 76-05, Technical University Berlin, FB 20.

[13] Goguen, J. A.: Systems and Minimal Realization, Proc. 1971 IEEE Conf. Dec. Contr., 42—46.

[14] —: Minimal Realization of Machines in Closed Categories, Bull. AMS 78 (1972), 777—783.

[15] Goguen, J. A.; Thatcher, J. W.; Wagner, E. G.; Wright, J. B.: Parallel Realization of Systems, Using Factorizations and Quotients in Categories, Research-Report 1975.

[16] HERRLICH, H.: Factorization of Morphisms $f: B \to FA$, Math. Z. **114** (1970), 180–186.
[17] HERRLICH, H.; STRECKER, G. E.: Category Theory, Allyn and Bacon Inc., Boston 1973.
[18] KALMAN, R. E.; FALB, P. L.; ARBIB, M. A.: Topics in Mathematical System Theory, McGraw-Hill, New York 1969.
[19] SUSSMANN, H. J.: Minimal Realizations and Canonical Forms for Bilinear Systems, Research Report (Rutgers Univ.) 1975.
[20] —: Existence and Uniqueness of Minimal Realizations of Nonlinear Systems, Math. Syst. Theory. **10** (1976/77), 263–284.
[21] TRNKOVÁ, V.: On Minimal Realizations of Behaviour Maps in Categorical Automata Theory, Com. Math. Univ. Carol. **15**, 3 (1974), 555–566.

Sheaf representations of algebras and transfer theorems for model theoretic properties

by HUGO VOLGER, Tübingen

0. Introduction

In the following we want to present a survey of some recent work of BURRIS, KEIMEL and WERNER on representations of algebras as algebras of sections of sheaves over boolean spaces. Our presentation will follow fairly closely the presentation in WERNER [17], although we will put the emphasis on te generalized FEFERMAN-VAUGHT theorem and the transfer theorems of model theoretic properties.

The situation to be considered will be the following. Every algebra A of a class $\mathfrak{K}$ can be presented as the algebra of global sections of a sheaf of algebras whose stalks belong to a subclass $\mathfrak{K}_0$ of $\mathfrak{K}$. A transfer theorem for a property P states that under certain conditions P holds for $\mathfrak{K}$, whenever P holds for $\mathfrak{K}_0$. A typical example, to be kept in mind, is the PIERCE representation [11] of a commutative regular ring R as a sheaf of fields over $B(R)^*$, the dual space of the boolean algebra of idempotents of R.

The reasons for restricting the attention to sheaves over boolean spaces are two-fold. For the class of discriminator varieties, which contains a large number of well known varieties, we have the representation theorem of KEIMEL and WERNER [8], which uses sheaves over boolean spaces. Moreover, for sheaves over boolean spaces a generalization of the theorem of FEFERMAN and VAUGHT [7] is available, which relates the properties of the algebra of global sections with those of the stalks. Thus it will be a basic tool in obtaining transfer theorems of model theoretic properties.

Although most of the results discussed here remain true for relational structures we will restrict ourselves to algebras for the sake of simplicity. In the first chapter we will introduce sheaves of algebras over boolean spaces and, in particular, boolean sheaves. We will try to pin-point the role of the ternary discriminator for these sheaves in the second chapter, where we will introduce the notion of a discriminator variety. Then the representation theorem of KEIMEL and WERNER [8] will be presented together with examples of discriminator varieties. The third chapter contains the generalized FEFERMAN-VAUGHT theorem which is basic for all the applications which follow. In the fourth chapter we will present the decidability result of BURRIS and WERNER [2] for residually finite discriminator varieties. The fifth chapter will contain results on ω-categoricity and related results.

It should be mentioned that boolean sheaves can also be described in a more algebraic way, which does not use topological spaces, as structures with values in a boolean algebra. The equivalence of the two approaches will be shown. Using the boolean-valued approach a completely algebraic proof of the FEFERMAN-VAUGHT theorem (cf. [14]) and the representation theorem can be given. Using a similar approach WEISPFENNING [16] obtained a number of transfer theorems for model completeness and quantifier elimination. Since he makes use of a different technique, we will not be able to present his results in this survey.

1. Sheaves over boolean spaces and boolean sheaves

Let X be a topological space. A *sheaf over* X is a topological space P together with a surjective map $\pi: P \to X$ which is a local homeomorphism, i.e., for every p in P there exists a neighbourhood U of p such that $\pi \mid U$ is a homeomorphism from U to $\pi(U)$. The inverse image $P_x = \pi^{-1}(x)$ of x in X is called the *stalk at* x. $\Gamma(X, P)$, the set of *global sections*, consists of continuous functions $\alpha: X \to P$ with $\pi\alpha = \mathrm{id}_X$. Thus $\Gamma(X, P)$ is a subset of $\Pi\langle P_x \mid x \in X\rangle$. Since P is a sheaf over X, the set $[\alpha_1 \equiv \alpha_2] = \{x \in X: \alpha_1(x) = \alpha_2(x)\}$ is open in P for α_1, α_2 in $\Gamma(X, P)$ and P_x is a discrete space in the induced topology for every x in X. Thus $\Gamma(X, P)$ should be considered as a continuous product of discrete spaces. If P is a T_2-space then $[\alpha_1 \equiv \alpha_2]$ is clopen for α_1, α_2 in $\Gamma(X, P)$.

The following examples will illustrate these definitions. If X is a discrete space, then a sheaf over X is a discrete space P together with a surjective map $\pi: P \to X$. $\Gamma(X, P)$ is in this case the product $\Pi\langle P_x \mid x \in X\rangle$. If A is a discrete space then the product space $X \times A$ together with the projection onto X is the constant sheaf for A over X. Every stalk is isomorphic to A and $\Gamma(X, X \times A)$ is isomorphic to $C(X, A)$, the continuous functions from X to A.

For a sheaf $\pi: P \to X$ we can construct a new sheaf $\pi_n: P^{(n)} \to X$, which is the n-th power of the given sheaf in an appropriate category, as follows. $P^{(n)}$ is the subspace of the points $\langle p_1, \dots, p_n\rangle$ in P^n with $\pi(p_1) = \dots = \pi(p_n)$ and π_n is defined by $\pi_n(\langle p_1, \dots, p_n\rangle) = \pi(p_1)$. Thus we have $P^{(n)}_x = (P_x)^n$ and $P^{(n)} = \bigcup \langle (P_x)^n : x \in X\rangle$ and $\Gamma(X, P^{(n)}) \cong (X, P)^n$, i.e., global sections of $P^{(n)}$ are n-tupels of global sections.

For the following let $\Delta = \{n_i \mid i \in I\}$ be a fixed finitary type and all algebras considered will be of type Δ. A *sheaf of Δ-algebras* over X is a sheaf $\pi: P \to X$ over X together with a continuous map $f_i: P^{(n_i)} \to P$ with $\pi f_i = \pi_{n_i}$ for every i in I. The restrictions $f_{i,x}: P_x^{n_i} \to P_x$ make every stalk P_x a Δ-algebra. Moreover, $\Gamma(X, P)$ can be made into a Δ-algebra where the operation $\bar{f}_i$ means composing with f_i, i.e., $\bar{f}_i(\alpha_1, \dots, \alpha_{n_i})(x) = f_{i,x}(\alpha_1(x), \dots, \alpha_{n_i}(x))$. In this way, $\Gamma(X, P)$ becomes a subalgebra of the product $\Pi\langle P_x \mid x \in X\rangle$. It should be mentioned that constant operations on P correspond to global sections.

In the following we will be interested in sheaves over boolean spaces, i.e., compact T_2-spaces with a base of clopen subsets. The most important property of boolean spaces is the *finite partition property* (FPP) which states: For every open covering $\langle U_i \mid i \in I\rangle$ of a boolean space X there exist a finite partition $\langle V_j \mid j = 1, \dots, k\rangle$ of clopen subsets of X such that there is a U_i with $V_j \in U_i$ for every j. Using the FPP one can show that for every sheaf $\pi: P \to X$, where X is boolean, $\Gamma(X, P)$ is a subdirect product of the stalks P_x for x in X, i.e., for every p in a stalk P_x there exists a global section α in $\Gamma(X, P)$ with $\alpha(x) = p$. As we have seen above, if P is T_2 then $[\alpha_1 \equiv \alpha_2]$ is in X^*, the dual algebra of clopen subsets.

In this case a more algebraic description of the algebras of global sections can be given. For a Δ-algebra A and a boolean space X the following three statements are equivalent:

(1) A is the algebra of global sections of a sheaf P of Δ-algebras of X where P is T_2.

(2) A is a subdirect product of Δ-algebras P_x for x in X such that $[a_1 \equiv a_2] = \{x \in X: a_1(x) = a_2(x)\} \in X^*$ for every a_1, a_2 in A and A satisfies the *patchwork principle* (PC), i.e., for every a_1, a_2 in A and N in X^* there exists an a in A with $N \subseteq [a \equiv a_1]$, $-N \subseteq [a \equiv a_2]$.

(3) A is a Δ-algebra together with a X^*-valued congruence relation $E: A \times A \to X^*$ such that $E(a_1, a_2) = 1$ implies $a_1 = a_2$ and which satisfies the condition (PC) if $[\equiv]$ is replaced by E.

This is shown by a circle of implications. The main difficulty lies in the step from (3) to (1). Here P_x is defined by $A/\equiv_x$, where $a_1 \equiv_x a_2$ iff $E(a_1, a_2) \in D_x$, the ultrafilter on X^* corresponding to x. Then the sets $W_{a,N} = \{a/\equiv_x \mid x \in N\}$ for a in A and N in X^* form a base of the topology on $P = \bigcup \langle A/\equiv_x \mid x \in X\rangle$, which makes P into a sheaf over X. Finally, one has to show that every global section is of the form $\hat{a}$, where $\hat{a}(x) = a/\equiv_x$ for x in X.

As a typical application of this fact we will consider a reduced product $\Pi \langle A_i \mid i \in I\rangle/F$ of Δ-algebras, which may be considered as a $2^I/F$-valued structure with $E(\alpha_1/F, \alpha_2/F) = \{i \in I: \alpha_1(i) = \alpha_2(i)\}/F$. The dual space $(2^I/F)^*$ consists of ultrafilters D on I which contain F, and the stalk at D will be the ultraproduct $\Pi \langle A_i \mid i \in I\rangle/D$ (cf. Ellerman [6]). This is also an example of the more general notion of a quotient with respect to a filter on X, which corresponds to restricting the sheaf to a closed subspace.

As another example which is described most easily in terms of (3) we consider a commutative ring R which is regular i.e. for every r in R there exists r^{-1} with $r^2r^{-1} = r$. Then $r^* = rr^{-1}$ belongs to $B(R)$, the boolean algebra of idempotents of R and we can define $E: R \times R \to B(R)$ by $E(r_1, r_2) = 1 - (r_1 - r_2)^*$. Moreover, for r_1, r_2 in R and e in $B(R)$ we have $(r - r_1)^* \leqq e$, and $(r - r_2)^* \leqq (1 - e)$ with $r = er_1 + (1 - e)\, r_2$ which yields the condition (PC). Noting that maximal ideals of R correspond to maximal ideals of $B(R)$, we have obtained in this way the representation given by Pierce [11].

For a sheaf $\pi: P \to X$ where X is boolean and P is T_2 let $Z(P)$ be the set $\{x \in X: |P_x| = 1\}$. Now, we may assume $|Z(P)| \leqq 1$, since we can arrive at this situation without changing the global sections by removing first $\operatorname{int}(Z(P))$ from $Z(P)$ and then contracting the rest to a point. This motivates the following definition: A *boolean sheaf of* Δ-algebras is a sheaf of Δ-algebras P over a boolean space X such that P is T_2 and $|Z(P)| \leqq 1$.

In a boolean sheaf P over X the topology on X is very much determined by the sets $[\alpha_1 \equiv \alpha_2]$ for α_1, α_2 in $\Gamma(X, P)$. More precisely, we can prove

$$X^* = \bigcup \langle\{[\alpha_1 \equiv \alpha_2], [\alpha_1 \not\equiv \alpha_2]\} : \alpha_1, \alpha_2 \in \Gamma(X, P)\rangle .$$

2. The representation theorem for discriminator varieties

In the following the ternary discriminator $t: A^3 \to A$ on a set A, which is defined by $t(a_1, a_2, a_3) = a_3$ if $a_1 = a_2$ and $t(a_1, a_2, a_3) = a_1$ if $a_1 \neq a_2$, will play a central role, which is due to the following result:

For a subdirect product A of Δ-algebras P_x for x in X, where X is a boolean space and $[a_1 \equiv a_2]$ is in X^* for a_1, a_2 in A, the following statements are equivalent:

(1) A satisfies the patchwork principle (PC) and

$$X^* = \bigcup \langle \{[a_1 \equiv a_2], [a_1 \not\equiv a_2]\} : a_1, a_2 \in A \rangle .$$

(2) There exists t: $A^3 \to A$ such that t_x: $P_x^3 \to P_x$ is the ternary discriminator on P_x for every x in X and

$$X^* = \bigcup \langle \{[a_1 \equiv a_2], [a_1 \not\equiv a_2]\} : a_1, a_2 \in A \rangle .$$

In the presence of (PC) t can be determined such that $[t(a_1, a_2, a_3) \equiv a_3] \geqq [a_1 \equiv a_2]$ and $[t(a_1, a_2, a_3) \equiv a_1] \geqq [a_1 \not\equiv a_2]$. Conversely, the element required in (PC) can be built up using t and related functions.

Clearly, the ternary discriminator is definable by a first-order formula. However, the full strength of (2) can be used, when the discriminator is definable by a term on all algebras P_x. This motivates the following definition. A variety $\mathfrak{V}$ of Δ-algebras is called a *discriminator variety* iff there exist a term $q(v_1, v_2, v_3)$ and a class $\mathfrak{A}$ which generates $\mathfrak{V}$, i.e., $\mathfrak{V} = V(\mathfrak{A})$ such that q is the ternary discriminator on every A in $\mathfrak{A}$.

The *representation theorem* of KEIMEL-WERNER [8], [17] states the following: If $\mathfrak{V} = V(\mathfrak{A})$ is a discriminator variety then every R in $\mathfrak{V}$ can be represented as the algebra of global sections of a boolean sheaf with stalks in $SP_U(\mathfrak{A})$, where $SP_U(\mathfrak{A})$ is the class of subdirect irreducible algebras of $\mathfrak{V}$.

The *duality theorem* of KEIMEL-WERNER [8] states the following: If $\mathfrak{V} = V(\mathfrak{A})$ is a discriminator variety then $\mathfrak{V}$ is dual to the category of boolean sheaves with stalks in $SP_U(\mathfrak{A})$. The proof relies on the following facts.

(a) The discriminator term q satisfies the equations $q(v_1, v_2, v_1) \equiv v_1$ and $q(v_1, v_2, v_2) \equiv v_1$, which imply that congruences of R in $\mathfrak{V}$ permute and distribute.

(b) Using JÓNSSON's characterization in [18] of subdirect irreducible algebras in congruence distributive varieties, one can show for R in $\mathfrak{V}$ with $|R| \geqq 2$: R subdirect irreducible iff $R \in SP_U(\mathfrak{A})$ iff q ternary discriminator on R iff R simple iff R directly indecomposable.

(c) Thus every R in $\mathfrak{V}$ is semisimple and Spec $R = \{\theta \in \mathrm{Con}(R)\colon \theta$ maximal or $\theta = R \times R\}$ determines a subdirect representation of R. With the help of the ternary discriminator it remains to be shown that $\bigcup \langle \{[r_1 \equiv r_2], [r_1 \not\equiv r_2]\} \mid r_1, r_2 \in R \rangle$ constitutes a boolean algebra, where $[r_1 \equiv r_2] = \{\theta \in \mathrm{Con}(R)\colon \langle r_1, r_2\rangle \in \theta\}$. Then (2) yields the representation theorem because of the algebraic characterization of boolean sheaves given in chapter 1.

(d) To obtain the duality theorem one has to show that the representation given above is functorial and that it is inverse to the global section functor.

For the decidability results we need a characterization of the residually finite discriminator varieties. The following statements are equivalent for a discriminator variety $\mathfrak{V}$:

(i) $\mathfrak{V}$ is *residually finite*, i.e., all subdirectly irreducible algebras are finite.

(ii) $\mathfrak{V}$ is residually small, i.e., there is up to isomorphism only a set of subdirectly irreducible algebras.

(iii) $\mathfrak{V}$ has up to isomorphism only finitely many simple algebras .

(iv) There exists a finite set $\mathfrak{A}$ of finite algebras such that $\mathfrak{V} = V(\mathfrak{A})$.

The basic example of a discriminator variety is of course the variety $\mathrm{BA} = V(\{2\})$ of boolean algebras on which the whole theory is based and which should be kept in mind as a test example. The following list of examples shows that the class of discriminator varieties is fairly large.

The first set of examples can be obtained as subvarieties of the variety $\mathfrak{H}$ of HEYTING algebras with an additional unary operation d which satisfies the following equations: $d(v) \wedge d^2(v) = 0$, $d(v) \vee d^2(v) = 1$ (i.e., $d(v)$ is complemented), $d(v_1) \wedge d(v_2) = d(v_1 \wedge \wedge d(v_2)) \wedge d(v_2)$, $d^2(v) \wedge v = d^2(v)$. If H in $\mathfrak{H}$ then $d^2(1) = 1$ and $d^2(a) = 0$ for $a \neq 1$. Thus the discriminator term can be defined by $q(v_1, v_2, v_3) = (v_3 \wedge d^2p(v_1, v_2)) \vee \vee (v_1 \wedge \neg d^2p(v_1, v_2))$ where $p(v_1, v_2) = (v_1 \vee v_2) \rightarrow (v_1 \wedge v_2)$. Spec H for H in $\mathfrak{H}$ is the dual space of the boolean algebra of complemented elements. — In this way one obtains as special cases the variety $\mathrm{BA} = V(\{2\})$ of boolean algebras, the variety $\mathrm{Post}_n = V(\{P_n\})$ resp. $\mathrm{Luk}_n = V(\{L_n\})$ of POST algebras resp. ŁUKASIEWICZ algebras of order n, the variety $\mathrm{Mon}_n = V(\{M_n\})$ of monadic algebras of degree n and the variety Cyl_n of cylindric algebras of dimension n.

A second set of examples can be obtained as subvarieties of the variety $\mathfrak{R}$ of rings with an additional unary operation g which satisfies the following equations: $g(v)\, g(v) = g(v)$, $g(v_1)\, v_2 = v_2 g(v_1)$ (i.e., $g(v)$ is a central idempotent), $g(v)\, v = v$, $g(v_1 g(v_2)) = g(v_1)\, g(v_2)$, $g(v_1 - v_1 g(v_2)) = g(v_1) - g(v_1)\, g(v_2)$. If R in $\mathfrak{R}$ is directly indecomposable then $g(0) = 0$, $g(a) = 1$ if $a \neq 0$. Thus the discriminator term can be defined by $q(v_1, v_2, v_3) = v_3 + (v_1 - v_3)\, g(v_1 - v_2)$. Spec R for R in $\mathfrak{R}$ is the dual space of $B(R)$, the boolean algebra of central idempotents. In this way one obtains as special cases the variety of strongly regular rings which is generated by the division rings, the variety of BAER*-rings which is generated by the domains, i.e., the rings without zero divisors, the variety of commutative pp. rings which is generated by the integral domains, the variety of commutative regular rings which is generated by the fields and the variety of n-rings which is generated by the Galois fields $\mathrm{GF}(q)$ with $n - 1 \equiv \equiv 0 \bmod q - 1$.

Another example is the variety of relatively complemented distributive lattices. Still another example is given by the variety of complementary semigroups of rank k.

3. The theorem of FEFERMAN and VAUGHT

The theorem of FEFERMAN and VAUGHT (cf. [7]) in its simplest form applied to Δ-algebras states that for every formula $\varphi(v_1, \dots, v_n)$ of the language L of Δ-algebras there exist finitely many formulas $\psi_1(v_1, \dots, v_n), \dots, \psi_m(v_1, \dots, v_n)$ and a formula $\Phi(y_1, \dots, y_m)$ of the language L_{BA} of boolean algebras such that for every product $\Pi\langle A_i | \, i \in I\rangle$ of Δ-algebras and n-tupel $\langle a_1, \dots, a_n\rangle$ in $\Pi\langle A_i \mid i \in I\rangle$ the formula $\varphi(a_1, \dots, a_n)$ holds in $\Pi\langle A_i \mid i \in I\rangle$ iff the formula $\Phi([\psi_1(a_1, \dots, a_n)], \dots, [\psi_m(a_1, \dots, a_n)])$ holds in 2^I, where $[\psi_j(a_1, \dots, a_n)]$ is the set $\{i \in I : A_i \models \psi_j(a_1(i), \dots, a_n(i))\}$. Thus the validity of the formula φ in the products is determined by the validity of the formulas $\psi_1, \dots, \psi_m$ in the factors.

The theorem of FEFERMAN and VAUGHT can be generalized to boolean sheaves of Δ-algebras which satisfy the condition (C) of COMER in [3]. For our purpose we need a more refined version of the theorem which can be applied to a larger class of sheaves at the cost of restricting the class of formulas to which it applies.

Let us say, that a boolean sheaf P of Δ-algebras over the boolean space X satisfies the *condition* (C_n) if for every formula $\varphi(v_1, \dots, v_n)$ of quantifier rank at most n and for $\alpha_1, \dots, \alpha_n$ in $\Gamma(X, P)$ the set $[\varphi(\alpha_1, \dots, \alpha_n)] = \{x \in X : P_x \models \varphi(\alpha_1(x), \dots, \alpha_n(x))\}$ is in X^*. (C_0) is clearly satisfied for every boolean sheaf. We say that it satisfies the condition (C_ω) if $[\varphi(\alpha_1, \dots, \alpha_n)]$ is in X^* for an arbitrary formula. In the context of X^*-valued Δ-algebras the condition (C_n) corresponds to a maximum condition (M_n) which states that the supremum which determines the truthvalue of a formula $\exists v \varphi(v, \alpha_1, \dots, \alpha_n)$, where φ has quantifier rank less than n, is taken on, i.e.

$$[\exists v \varphi(v, \alpha_1, \dots, \alpha_n)] = \sup \{[\varphi(\alpha, \alpha_1, \dots, \alpha_n)] : \alpha \in \Gamma(X, P)\}$$
$$= [\varphi(\alpha_0, \alpha_1, \dots, \alpha_n)] \text{ for some } \alpha_0 \text{ in } \Gamma(X, P).$$

Let $\mathfrak{K}$ be a class of Δ-algebras.

Definition: $\Gamma^n(\mathfrak{K})$ resp. $\Gamma^\omega(\mathfrak{K})$ denotes the class of Δ-algebras which can be represented as algebras of global sections of a boolean sheaf of algebras from $\mathfrak{K}$ which satisfies (C_n) resp. (C_ω). This yields a decreasing chain of subclasses of $\Gamma^0(\mathfrak{K})$.

Now the FEFERMAN-VAUGHT theorem for the operator Γ^n can be formulated as follows: For every formula $\varphi(v_1, \dots, v_n)$ with quantifier rank at most n there exist finitely many formulas $\psi_1(v_1, \dots, v_n), \dots, \psi_m(v_1, \dots, v_n)$ of quantifier rank at most n and a formula $\Phi(y_1, \dots, y_m)$ of the language L_{BA} such that for every boolean sheaf P of Δ-algebras over X which satisfies (C_n) and $\alpha_1, \dots, \alpha_n$ in $\Gamma(X, P)$ the following are equivalent:

(i) $\Gamma(X, P) \models \varphi(\alpha_1, \dots, \alpha_n)$,

(ii) $X^* \models \Phi([\psi_1(\alpha_1, \dots, \alpha_n)], \dots, [\psi_m(\alpha_1, \dots, \alpha_n)])$.

The sequence $\langle \Phi_i, \psi_1, \dots, \psi_m \rangle$ which has to be associated with φ can be taken from FEFERMAN-VAUGHT [7], and their proof can be adapted to this more general situation. Using boolean-valued Δ-algebras rather than boolean sheaves ot Δ-algebras it becomes quite clear that the patchwork condition (PC) and the maximum condition (M_n) is all what is required to make it work.

As an immediate corollary we have the fact that Γ^n preserves elementary equivalence up to quantifier rank n, i.e., if P and Q are boolean sheaves over the same space X which satisfy (C_n) then in $\Gamma(X, P)$ and $\Gamma(X, Q)$ the same sentences of quantifier rank at most n are true, whenever this holds for all stalks. The constant sheaf $X \times A$ over a boolean space X satisfies (C_ω), since α in $C(X, A) \cong (X, X \times A)$ has only finitely many values and

$$[\varphi(\alpha_1, \dots, \alpha_n)] = \bigcup \langle \alpha_1^{-1}(a_1) \cap \dots \cap \alpha_n^{-1}(a_n) : A \models \varphi(a_1, \dots, a_n) \rangle .$$

As another corollary we have that fact, that $\Gamma(X_1, X_1 \times A_1)$ is elementarily equivalent to $\Gamma(X_2, X_2 \times A_2)$ whenever A_1 is elementarily equivalent to A_2 and the same holds for X_1^* and X_2^*. This is a result on bounded boolean powers (cf. [1]) since $\Gamma(X, X \times A) \cong A^*[X^*]$, the bounded boolean power of A with respect to X^*.

Let $\mathfrak{A}$ be a class of Δ-algebras. The theory $\mathrm{Th}(\mathfrak{A})$ is called *complete* if either a sentence of L or its negation belongs to $\mathrm{Th}(\mathfrak{A})$, i.e., all algebras in $\mathfrak{A}$ are elementarily equivalent. Thus $\mathrm{Th}(\mathfrak{A})$ is complete iff there exists an A in $\mathfrak{A}$ with $\mathrm{Th}(\mathfrak{A}) = \mathrm{Th}(A)$. Since Γ^ω preserves elementary equivalence, we obtain the following *transfer theorem for completeness for* Γ^ω: If $\mathfrak{A}$ is a class of Δ-algebras and $\mathfrak{B}$ is a class of boolean algebras with $\mathrm{Th}(\mathfrak{A}) = \mathrm{Th}(A)$ and $\mathrm{Th}(\mathfrak{B}) = \mathrm{Th}(B)$ for some A in $\mathfrak{A}$ and B in $\mathfrak{B}$ then $\mathrm{Th}(\Gamma^\omega_{\mathfrak{B}}(\mathfrak{A})) =$

$= \mathrm{Th}(\Gamma(B^*, B^* \times A))$, where $\Gamma^{\omega}_{\mathfrak{B}}(\mathfrak{A})$ is the subclass of $\Gamma^{\omega}(\mathfrak{A})$ of those algebras $\Gamma(X, P)$ with X^* in $\mathfrak{B}$. For example, if $\mathfrak{A}$ is the class of algebraically closed fields of characteristic zero and if $\mathfrak{B}$ is the class of atomless boolean algebras, then $\Gamma^{\omega}_{\mathfrak{B}}(\mathfrak{A})$ is the class of algebraically closed commutative regular rings with an atomless algebra of idempotents. Since $\mathrm{Th}(\mathfrak{A})$ and $\mathrm{Th}(\mathfrak{B})$ are complete, $\mathrm{Th}(\Gamma^{\omega}_{\mathfrak{B}}(\mathfrak{A}))$ is complete.

4. Decidability results for residually finite discriminator varieties

As in [7] the generalized Feferman-Vaught theorem yields a transfer theorem for decidability for the operator Γ^{ω}. Using the same device as in [3] which replaces the formula Φ by a sentence $\hat{\Phi}$, we obtain the following *transfer theorem for decidability* (cf. Comer [3]): If $\mathfrak{K}$ is a class of Δ-algebras and $\mathfrak{B}$ is a class of boolean algebras such that the theories $\mathrm{Th}(\mathfrak{K})$ and $\mathrm{Th}(\mathfrak{B})$ are decidable, then the theory $\mathrm{Th}(\Gamma^{\omega}_{\mathfrak{B}}(\mathfrak{K}))$ is decidable, where $\Gamma^{\omega}_{\mathfrak{B}}(\mathfrak{K})$ is the subclass of those algebras $\Gamma(X, P)$ in $\Gamma^{\omega}(\mathfrak{K})$ for which X^* is in $\mathfrak{B}$.

This result does not apply to discriminator varieties, since they are of the form $\Gamma^0(S(\mathfrak{A}))$ for some $\mathfrak{A}$. In order to obtain a decidability result for Γ^0 one has to use different methods. Modifying a method developed by Comer in [4] for monadic algebras Burris and Werner [2] proved the following *theorem*: If $\mathfrak{B}$ is a residually finite discriminator variety then the theory $\mathrm{Th}(\mathfrak{B})$ is decidable. More general, if $\mathfrak{A}$ is a finite set of finite algebras then the theory $\mathrm{Th}(\Gamma^0 S(\mathfrak{A}))$ is decidable.

The proof consists of the following steps:

(1) For every finite set of finite algebras $\mathfrak{A}$ there exists a partial algebra A which amalgamates $\mathfrak{A}$. A is said to amalgamate $\mathfrak{A}$ if the following holds:

(i) $S(\mathfrak{A}) \subseteq S(A)$ up to isomorphism,
(ii) every boolean sheaf over a boolean space X with countable base with stalks in $S(\mathfrak{A})$ is isomorphic to a subsheaf of the constant sheaf of A over X.

(2) Let k be the number of isomorphism types of algebras in $S(\mathfrak{A})$. Then there exists a formula $\Lambda(Y_1, \dots, Y_k)$ in the language of boolean algebras with ideals, where the Y_j's are variables for ideals, such that the following holds: The subsheaves of the constant sheaf of A over X with stalks in $S(\mathfrak{A})$ correspond to k-tupels $\langle X^*, I_1, \dots, I_k\rangle$ such that $\Lambda(I_1, \dots, I_k)$ holds in X^*. The ideal I_j is the dual of the open set X_j of those x in X whose stalk contains the algebra A_j.

(3) Let $B(\mathfrak{A})$ be the class of the sequences $\langle B, I_1, \dots, I_k\rangle$ such that $\Lambda(I_1, \dots, I_k)$ holds in B. Then the theory of $B(\mathfrak{A})$ is decidable, since it is a finitely axiomatizable extension of the theory of boolean algebras with k specified ideals which is decidable by a result of Rabin in [12].

(4) Let m be the number of elements of A the amalgamating partial algebra of $\mathfrak{A}$. There exists a formula $\Delta(y_1, \dots, y_m, Y_1, \dots, Y_k)$ in the language of boolean algebras with k specified ideals and there exist formulas $\delta_1(v), \dots, \delta_m(v)$ of the language for $\mathfrak{A}$ such that the following holds: If P is a subsheaf of the constant sheaf A over the boolean space X with stalks in $S(\mathfrak{A})$ then for every α in $\Gamma(X, X \times A)$ we have

$$\alpha \text{ in } \Gamma(X, P) \quad \text{iff} \quad X^* \models \Delta(\delta_1(\alpha), \dots, \delta_m(\alpha), I_1, \dots, I_m),$$

where $I_1, \dots, I_m$ are the ideals determined in (3). This shows that the subalgebra $\Gamma(X, P)$ of $\Gamma(X, X \times A)$ can be obtained by a relativization with respect to $\langle \Delta; \delta_1, \dots, \delta_m \rangle$.

(5) Now a relativized version of the FEFERMAN-VAUGHT theorem for Γ^ω yields the required result, since the theory of $S(\mathfrak{A})$ is decidable.

Clearly the difficult part of the proof is step (1), which involves a complicated induction along the embeddability relation of the algebras in $S(\mathfrak{A})$.

5. Results on ω-categoricity ond other model theoretic properties

Let $\mathfrak{A}$ be a class of Δ-algebras. The theory $\mathrm{Th}(\mathfrak{A})$ is called *ω-categorical* if any two countable algebras in $\mathfrak{A}$ are isomorphic. If $\mathrm{Th}(\mathfrak{A})$ is ω-categorical then $\mathrm{Th}(\mathfrak{A})$ is complete iff either all algebras in $\mathfrak{A}$ are infinite or all algebras in $\mathfrak{A}$ are finite.

RYLL-NARDZEWSKI has characterized in [13] complete ω-categorical theories. For every n let $F_n(L)$ be the set of formulas whose free variables are among the variables $v_0, \dots, v_{n-1}$, and let $S_n(\mathfrak{A})$ be the boolean algebra which can be obtained by identifying two formulas in $S_n(\mathfrak{A})$ if they are equivalent with respect to $\mathrm{Th}(\mathfrak{A})$. For types Δ which are countable the theorem states that a complete theory $\mathrm{Th}(\mathfrak{A})$ is ω-categorical iff $S_n(\mathfrak{A})$ is finite for every n in ω.

WASKIEWICZ [15] proved the following *transfer theorem for ω-categoricity* for Γ^ω: If $\mathrm{Th}(\mathfrak{A}) = \mathrm{Th}(A)$ and $\mathrm{Th}(\mathfrak{B}) = \mathrm{Th}(B)$ are ω-categorical then $\mathrm{Th}(\Gamma^\omega_\mathfrak{B}(\mathfrak{A})) = \mathrm{Th}(\Gamma(B^*, B^* \times A))$ is ω-categorical. — An application of the FEFERMAN-VAUGHT theorem yields $|S_n(\Gamma(B^*, B^* \times A))| \leqq |S_m(B)| \leqq \omega$, where $m = |S_n(A)| < \omega$. The theory $\mathrm{Th}(B)$ of a boolean algebra B is ω-categorical iff B has at most finitely many atoms. Therefore, $\mathrm{Th}(A_1 \times A_2)$ is ω-categorical, whenever $\mathrm{Th}(A_1)$ and $\mathrm{Th}(A_2)$ are ω-categorical. WASKIEWICZ [15] also proved the following *stronger result on transfers of ω-categoricity*: If $\mathfrak{A} = \mathfrak{A}_1 \cup \dots \cup \mathfrak{A}_l$ and $\mathrm{Th}(\mathfrak{A}_j) = \mathrm{Th}(A_j)$ for $j = 1, \dots, l$ and $\mathrm{Th}(X^*)$ are ω-categorical and if $\Gamma(X, P) \in \Gamma^\omega(\mathfrak{A})$ then $\mathrm{Th}(\Gamma(X, P))$ is ω-categorical. — As above, one shows $|S_n(\Gamma(X, P))| \leqq |S_m(X^*)| < \omega$, where $m = |S_n(A_1)| \cdot \dots \times |S_n(A_l)| < \omega$.

Now let $\mathfrak{B} = V(\mathfrak{A})$ be a residually finite discriminator variety. Thus $S(\mathfrak{A})$ is a finite set of finite algebras. Clearly, $\mathrm{Th}(\mathfrak{A})$ is ω-categorical for every A in $S(\mathfrak{A})$, since A is finite. Using the fact, mentioned in the previous chapter, that every R in $\mathfrak{B}$ with $\mathrm{Spec}(R)$ countable has a representation as a subsheaf of a constant sheaf, we obtain the following result on *ω-categoricity in residually finite discriminator varieties*: Let $\mathfrak{B}$ be a residually finite discriminator variety and let R be an algebra in $\mathfrak{B}$ with $\mathrm{Spec}(R)$ countable then the following holds:

(1) If $\mathrm{Th}(\mathrm{Spec}(R)^*)$ is ω-categorical, then $\mathrm{Th}(R)$ is ω-categorical.

(2) If $\mathrm{Th}(\mathrm{Spec}(R)^*, I_1, \dots, I_k)$ is ω-categorical, then $\mathrm{Th}(R)$ is ω-categorical, where $\langle \mathrm{Spec}(R)^*, I_1, \dots, I_k \rangle$ is the sequence associated with R, as indicated in the previous chapter.

The first statement follows from WASKIEWICZ's result mentioned above. The second statement, which implies the first, follows from a result of MACINTYRE and ROSENSTEIN in [10]. This relies on the fact that R can be interpreted in $\langle \mathrm{Spec}(R)^*, I_1, \dots, I_k \rangle$.

As an example, one obtains COMER's result in [5], that $\mathrm{Th}(\Gamma(B_0^*, B_0^* \times M_m))$ is ω-categorical, where B_0 is a countable atomless boolean algebra and M_m is the simple monadic algebra with 2^m elements.

In the rest of this chapter we will mention briefly a few more transfer theorems on model theoretic properties. Extending a result of MACINTYRE in [9], BURRIS and WERNER [2] proved a transfer theorem for model completeness for the operator $\Gamma^0_{\mathfrak{B}_0}$. $\mathfrak{B}_0$ is the class of atomless boolean algebras and it corresponds to the class of boolean spaces without isolated points. The basic tool is MACINTYRE's lemma for sheaves over boolean spaces without isolated points (cf. 3.7. in [17]). Moreover, having characterized the existentially closed members of a discriminator variety, BURRIS and WERNER [2] can show that a discriminator variety $\mathfrak{B} = V(\mathfrak{A})$ has a model companion whenever $\mathfrak{A}$ has a model companion.

References

[1] S. BURRIS: Boolean powers, Algebra univ. **5** (1975), 341–360.

[2] S. BURRIS; H. WERNER: Sheaf constructions and their elementary properties. I, II, preprints (1975), TH Darmstadt.

[3] S. D. COMER: Elementary properties of structures of sections, Bol. Soc. Math. Mexicana. (2) **19** (1974), 78–85.

[4] S. D. COMER: Monadic algebras with finite degree, Algebra univ. **5** (1975), 313–327.

[5] S. D. COMER: Complete and model-complete theories of monadic algebras, Colloq. Math. **34** (1976), 183–190.

[6] D. ELLERMAN: Sheaves of structures and generalized ultraproducts, Ann. of Math. Logic **7** (1974), 163–195.

[7] S. FEFERMAN; R. L. VAUGHT: The first order properties of products of algebraic systems, Fund. Math. **47** (1959), 57–103.

[8] K. KEIMEL; H. WERNER: Stone-duality for varieties generated by quasi-primal algebras, Memoirs AMS **148** (1975), 59–85.

[9] A. MACINTYRE: Model completeness for sheaves of structures, Fund. Math. **81** (1973), 73–89.

[10] A. MACINTYRE; J. G. ROSENSTEIN: $\aleph_0$-categoricity for rings without nilpotent elements and for boolean structures, preprint 1973.

[11] R. S. PIERCE: Modules over commutative regular rings, Memoirs AMS **70** (1967).

[12] M. O. RABIN: Decidability of second order theories and automata on infinite trees, Trans. AMS **141** (1969), 1–35.

[13] C. RYLL-NARDZEWSKI: On the categoricity in power, Bull. Acad. Sci. Polon. **7** (1959), 545 bis 548.

[14] H. VOLGER: The Feferman-Vaught theorem revisited, Colloq. Math. **36** (1976), 1–11.

[15] J. WASKIEWICZ: ω_0-categoricity of generalized products, Colloq. Math. **27** (1973), 1–5.

[16] V. WEISPFENNING: Model completeness and elimination of quantifiers for subdirect products of structures, J. Algebra. **36** (1975), 252–277.

[17] H. WERNER: Algebraic representation and model theoretic properties of algebras with the ternary discriminator, preprint 1975, TH Darmstadt.

[17a] H. WERNER: Discriminator-Algebras. Akademie-Verlag, Berlin 1978.

[18] B. JÓNSSON, Algebras whose congruence lattices are distributive, Math. Scand. **21** (1967), 110–121.

Bemerkungen zur Kompliziertheit der Berechnungen auf stochassitchen Automaten[1)]

von B. A. Trachtenbrot, Nowosibirsk

(Übersetzung aus dem Russischen von R. Peters)

§ 1. Einführung

Wie K. de Leeuw, E. F. Moore und andere [1] zeigten, berechnet eine Maschine, welche die Ausgänge eines Bernoulli-Generators D_p (p: Wahrscheinlichkeit der Eins, $1 - p$: Wahrscheinlichkeit der Null) aufnimmt, eine bezüglich p effektive Funktion. Folglich führt das Hinzuschalten eines Generators mit den effektiven Parametern p in den Aufbau der Maschine nicht aus der Klasse der rekursiven Funktionen heraus. Kann dies bezüglich der Geschwindigkeit des Erhaltens eines Resultates oder irgendeines anderen Maßes der Kompliziertheit der Berechnung nützlich sein? Bestimmte Überlegungen über eine annehmbare Lösung „durchmusterter" Aufgaben in der Klasse der Algorithmen mit zufälligen Akten der Auswahl wurden von S. W. Jablonski in [2] ausgesprochen. Später zeigte J. M. Barzdin [3], daß bei einer spezifischen Interpretation der Arbeit einer Wahrscheinlichkeitsmaschine eine Beschleunigung des Rechenprozesses im Vergleich zu einer determinierten Maschine tatsächlich gilt. Doch in der Form, die am natürlichsten erscheint, sind ähnliche Resultate noch nicht bekannt. Einige Beobachtungen des Autors, die sich auf diese Frage beziehen, werden in dem vorliegenden Artikel dargelegt.

Zu Beginn vereinbaren wir die folgenden Termini und Bezeichnungen.

Unter einer Maschine wollen wir eine mit einem Band versehene Turing-Maschine verstehen, welche gleichzeitig mit den Bandsymbolen bei jedem Takt ein Symbol (0 oder 1) aufnimmt, welches vom Generator D erzeugt wurde. Formal bedeutet das (unter der Voraussetzung, daß nur ein Lesekopf vorhanden ist), daß der linke Teil jedes Kommandos ein Tripel der Art: q (laufender Zustand), s (Bandsymbol), σ (Ausgang des Generators) ist. Damit ist für eine beliebige Ausgangskonfiguration K und für ein beliebiges duales Wort $\xi = \xi(1)\, \xi(2) \dots \xi(l)$ die Konfiguration $N(K, \xi)$ definiert, in der sich die Maschine nach l Takten befindet. Die Maschine N wird eine p-Maschine, wenn dem Generator D die Wahrscheinlichkeit p der Ausgabe der Eins (d. h. $D = D_p$) zugeschrieben wird.

Mit einer p-Maschine N wird die Zufallsgröße $\tau_N(x)$, die Arbeitszeit, assoziiert, wenn die Maschine mit dem Ausgangswort x beginnt; das Ereignis „$N(x) = y$", wenn N das Wort x in das Wort y umbildet und ähnliches mehr.

Wir werden sagen, daß die p-Maschine N das Prädikat Γ (im allgemeinen partiell) mit der Zuverlässigkeit Δ berechnet, wenn für ein beliebiges x aus dem Definitionsbereich dieses Prädikates

$$\text{wahr}\,\{N(x) = \Gamma(x)\} > \Delta$$

1) Diese Arbeit erschien im Original in: [6a]; eine genauere Übersetzung als „stochastische Automaten" wäre „Zufalls-" oder „Wahrscheinlichkeitsmaschinen" (Anm. d. Herausg.)

12*

ist. Weiterhin werden wir $\Delta \geqq \frac{1}{2}$ annehmen, was die Einzigkeit des Prädikates garantiert, welches die gegebene Maschine berechnen kann.

Als Grundmaß der zeitlichen Kompliziertheit wird in diesem Artikel die zeitliche Δ-Signalisierende $t_N(\Delta, x)$ verwendet, welche wir so definieren: $t_N(\Delta, x)$ ist das kleinste a derart, daß

$$[\text{wahr}\ \{N(x) = \Gamma(x)\ \&\ \tau_N(x) \leqq a\} > \Delta]$$

ist.

Es ist klar, daß $t_N(\Delta, x)$ überall im Definitionsbereich des Prädikates Γ existiert, sobald es die Maschine N mit der Zuverlässigkeit Δ berechnet. Andere mögliche Maße der zeitlichen Kompliziertheit sind die maximale und die mittlere Arbeitszeit der mit dem Argument x die Arbeit beginnenden Maschine N (natürlich in den Fällen, wo diese Größe endlich ist).

Entsprechende Bezeichnungen sind

$$t_N^{\max}(x);\ t_N^{\text{mittl}}(x)\,.$$

Zum Schluß bemerken wir, daß man eine determinierte Maschine als Spezialfall einer p-Maschine betrachten kann, wenn nämlich $p = 1$ oder $p = 0$ ist. Dabei fallen $t(\Delta, x)$, $t^{\max}(x)$ und $t^{\text{mittl}}(x)$ mit der gewöhnlichen zeitlichen Signalisierenden zusammen.

Wir nehmen an, daß wir über irgendeine obere Abschätzung für die Δ-Signalisierende des Prädikates Γ in der Klasse der Maschinen mit dem Generator $D_{p'}$ verfügen. Welche obere Abschätzung kann beim Übergang zum Generator $D_{p''}$ garantiert werden? Gewisse Abschätzungen, die sich auf den Übergang von $p' = \frac{1}{2}$ zu $p'' = 1$ und von einem beliebigen p' zu $p'' = \frac{1}{2}$ beziehen, sind in den Sätzen 2, 5 bzw. 8 enthalten. Man erhält sie auf dem Wege der Modellierung des $D_{p'}$ gebrauchenden Rechenprozesses mittels einer p''-Maschine. Beim Beweis werden neben gewöhnlicher Programmiertechnik natürlich auch wahrscheinlichkeitstheoretische Abschätzungen, die mit dem Bernoulli-Test verbunden sind, verwendet. Die genannten Sätze geben gewisse Vorstellungen darüber, mit welcher maximalen Beschleunigung des Rechenprozesses man auf Kosten der Auswahl des Parameters p rechnen kann. Leider ist es uns nicht einmal gelungen, die folgende Frage zu lösen.

Existiert ein überall definiertes Prädikat Γ, das auf einer $\frac{1}{2}$-Maschine mit der Zuverlässigkeit $\Delta > \frac{1}{2}$ und in der Zeit $t(\Delta, x)$, die hinsichtlich der Ordnung kleiner als die Zeit einer beliebigen determinierten Berechnung ist, berechenbar ist?[1]

Dieser Umstand zeigt, wieweit wir noch von einer Begründung der Endgültigkeit der Abschätzungen aus den Sätzen 2, 5 und 8 oder wenigstens von ihrer wesentlichen Verbesserung entfernt sind.

Trotzdem gelingt es, bei gewissen zusätzlichen Beschränkungen zu klären, was und was man nicht mit Hilfe einer Wahrscheinlichkeitsberechnung erreichen kann. Im § 2 werden in bezug auf Berechnungen auf $\frac{1}{2}$-Maschinen sogenannte $\exists$-Algorithmen betrachtet, die eine Beschleunigung des Rechenprozesses mit der Zuverlässigkeit $\Delta = \frac{1}{2}$ und für nicht überall definierte Prädikate sogar mit der Zuverlässigkeit $\Delta > \frac{1}{2}$ gewährleisten (Sätze 2 und 3).

(Es bedeute $\lesssim$ „nicht größer bezüglich der Ordnung" und $\Phi_p(n)$ „Zeit, die für die Berechnung der ersten n Ziffern in der Dualzerlegung der Zahl p aufgewendet wurde".)

[1]) Im Jahre 1975 hat R. W. Freiwald neue Resultate in dieser Richtung erzielt.

A priori ist klar, daß eine Beschleunigung der Berechnung des Prädikates Γ auf Kosten einer adäquaten Auswahl des Parameters p, nämlich auf Kosten einer entsprechenden Kodierung des Prädikates Γ mittels der Dualzerlegung der Zahl p, erhalten werden kann. Im § 3 werden genauere Vorstellungen über den Effekt, den man auf diesem Wege erreichen kann, gegeben. Es erweist sich, daß unzureichend lange Berechnungen nicht gestatten, eine in p enthaltene nützliche Information zu entnehmen (Sätze 6 und 7), und die Anwendung eines unzureichend komplizierten p liefert keine wesentlichen Vorteile im Vergleich zum Wert $p = \frac{1}{2}$ (Folgerung aus Satz 7).

Wie schon bemerkt wurde, wollen wir in dieser Arbeit hauptsächlich aufklären, wie die Auswahl des Parameters p eines der Maschine beigefügten BERNOULLI-Generators D_p die Geschwindigkeit der Berechnung beeinflußt. Wir setzen jetzt voraus, daß ein beliebiges (möglicherweise sogar nichteffektives) p_0 fixiert sei und betrachten nur Berechnungen in der Klasse der p_0-Maschinen. Wie beeinflußt die Größe des erklärten Zuverlässigkeitsniveaus Δ die Geschwindigkeit der Berechnung und wie ist der Zusammenhang zwischen der mittleren Zeit und der Δ-Signalisierenden ?[1] Der unten formulierte Satz ist sowohl in der Klasse der Maschinen mit einem Kopf als auch in der Klasse der Maschinen mit mehreren Köpfen richtig.

Satz 1. *Möge die p_0-Maschine N das Prädikat Γ mit der Zuverlässigkeit $\Delta \geqq \frac{1}{2}$ berechnen.*

Dann gilt:

(i) *Bei $\Delta = \frac{1}{2}$ gibt es eine p_0-Maschine M, die Γ mit der Zuverlässigkeit Δ berechnet, so daß gilt*

$$t_M(\Delta, x) \leq t_N(\Delta, x) \;\&\; t_M^{\text{mittl}}(x) \leqq \text{const}(p_0);$$

(ii) *bei $\Delta > \frac{1}{2}$, wie auch immer $\Delta' > \Delta$ sei, gibt es eine p_0-Maschine M, die Γ mit der Zuverlässigkeit Δ' berechnet, so daß gilt*

$$t_M(\Delta', x) \leqq t_M^{\text{mittl}}(x) \leq t_N^{\text{mittl}}(x)\,.$$

Beweis (I). Die Maschine arbeitet so: Bei geraden Takten $2, 4, 6, \ldots, 2v, \ldots$ modelliert sie die entsprechenden Schritte der Maschine N mit den Nummern $1, 2, 3, \ldots, v, \ldots$.

Dies geschieht solange, wie bei den ungeraden Takten ausnahmslos Nullen oder ausnahmslos Einsen eintreffen. Beim ersten ungeraden m mit $\xi(m) \neq \xi(m-2)$ beendet die Maschine M die Arbeit und liefert das Resultat $\xi(m)$.

(II). Eine Verbesserung der Zuverlässigkeit (wir gehen zu $\Delta' > \Delta$ über) wird dadurch erreicht, daß M mehrfach wiederholte Berechnungen auf N modelliert. Es ist klar, wenn $t_N^{\text{mittl}}(x)$ existiert, so existiert auch $t_M^{\text{mittl}}(x) \leqq \text{const}(\Delta, \Delta') \cdot t_N^{\text{mittl}}(x)$. Um eine Abschätzung der Δ'-Signalisierenden zu erhalten, wählen wir die Konstante $c > 0$ und die Maschine M so, daß gilt: $\Delta' + 1/c < 1$; die Maschine M berechnet Γ mit der Zuverlässigkeit $\Delta'' = \Delta' + 1/c$ und außerdem

$$t_M^{\text{mittl}}(x) \leqq \text{const}\,(\Delta', \Delta'') \cdot t_N^{\text{mittl}}(x)\,.$$

Wegen

$$\text{wahr}\,\{\tau_M(x) > c \cdot t_M^{\text{mittl}}(x)\} < \frac{1}{c}\,,$$

[1]) Ein Vergleich mit der maximalen Zeit ist trivial.

gilt für die Maschine M die Behauptung: M berechnet Γ mit der Zuverlässigkeit $\Delta'' - 1/c = \Delta'$ und mit der Abschätzung

$$t_M(\Delta', x) \leqq c \cdot t_M^{\text{mittl}}(x) \leqq c \cdot \text{const}\,(\Delta', \Delta'') \cdot t_N^{\text{mittl}}(x)\,.$$

Bemerkungen. (1). Also kann bei $\Delta = \frac{1}{2}$ die Abschätzung der mittleren Zeit nicht als Grundlage für eine nichttriviale Klassifikation der Prädikate bezüglich ihrer Kompliziertheit dienen, während bei $\Delta > \frac{1}{2}$, da die mittlere Zeit hinsichtlich der Ordnung nicht kleiner als die Δ-Signalisierende ist, schon (siehe Satz 2) eine nichttriviale Klassifikation möglich ist. Da wir an Abschätzungen mit einer Genauigkeit bis auf die Ordnung interessiert sind, brauchen wir uns dabei um die Größe Δ nicht zu sorgen, sofern nur $\Delta > \frac{1}{2}$ ist.

(2). Sei $\Delta > \frac{1}{2}$. In der Klasse der Maschinen mit mehreren Köpfen garantiert die Modellierung wiederholter Berechnungen der Maschine N mittels einer entsprechenden Maschine M auch für beliebiges Δ' die Abschätzung:

$$t_M(\Delta', x) \leqq t_N(\Delta, x)\,.$$

In der Klasse der Maschinen mit einem Kopf gelang es uns jedoch nur, eine schwächere Abschätzung zu erhalten:

$$t_M(\Delta', x) \leqq t_N(\Delta, x) \cdot \log t_N(\Delta, x)\,.$$

Es bleibt ungeklärt, ob hier der logarithmische Faktor wesentlich ist.

Im weiteren wird dort, wo es nötig ist, klar gesagt, ob über Maschinen mit einem Kopf oder Maschinen mit mehreren Köpfen gesprochen wird.

§ 2. 1/2-Maschinen

Satz 2. *Möge eine einköpfige $\frac{1}{2}$-Maschine N das Prädikat Γ mit der Zuverlässigkeit $\Delta \geqq \frac{1}{2}$ berechnen. Dann gibt es eine einköpfige determinierte Maschine L, die Γ berechnet und für die*

$$t_L(x) \leqq 2^{t_N(\Delta, x)} \cdot \log^2 t_N(\Delta, x)$$

ist.

Beweis. Die Maschine L geht nacheinander alle dualen Wörter der Länge 1, der Länge 2 usw. durch, bis zuerst ein solches ν ermittelt wird, das der Bedingung: unter den Wörtern ξ der Länge ν gibt es mehr als $(1 - \Delta) \cdot 2^\nu$ derart, daß $N(\xi, x)$ ein und derselbe Wert σ ist, genügt. Offensichtlich ist σ ($\sigma = 1, 0$) auch Wahrheitswert von $\Gamma(x)$, und die Größe ν fällt mit dem Wert $t_N(\Delta, x)$ zusammen. Eine grobe Abschätzung dieser Wahl führt unmittelbar zu der oberen Grenze $\asymp 2^\nu \cdot \nu^2$. Bei einer etwas verbesserten Wahl gelingt es, diese Grenze auf $\asymp 2^\nu \cdot \log^2 \nu$ herabzudrücken.

Bemerkung. In der Klasse der Maschinen mit mehreren Köpfen kann man die Abschätzung

$$t_L(x) \leqq 2^{t_N(\Delta, x)} \cdot \log t_N(\Delta, x)$$

garantieren.

Folglich kann man bezüglich der Anwendung eines Generators $D_{1/2}$ nur eine logarithmische Verkleinerung der Rechenzeit (sogar wenn man sich mit der Zuverlässigkeit

$\Delta = \frac{1}{2}$ zufrieden gibt!) erreichen. Ist in der Tat ein maßgeblicher Vorteil erreichbar? Die Technik der verallgemeinerten endlichen Wahrscheinlichkeitsautomaten von TURAKAINEN [4] erlaubt, wie A. W. WEISSER bemerkte, Prädikate zu finden, für die mit der Zuverlässigkeit $\Delta = \frac{1}{2}$ eine Berechnung mit Zeitvorteil von der Ordnung einer Quadratwurzel möglich ist. So ist zum Beispiel das Erkennen der Symmetrie eines dualen Wortes mit einer linearen Abschätzung der Zeit (sogar der maximalen!) auf einer $\frac{1}{2}$-Maschine mit der Zuverlässigkeit $\Delta = \frac{1}{2}$ möglich. Dagegen ist nach einem bekannten Theorem von J. M. BARZDIN ein determiniertes Erkennen der Symmetrie in einer Zeit, bezüglich der Ordnung kleiner als n^2, unmöglich. Was die Beschleunigung des Rechenprozesses bei garantierter Zuverlässigkeit $\Delta > \frac{1}{2}$ betrifft, so ist es hier angebracht, den Algorithmus (darunter auch einen Wahrscheinlichkeitsalgorithmus), der das partielle Prädikat Γ berechnet, als trennenden Algorithmus der Mengen $\{x \mid \Gamma(x) = 0\}$ und $\{x \mid \Gamma(x) = 1\}$ zu deuten. Eben in dieser Terminologie führte I. N. KOWALENKO [5] erstmals eine vergleichende Abschätzung der Anzahl der Zustände für einen Wahrscheinlichkeitsautomaten und für einen determinierten endlichen Automaten durch, die ein und dasselbe Paar von Mengen trennen. Jedoch gestattet ein gewisser Umbau des Beispiels von I. N. KOWALENKO das unten beschriebene Paar von Mengen L_0 und L_1, das mit der Zuverlässigkeit $\Delta > \frac{1}{2}$ trennbar ist, mit einer linearen Abschätzung der Δ-Signalisierenden zu erhalten, obwohl jede ihrer determinierten Trennungen bezüglich der Ordnung nicht weniger als $n \cdot \log n$ Schritte erfordert.

Die Mengen L_0 und L_1 bestehen aus den Wörtern x im Alphabet $\{0, 1, 2\}$ mit dem folgenden Aussehen:

$$x = A\xi(1)\, A\xi(2) \ldots A\xi(2^k)\,, \qquad k = 1, 2, 3, \ldots,$$

wobei $A = 22 \ldots 2$ (k-mal) ist.

Dabei ist $x \in L_0$ ($x \in L_1$), wenn unter den $\xi(i)$ die Anzahl der Nullen (Anzahl der Einsen) $\geqq 2/(3 \cdot 2^k)$ ist. Wir gehen zur Beschreibung einer gewissen, häufig anzutreffenden Situation über, in der vermutlich Erscheinungen, analog denen, die eben erst aufgezeigt wurden, gelten.

∃-Algorithmen

Wir legen folgende Bezeichnungen fest:

I — Menge aller dualen Wörter. Für jedes $x \in I$:

$l(x)$ — Länge des Wortes x.

$\mu(x)$ — Länge des maximalen Anfangsstückes des Wortes x, das ausnahmslos aus Nullen besteht.

$$\nu(x) \underset{\text{def}}{=} l(x) - \mu(x)\,.$$

$n(x)$ — natürliche Zahl, deren Dualdarstellung x ist. Es ist $n(x) \geqq 2^{\nu(x)-1}$.

$x < y \underset{\text{def}}{=} l(x) = l(y)$ & das Wort x steht lexikografisch vor dem Wort y.

$R(x, y)$ — zweistelliges Prädikat mit Argumenten in I.

$\varphi(x)$ — einstellige Funktion aus I in I.

Neben dem gewöhnlichen beschränkten Quantor der Art $\underset{y<\varphi(x)}{\exists y}$ werden wir für jedes $0 < p < 1$ den Quantor $\underset{y<\varphi(x)}{\exists^p}$ verwenden, der folgendes bedeutet: „Unter den Wörtern, die dem Wort $\varphi(x)$ vorangehen, existieren nicht weniger als $p \cdot n(\varphi(x))$ derart, daß ...", anders gesagt, „unter den Wörtern $< \varphi(x)$ ist der Anteil derer, welche ... nicht kleiner als p".

Wir betrachten jetzt das überall definierte einstellige Prädikat Γ:

$$\Gamma(x) \underset{\text{def}}{=} \underset{y<\varphi(x)}{\exists} yR(x, y) ,$$

und auch für jedes $0 < p < 1$ das partielle Prädikat Γ_p, welches durch die Bedingung

$$\underset{y<\varphi(x)}{\exists^p} yR(x, y) \rightarrow \Gamma_p(x) = 1; \quad \overline{\underset{y<\varphi(x)}{\exists} yR(x, y)} \rightarrow \Gamma_p(x) = 0$$

(anders gesagt, Γ_p trennt die Mengen $M_0 = \left\{x \mid \overline{\underset{y<\varphi(x)}{\exists} yR(x, y)}\right\}$ und $M_p = \left\{x \mid \underset{y<\varphi(x)}{\exists^p} yR(x, y)\right\}$) definiert sei. Es ist klar, daß man diese Prädikate mit der Methode der Durchmusterung berechnen kann; zum Beispiel kann man den Wert von $\Gamma(x)$ durch aufeinanderfolgende Überprüfung der Werte $R(x, y_0)$, $R(x, y_1)$, ... , $R(x, \varphi(x))$[1]) bis zur ersten Ermittlung eines y_i mit $R(x, y_i) = 1$, oder bis zur Aufklärung, daß ein solches y_i nicht existiert, finden. Als Alternative zur Durchmusterung werden wir folgende Prozedur betrachten: Für ein zufällig ausgewähltes $y_i < \varphi(x)$ berechnen wir $R(x, y_i)$. Ist $R(x, y_i) = 1$, so ist damit der Wahrheitswert von $\Gamma(x)$ gefunden; andernfalls wollen wir den Wert von $\Gamma(x)$ vermuten, indem wir uns nur vom Resultat eines gewissen zufälligen Auswahlaktes leiten lassen (ohne zusätzliche Untersuchung des Prädikates R). $\exists(r)$-Algorithmen und $\exists$-Algorithmen sind eine Realisierung dieser Ideen, bei denen die Richtigkeit des Resultates mit einer Zuverlässigkeit $\Delta = \frac{1}{2}$ für die Berechnung von $\Gamma(x)$ und mit einer Zuverlässigkeit von $\Delta > \frac{1}{2}$ für die Berechnung von $\Gamma_p(x)$ garantiert wird. Die genaue Formulierung eines $\exists(r)$-Algorithmus ist die folgende:

A. Berechnen des Wertes $y = \varphi(x)$. Übergang zu **B**.

Bemerkung. Die Parameter $l(y)$, $\mu(y)$, $\nu(y)$ des erhaltenen Wortes y bezeichnen wir einfach durch l bzw. μ bzw. ν.

B. Es ist das Wort

$$\xi = \xi(1) \dots \xi(l)\, \xi(l+1)\, \xi(l+2)$$

zu bilden, wobei $\xi(1) = \xi(2) = \dots = \xi(\mu) = 0$ ist; $\xi(j)$ bei $j > \mu$ ist der Ausgang des Generators $D_{1/2}$. Übergang zu **C**.

Bemerkung. Das Anfangsstück der Länge l des Wortes ξ bezeichnen wir mit ξ'. Folglich ist

$$\xi = \xi' \cdot \xi(l+1)\, \xi(l+2) .$$

C. Überprüfen: $\xi < \varphi(x)$? Wenn ja, so **E**, sonst **D**.
D. Resultat $= \xi(l+1)$. Berechnung beendet.
E. Überprüfen: $R(x, \xi') = 1$? Wenn ja, so **F**, sonst **G**.
F. Resultat $= 1$. Berechnung beendet.
G. Überprüfen: $\xi(l+1) = 0$? Wenn ja, so **H**, sonst **I**.
H. Resultat $= 0$. Berechnung beendet.
I. Überprüfen: $\xi(l+2) = 1$? Wenn ja, so **J**, sonst **K**.
J. Resultat $= 1$. Berechnung beendet.

[1]) Hier ist y_i die Dualdarstellung der Zahl i.

K. Überprüfen: $\xi(1) \ldots \xi(l) < \underbrace{0 \ldots 0}_{\mu} \underbrace{0 \ldots 01}_{r} \underbrace{0 \ldots 0}_{\nu - r}$?[1]) Wenn ja, so **L**, sonst **M**.

L. Resultat = 0. Berechnung beendet.

M. Resultat = 1. Berechnung beendet.

Was einen $\exists$-Algorithmus betrifft, so unterscheidet er sich von einem $\exists(r)$-Algorithmus nur dadurch, daß in ihm der Schritt **K** das folgende Aussehen hat:

K. Überprüfen: $\xi(1) \ldots \xi(l) < \underbrace{0 \ldots 01}_{l}$? Wenn ja, so **L**, sonst **M**.

Bemerkung. Im wesentlichen ist ein $\exists$-Algorithmus ein $\exists(\nu)$-Algorithmus. Aber bei einer solchen Deutung ist der Parameter ν eine veränderliche Größe, die von $\varphi(x)$ — und das bedeutet letztendlich vom Argument x — abhängt. Währenddessen nimmt ein $\exists(r)$-Algorithmus an, daß der Wert r fixiert worden sei.

Leicht ist die Richtigkeit der folgenden Behauptung zu überprüfen.

Satz 3. (I). *Ein $\exists$-Algorithmus berechnet das Prädikat Γ mit der Zuverlässigkeit* $\Delta = \frac{1}{2}$;

(II). *Für jedes $0 < p < 1$ gibt es ein solches $\Delta > \frac{1}{2}$ und ein solches r, daß der $\exists(r)$-Algorithmus das Prädikat Γ_p mit der Zuverlässigkeit Δ berechnet.*

Der Beweis wird zurückgeführt auf die Ermittlung der Anzahl $h(x)$ von sogenannten wahren Wörtern, d. h. von Wörtern ξ (unter allen $2^{\nu+2}$ Wörtern, welche während der Etappe **B** entstehen können), die ein richtiges Resultat des Algorithmus gewährleisten. Für jedes x bezeichnen wir mit $k(x)$ die Anzahl der $y < \varphi(x)$, für die $R(x, y) = 1$ ist.

(I). Sei $k(x) = 0$, d. h. $\Gamma(x) = 0$. Dann ist beim $\exists$-Algorithmus: $h(x) = 2^{\nu+1}$ (Wörter ξ, für die $\xi(l + 1) = 0) + 1$ (Wort $\underbrace{0 \ldots 01}_{l}0$).

Sei $k(x) > 0$, d. h. $\Gamma(x) = 1$. Dann ist beim $\exists$-Algorithmus: $h(x) \geqq 4 \cdot k(x)$ (Wörter ξ, für die $R(x, \xi') = 1) + (\varphi(x) - k(x))$ (Wörter $\xi'01$, wobei $R(x, \xi') = 0) + (\varphi(x) - k(x) - 1)$ (Wörter $\xi'00$, $R(x, \xi') = 0) + 2[2^\nu - \varphi(x)]$ (Wörter $\xi'\xi(l + 1) \cdot \xi(l + 2)$, wobei $\varphi(x) \nless \xi') = 2^{\nu+1} + 2k(x) - 1 \geqq 2^{\nu+1} + 1$.

Folglich ist in beiden Fällen der Anteil der wahren Wörter größer als $\frac{1}{2}$.

(II). In diesem Falle wird die Rechnung analog durchgeführt, nur anstelle des Summanden (bzw. Subtrahenden) $1 (= 2^{\nu-\nu})$ erscheint $2^{\nu-r}$. Außerdem muß man berücksichtigen, daß im Sinne unserer Bezeichnung $\varphi(x) \geqq 2^{\nu-1}$ ist.

Also sei $k(x) = 0$, d. h. $\Gamma_p(x) = 0$. Dann ist beim $\exists(r)$-Algorithmus

$$h(x) = 2^{\nu+1} + 2^{\nu-r}.$$

Sei $k(x) > p \cdot n(\varphi(x))$, d. h. $\Gamma_p(x) = 1$. Dann ist beim $\exists(r)$-Algorithmus

$$h(x) \geqq 2^{\nu+1} + 2k(x) - 2^{\nu-r} \geqq 2^{\nu+1} + 2p \cdot n(\varphi(x)) \geqq 2^{\nu+1} + p \cdot 2^\nu - 2^{\nu-r}.$$

Darum ist die Wahrscheinlichkeit einer richtigen Antwort nicht kleiner als $\frac{1}{2} + \frac{1}{2^{r+2}}$ bzw. $\frac{1}{2} + \frac{p}{4} - \frac{1}{2^{r+2}}$. Hieraus ist ersichtlich, daß die Wahl von r: $\frac{p}{16} < \frac{1}{2^{r+2}} \leqq \frac{p}{8}$ eine Zuverlässigkeit $\Delta = \frac{1}{2} + \frac{p}{16}$ gewährleistet.

[1]) D. h., $n(\xi') < 2^{\nu-r}$?

Mögen R und φ determinierte Berechnungen mit den Signalisierenden R^* bzw. φ^* zulassen. In einem derartigen Fall ist klar, daß die Realisierung eines $\exists$-Algorithmus oder eines $\exists(r)$-Algorithmus auf einer $\frac{1}{2}$-Maschine mit der Abschätzung

$$t^{\max}(x) \leq \varphi^*(x) + \max_{y<\varphi(x)} R^*(x, y) + l(x)$$

vor sich geht, wobei der Summand $l(x)$ den Aufwand auf allen von **A** und **E** verschiedenen Etappen berücksichtigt. Währenddessen erfordert die Berechnung von Γ oder Γ_p mit einer der Methode der Durchmusterung ähnlichen Methode eine Zeit $\asymp \varphi^*(x) + \sum_{y<\varphi(x)} R^*(x, y)$, die sich als bedeutend größer erweisen kann. Darum kann man von einem $\exists$-Algorithmus ($\exists(r)$-Algorithmus) im Vergleich zu determinierten Berechnungen einen Zeitvorteil erwarten.

Beispiel 1. Möge $R(x, y)$ folgendes bedeuten: Im Wort x ist der Buchstabe mit der Nummer $n(y)$ verschieden vom symmetrisch zu ihm gelegenen Buchstaben mit der Nummer $l(x) - n(y)$. Sei $\varphi(x)$ die Dualdarstellung der Zahl $l(x) + 1$. Dann ist $\Gamma(x)$ das Prädikat „x ist nichtsymmetrisches Wort"; M_0 ist die Menge aller symmetrischen Wörter; M_p ist die Menge aller p-symmetrischen Wörter, d. h. die Menge aller der Wörter $x = x(1) \ldots x(l)$, für deren jedes der Anteil der von $x(l-1)$ verschiedenen Buchstaben $x(i)$ nicht kleiner als p ist. Es ist offensichtlich, daß man in dem betrachteten Beispiel für Maschinen mit einem Kopf

$$\varphi^*(x) \leq l(x) \cdot \log l(x), \quad R^*(x, y) \leq l(x) \cdot \log l(x) \qquad (\text{bei} \quad y \leq \varphi(x))$$

wählen kann.

Folglich ist in der Klasse der $\frac{1}{2}$-Maschinen mit einem Kopf eine Bestimmung der Symmetrie mit der Zuverlässigkeit $\Delta = \frac{1}{2}$ und auch die Teilung der Mengen M_0 und M_p mit der Zuverlässigkeit $\Delta > \frac{1}{2}$ in der Zeit $t^{\max}(x) \asymp l(x) \cdot \log l(x)$ möglich. Das ist bezüglich der Ordnung besser, als die mögliche Zeit einer determinierten Berechnung, da der bekannte Beweis von J. M. BARZDIN über die Kompliziertheit einer TURINGschen Symmetriebestimmung für den Beweis der folgenden Behauptung, in der H die Entropie bedeutet, adaptiert werden kann.

Satz 4. *Sei $0 < p < 1$ und $H(2p) < 1$. Dann gilt für jede mit einem Kopf versehene* TURING-*Maschine L, welche M_0 und M_p teilt,*

$$\max_{l(x)=n} t_L(x) \geqq O(n^2)\,.$$

Erläuterung. Eine solche Auswahl des Parameters p ist einzig und allein durch das Bedürfnis nach einer guten oberen Abschätzung für die Größe $v(l, s)$ bedingt, die nach Definition gleich dem Volumen einer Kugel mit dem Radius s im Raum der dualen Wörter der Länge l (HAMMINGsche Metrik!) ist. Es ist nämlich bei $n \leqq l \leqq 2n$ und $n \to \infty$ $v(l, pn) \leqq 2^{H(2p)\,l + \log l}$. Folglich ist durch die Bedingung $H(2p) < 1$

$$v(l, pn) \leqq 2^{(1+\varepsilon)\,l}$$

garantiert.

Beispiel 2. Möge $R(x, y)$ bedeuten: „Die Zahl $n(y)$ ist nichttrivialer Teiler der Zahl $n(x)$", und $\varphi(x)$ sei die duale Darstellung des ganzen Teiles der Zahl $\sqrt{n(x)} + 1$. Da die Berechnung von R und φ mit quadratischer Abschätzung möglich ist, so ist auch die Berechnung des Prädikates „x ist Dualdarstellung einer Primzahl" mit der Zuverlässigkeit $\Delta = \frac{1}{2}$ mittels einer $\frac{1}{2}$-Maschine in der Zeit

$$t^{\max}(x) \leqq l^2(x)$$

möglich.

Dagegen ist für kein natürliches k ein Verfahren zur determinierten TURING-Berechnung dieses Prädikates mit einer Abschätzung der Zeit $\asymp l^k(x)$ bekannt. Wenn die TURING-Bestimmung der Primzahleigenschaft tatsächlich eine zeitliche, eine beliebige Potenzfunktion übertreffende Kompliziertheit hat, so würde sich schon an diesem Beispiel die Diskrepanz zwischen den Zeiten einer determinierten und einer Wahrscheinlichkeitsberechnung als nahe der maximal möglichen erweisen (in Übereinstimmung mit Satz 2). Es wäre interessant, eine der Wahrheit nahekommende Kandidatur für die Diskrepanz derselben Ordnung in der Situation des Trennens der Mengen M_0 und M_p mit der Zuverlässigkeit $\Delta > \frac{1}{2}$ aufzuzeigen.

§ 3. p-Maschinen (beliebiges p)

Eine Dualzerlegung der Zahl $p = 0, p(1)\,p(2) \ldots p(n) \ldots$ kann man als Prädikat p betrachten, welches auf der Menge der natürlichen Zahlen definiert ist. Allgemein gesagt, kann dieses Prädikat auch nichteffektiv sein. Wenn p effektiv ist (und dafür eine der möglichen TURING-Berechnungen fixiert ist!), vereinbaren wir, daß als Maß seiner Kompliziertheit die zusammengefaßte Signalisierende $\Phi_p(n)$ verwendet wird, die nach Definition gleich ist der Zeit, die für die Berechnung der ersten n Werte $p(1), \ldots, p(n)$ des Prädikates p verwendet wurde. Der Fall, wo p eine dual-rationale Zahl ist, unterscheidet sich, wie man leicht einsieht, nicht wesentlich vom Fall $p = \frac{1}{2}$, und darum werden wir unsere Aufmerksamkeit, wenn wir über beliebiges p sprechen, in der Tat auf jene Situation konzentrieren, wo $p(n)$ unendlich oft sowohl den Wert 1 als auch den Wert 0 annimmt. Wie bekannt ist [1], kann man die Arbeit einer p-Maschine N mittels einer determinierten Maschine L modellieren, wobei L Zugang zum Prädikat p hat, das als Orakel genommen wurde. (Kürzer: Mittels einer Maschine mit p-Orakel.) Die zeitliche Signalisierende $t_L(x)$ einer solchen Maschine berechnet die Zeit für die Vorbereitung der Fragen an das Orakel und für ihre Bearbeitung; es gilt allerdings, daß Antworten auf Fragen “$p(n) = ?$“ augenblicklich, ohne zusätzlichen Zeitaufwand, erhalten werden. Da wir das Orakel p als Prädikat eines natürlichen Argumentes n betrachten, so ist es vorteilhaft, auch das Prädikat, welches von einer p-Maschine berechnet wird, als von einem natürlichen Argument abhängig zu betrachten. Darüber hinaus werden wir überall in diesem Paragraphen den Standpunkt einnehmen, daß das natürliche Argument in einer unären Darstellung gegeben ist. Abschätzungen für andere Kodierungsverfahren kann man leicht aus denjenigen erhalten, die unter den genannten Einschränkungen erhalten werden.

Gleichzeitig mit der zeitlichen Signalisierenden $t_L(n)$ ist es nützlich, die Arbeit einer Maschine mit p-Orakel auch durch die zulässige Größe der Fragen, mit welchen sich die Maschine L an das Orakel wenden kann, zu charakterisieren. Als ein solches Maß betrachten wir die Funktion $\Phi_L(n)$, die nach Definition gleich der größten der natürlichen Zahlen r ist, bei denen das Orakel auf die Frage „$p(r) = ?$“ antworten muß, wenn die Maschine mit dem Argument n in Betrieb genommen wurde.

Bei den unten aufgeführten Abschätzungen tritt hin und wieder die Funktion h (möglicherweise mit Indizes) auf, die den folgenden Bedingungen genügt:

a) h ist eine nichtfallende und nicht beschränkte Funktion.
b) h ist eine effektive Funktion.

Die Menge aller dieser Funktionen bezeichnen wir durch H.

Tatsächlich interessieren uns Abschätzungen unter Zuhilfenahme langsam wachsender Funktionen. Es ist aber bekannt, daß jede Funktion $h \in H$ durch eine Funktion $h' \in H$, welche hinreichend einfach berechenbar ist, minoriert werden kann; $h'(n)$ ist zum Beispiel auf einer TURING-Maschine mit einem Kopf in der Zeit $\asymp n \cdot \log n$ berechenbar. Indem wir das oben Gesagte berücksichtigen, werden wir dort, wo es notwendig ist, ohne besonderen Hinweis den Standpunkt einnehmen, daß die betrachtete Funktion aus H mit logarithmischer Verzögerung berechenbar ist (d. h., in der Zeit $\asymp n \cdot \log n$).

Satz 5. (1). *Möge die p-Maschine N das Prädikat Γ mit der Zuverlässigkeit $\Delta > \frac{1}{2}$ berechnen. Dann gibt es für eine beliebige Funktion $h \in H$ eine Maschine L mit p-Orakel, die ebenfalls Γ berechnet und für die gleichzeitig die Abschätzung*

$$t_L(n) \leqq 2^{t_N(\Delta, n)} \cdot t_N^2(\Delta, n) ,$$

$$\varphi(n) \leqq \log_2 \sqrt{t_N(\Delta, n)} + h(n)$$

gilt.

(2). *Wenn unter Beibehaltung der vorhergegangenen Voraussetzungen p eine effektive Zahl ist, so gibt es eine Γ berechnende determinierte Maschine M, für die*

$$t_M(n) \leqq 2^{t_N(\Delta,n)} \cdot t_N^2(\Delta, n) + \Phi_p(\log_2 \sqrt{t_N(\Delta, n)} + h(n))$$

gilt.

Beweis. Klar ist, daß (2) eine unmittelbare Folge aus (1) ist. Die Begründung der Behauptung (1) ergibt sich aus folgenden Fakten:

A. Wahrscheinlichkeitstheoretische Abschätzung. Es werden im Raum I^ν aller dualen Wörter der Länge ν die Maße μ und μ_ν betrachtet, die von den BERNOULLI-Generatoren D_p bzw. D_{p_ν} induziert werden. Wir werden die Nähe dieser Maße bei fixiertem p und $p_\nu \to p$, d. h., die Größe

$$\varepsilon_\nu \underset{\text{def}}{=} \sup_{D \subseteq I^\nu} |\mu(D) - \mu_\nu(D)| ,$$

abschätzen.

Standardberechnungen ergeben die Richtigkeit der Implikation $|p_\nu - p| = = \nu\,(1/\sqrt{\nu}\,) \to \lim \varepsilon_\nu = 0$.

Als Näherungen p_ν kann man die Ausdrücke $\tilde{p}(s(\nu)) \underset{\text{def}}{=} 0, p(1) \ldots p(s(\nu))$ der dualen Zerlegung der Zahl $p = 0, p(1) \ldots p(n) \ldots$ nehmen. Es ist zum Beispiel hinreichend, daß für eine gewisse Funktion $h \in H$ die Gleichung

$$s(\nu) = \text{ganzer Teil von } [\tfrac{1}{2} \log \nu + h(\nu)] \qquad (*)$$

erfüllt ist.

Wichtig ist zu bemerken, daß bei einer solchen Definition der Funktion $s(\nu)$ nicht nur $\lim \varepsilon_\nu = 0$ garantiert wird, sondern daß auch die Funktion $s(\nu)$ selbst mit logarithmischer Verzögerung berechenbar ist.

B. Modellierung. Wir kehren nun zur p-Maschine N, die Γ mit der Zuverlässigkeit $\Delta > \frac{1}{2}$ ausrechnet, zurück und nehmen als Näherungen p_ν für ihren Parameter p die soeben definierten Größen $\tilde{p}(s(\nu))$. Es sei $\varepsilon \underset{\text{def}}{=} \frac{(\Delta - \frac{1}{2})}{2}$ und c eine solche Konstante, daß

$$\nu > c \to \varepsilon_\nu < \varepsilon .$$

Unten wird beschrieben, wie die Maschine L die Arbeit der p-Maschine N für die Werte des Argumentes n modelliert, für die $t_N(\Delta, n) > c$ ist. (Das Umstellen der Arbeit von L für den Fall $t_N(\Delta, n) \leqq c$ unter Beachtung der nötigen Abschätzungen ist trivial!) Also zerfällt die Arbeit der Maschine L in Zyklen $Z_1, Z_2, \ldots, Z_\nu, \ldots$.

Im Zyklus Z_ν geschieht das folgende:

a) Berechnung (mit logarithmischer Verzögerung) von $s(\nu)$ nach der Formel (*);

b) „Heraussondern“ von $p(1), \ldots, p(s(\nu))$ aus dem Werte-Orakel. Damit wird eine dual-rationale Näherung p_ν des Parameters p fixiert;

c) Modellieren der Arbeit der Maschine N in dualen Wörtern ξ der Länge ν, die vom Generator ausgegeben werden. Damit wird die Teilmenge der Wörter P_1 (P_0) ausgesondert, die das Resultat 1 (0) begünstigen;

d) Berechnung von $\mu_\nu(P_0)$ und $\mu_\nu(P_1)$;

e) Wenn eine der Mengen P_i schon ein hinreichend großes Maß hat, so wird die Berechnung beendet. Bei $\mu_\nu(P_0) > 1/(2 - \varepsilon)$ wird nämlich das Resultat 0 und bei $\mu_\nu(P_1) > 1/(2 - \varepsilon)$ das Resultat 1 fixiert. Andernfalls Übergang zum Zyklus $Z_{\nu+1}$.

Es ist hinreichend offensichtlich, daß die Modellierung gerade bei $\nu = t_N(\Delta, n)$ mit der Ausgabe des richtigen Resultates und der aufgezeigten Abschätzung für $\Phi_L(n)$ beendet wird. Was $t_L(n)$ betrifft, so bestätigt sogar eine hinreichend grobe Analyse der Etappen **c)**—**d)** die Richtigkeit der aufgezeigten Abschätzung. (Höchstwahrscheinlich wird diese Abschätzung sogar noch etwas verbessert werden können. Es kann sogar sein, daß für t_L eine gleiche Abschätzung gilt wie in Satz 2.)

Beim Vergleich von Funktionen bezüglich ihres Wachstums werden wir neben einer Beziehung vom Typ $f(n) = o(g(n))$ auch eine stärkere Beziehung $f(n) \underset{\text{eff}}{=} o(g(n))$ verwenden, die folgendes bedeutet: Es existiert eine effektive, gegen ∞ strebende Funktion $r(n)$ derart, daß $r(n) \leqq g(n)/f(n)$ ist. Weiter unten wird die nachstehende Folgerung nützlich sein.

Folgerung aus Satz 5. *Wie auch die Funktion $\lambda(n) \underset{\text{eff}}{=} o(4^n)$ beschaffen ist, es gibt ein Paar von Funktionen $h_1, h_2 \in H$, für das die folgende Behauptung gilt: Möge die p-Maschine N das Prädikat Γ mit der Zuverlässigkeit $\Delta > \frac{1}{2}$ berechnen, wobei $t_N(\Delta, n) \leqq \leqq \lambda(n)$ ist. Dann gilt für eine entsprechende Maschine L mit Orakel die Abschätzung*

$$t_L(n) \leqq 2^{4^{n-h_1(n)}}, \qquad \Phi_L(n) \leqq n - h_2(n) .$$

Es ist natürlich zu erwarten, daß man für jedes beliebige Prädikat Γ, bei Wahl eines entsprechenden Wortes des Parameters p, seine hinreichend schnelle Wahrscheinlichkeitsberechnung organisieren kann. Man kann z. B. wie in [1] gezeigt vorgehen, d. h., als p kann Γ selbst gewählt werden, während der Wahrscheinlichkeitsmaschine die angenäherte Berechnung der reellen Zahl p auf dem Wege der Häufigkeitsanalyse der Ausgabe von Einsen durch den D_p-Generator übertragen wird. Wenn jedoch in der dualen Zerlegung der Zahl p eine Serie aus Einsen oder Nullen mit recht schnell wachsender Länge angetroffen wird, dann wird auch die Zeit, die zur Führung der Statistik notwendig ist, sehr schnell wachsen. Die Ursache besteht darin, daß wir nicht einfach gute Näherungen für die Zahl p, sondern Näherungen mit Fehlern brauchen, die durch Bewahrung einer gewissen Anzahl von dualen Ziffern aus der Zerlegung von p und durch Verwerfung der übrigen erhalten werden. Diese Schwierigkeiten kann man beispielsweise folgendermaßen umgehen: Für p wählen wir eine Zahl, deren Dualzerlegung sich aus der dualen Folge

$$\Gamma(1)\,\Gamma(2)\,\Gamma(3)\,\ldots\,\Gamma(n)\,\ldots,$$

mittels Einfügen einer Null zwischen $\Gamma(1)$ und $\Gamma(2)$, einer Eins zwischen $\Gamma(4)$ und $\Gamma(5)$, einer Null zwischen $\Gamma(9)$ und $\Gamma(10)$ usw. ergibt, d. h., indem wir abwechselnd Nullen und Einsen zwischen $\Gamma(\nu^2)$ und $\Gamma(\nu^2 + 1)$ einsetzen. Schon eine solche Wahl von p gewährleistet die Richtigkeit der folgenden Abschätzung.

Satz 6. *Wie auch das Prädikat Γ sein möge, es existiert eine Maschine N, welche bei Hinzuschaltung eines entsprechenden D_p-Generators dieses Prädikat mit der Zuverlässigkeit $\Delta > \frac{1}{2}$ und so berechnet, daß*

$$t_N(\Delta, n) \leqq 4^{n(1+o(n))}$$

ist.

Für gewisse Prädikate kann man natürlich auch schnellere, diese Prädikate berechnende p-Maschinen auswählen. Indem wir zum Beispiel den Satz 6 verwenden, ist es nicht schwer zu zeigen, daß beliebig komplizierte Prädikate (bei determinierten Berechnungen) und sogar nichteffektive Prädikate existieren, welche auf entsprechenden p-Maschinen mit der Abschätzung $t(\Delta, n) \leqq n \cdot \log n$ berechenbar sind. Es bleibt ungeklärt, ob diese Behauptung ihre Gültigkeit behält, wenn die Forderung „$\leqq n \cdot \log n$" durch die härtere Forderung „$\leqq n$" ersetzt wird. In diesem Zusammenhang ist es angebracht, an das bekannte Resultat von M. O. Rabin [7] über die Darstellung von Ereignissen mittels endlicher Wahrscheinlichkeitsautomaten mit isoliertem Schnittpunkt zu erinnern. In unserer Terminologie lautet dieses Resultat so: Bei beliebigem p ist jedes Prädikat, welches auf einer p-Maschine mit der Zuverlässigkeit $\Delta > \frac{1}{2}$ und in der Zeit $\tau^{\max}(n) \leqq n$ berechenbar ist, ein endlich-automatiges. Anders gesprochen, bei sehr schnellen Wahrscheinlichkeitsberechnungen wird die Information, die im Parameter p enthalten ist, nicht hinreichend gut ausgenutzt werden können.

Ein anderes Echo auf diese Erscheinung ist der folgende Satz.

Satz 7. *Sei $\lambda(n) \underset{\text{eff}}{=} o(4^n)$. Dann existiert ein effektives Prädikat $\Gamma_\lambda(n)$, das auf einer gewissen determinierten Maschine in der Zeit $2^{o(4^n)}$ berechenbar ist, aber für kein p mit der Zuverlässigkeit $\Delta > \frac{1}{2}$ auf einer p-Maschine in der Zeit*

$$t(\Delta, n) \leqq \lambda(n)$$

berechenbar ist.

Beweis. Wegen Folgerung aus Satz 5 eignet sich als Γ_λ ein beliebiges Prädikat Γ, das den folgenden Bedingungen genügt:

A. Wenn irgendeine Maschine L mit Orakel Γ mit der Abschätzung $\varphi_L(n) \leqq$ $\leqq n - h_2(n)$ berechnet, so gilt

$$\exists_n^\infty [t_L(n) > 2^{4^{n-h_1(n)}}].$$

B. Γ ist determiniert in der Zeit $2^{o(4^n)}$ berechenbar.

Ein solches Prädikat Γ_λ kann man durch Vervollkommnung einer Diagonalkonstruktion von D. Hodshajew [8] ermitteln.

Für ein beliebiges Paar effektiver Funktionen f und g (es wird $g(n) \to \infty$ vorausgesetzt) wird nämlich das Prädikat Γ_{fg} konstruiert, das der Bedingung genügt

A'. Wenn irgendeine Maschine L mit Orakel Γ_{fg} mit der Abschätzung $\varphi_L(n) \leqq \leqq n - g(n)$ berechnet, so gilt

$$\exists^\infty n[t_L(n) > f(n)] .$$

Weiter, wenn f und g hinreichend einfach determiniert berechenbar sind, und zwar, wenn f in der Zeit $\leqq f(n) \cdot \log f(n)$, $g \in H$, berechenbar ist, so ist auch die Bedingung

B'. $\Gamma_{fg}(n)$ ist determiniert berechenbar in der Zeit $\leqq (\text{const})^n \cdot f(n) \cdot \log f(n)$, erfüllt.

Wenn jetzt als f und g die Funktionen $2^{4^n - h_1(n)}$ bzw. $h_2(n)$ genommen werden, so nehmen **A'** bzw. **B'** gerade die Form **A** bzw. **B** an. Es bleibt noch übrig zu bemerken, daß bei einer derartigen Auswahl das Prädikat Γ_{fg} nur von der Wahl von λ abhängt, d. h., es ist das gewünschte Prädikat Γ_λ.

Zum Schluß dieses Paragraphen werden wir die Rechenzeit auf p-Maschinen bei beliebigen Werten des Parameters p mit der Rechenzeit auf $\frac{1}{2}$-Maschinen vergleichen. Dazu schätzen wir die Arbeitszeit einer $\frac{1}{2}$-Maschine N ab, die mit einer Zuverlässigkeit $\Delta' > \frac{1}{2}$ den Berechnungsprozeß des Prädikates Γ mit der Zuverlässigkeit $\Delta > \frac{1}{2}$ mittels einer gegebenen p-Maschine N modelliert. Faktisch erweist sich Δ' etwas kleiner als Δ. Dieser Umstand ist aber, wenn wir Satz 1 berücksichtigen, unwesentlich.

Satz 8. *Möge eine p-Maschine N mit einem Kopf das Prädikat Γ mit der Zuverlässigkeit $\Delta > \frac{1}{2}$ berechnen. Dann gibt es eine $\frac{1}{2}$-Maschine N' mit einem Kopf, welche Γ mit einer gewissen Zuverlässigkeit $\Delta' > \frac{1}{2}$ berechnet und zwar so, daß gilt*

$$t_N(\Delta', n) \leqq t_N(\Delta, n) \cdot \log t_N(\Delta, n) + \Phi_p(\log t_N(\Delta, n) + \text{const}) .$$

Bemerkungen. (1). In der Klasse der Maschinen mit mehreren Köpfen kann man die bessere Abschätzung

$$t_N(\Delta', n) \leqq t_N(\Delta, n) + \Phi_p(\log t_N(\Delta, n) + \text{const})$$

garantieren.

(2). In Analogie zu Satz 5 könnte man den Begriff der $\frac{1}{2}$-Maschine mit Orakel einführen und für sie getrennt eine Abschätzung der Zeit für die Zurückführbarkeit und eine Abschätzung der Fragenlänge ermitteln.

Beweis von Satz 8. Die Maschine N' modelliert jede an den D_p-Generator gerichtete Anfrage der Maschine N mittels eines Vergleichszyklus und kopiert im übrigen die Arbeit der Maschine N. Im Verlaufe eines solchen Zyklus werden die Ausgänge des $\frac{1}{2}$-Generators $\xi(1), \xi(2), \ldots$ der Reihe nach mit den Ziffern $p(1), p(2), \ldots$ der Zerlegung der Zahl p verglichen und zwar solange, bis zuerst ein solches i gefunden ist, für das $\xi(i) \neq p(i)$ ist. Klar ist, daß wahr $\{p(i) = 1\} = p$, wahr $\{p(i) = 0\} = 1 - p$ ist. Der zweite Summand in der angeführten Abschätzung entspricht dem folgenden einfachen Fakt. Möge $R(v, c)$ die Wahrscheinlichkeit dafür bedeuten, daß in v aufeinanderfolgenden Vergleichszyklen mehr als $\log v + c$ Ziffern der Zerlegung von p

verwendet werden. Dann gilt $\forall \varepsilon \exists c \forall v$: $v > c \rightarrow R(v, c) < \varepsilon$. Der erste Summand spiegelt das Kopierungsverfahren der eigentlichen Rechenschritte der Maschine N wider.

Folgerung aus Satz 8. *Sei* $\Phi_p(n) = o(2^n)$. *Dann ist in der Klasse der Maschinen mit einem Kopf*

$$t_{N'}(\Delta', n) \leq t_N(\Delta, n) \cdot \log t_N(\Delta, n)$$

und in der Klasse der Maschinen mit mehreren Köpfen

$$t_{N'}(\Delta', n) \leq t_N(\Delta, n) .$$

Anders gesagt, die Verwendung eines „unkomplizierten" D_p-Generators ($\Phi_p(n) = o(2^n)$) kann keinen wesentlichen Gewinn an Rechengeschwindigkeit im Vergleich zu einem einfachen $\frac{1}{2}$-Generator geben.

Literatur

[1] DE LEEUW, K.; MOORE, E. F.; SHANNON, C. E.; SHAPIRO, N.: Computability by probabilistic machines. Automata Studies, Annals of mathematical studies, No. 34, Princeton U. Press, Princeton, N.Y., 1956.

[2] Яблонский, С. В.: Об алгоритмических трудностях синтеза минимальных контактных схем. Проблемы кибернетики, 1959, No. 2.

[3] Барздинь, Я. М.: О вычислимости на вероятностных машинах. Докл. АН СССР 185 (1969), No. 4.

[4] TURAKAINEN, P.: On languages representable in rational probabilistic automata. Acad. Sci. Fen., 1969, A1, No. 439.

[5] Коваленко, И.Н.: Замечание о сложности вероятностных и детерминированных конечных автоматов. Кибернетика, 1965, No. 2, 35—36.

[6] Трахтенброт, Б. А.: Оптимальные вычисления и частотное явление яблонского. Алгебра и логика, Семинар 4 (1965), No. 5, 79—93.

[6a] Трахтенброт, Е. А.: Замечания о сложности вычислений на вероятностных машинах. Теория Алгорифмов и математическая лосика (Посвяицается Андрею Андреевичу Маркову кего семидеся чилетию), редакц. Б. А. Кишнера; Н. М. Нагорного, 159—176, высисистелсный центр., Акад. Наук СССР, Москва, 1974.

[7] RABIN, M. O.: Probabilistic automata. Information and Control. 6 (1963), No. 3.

[8] Ходжаев, Дж.: О сложностпе вычислений на машинах Тьюринга с оракулом. Автореферат диссертации, Ташкент, 1970.

In der gleichen Reihe sind bisher erschienen

Band 1

Universale Algebren und Theorie der Radikale

herausgegeben von H. J. Hoehnke

1976. VI, 85 Seiten, 18,– M, Bestell-Nr. 7622237 (2174/1)

Band 2

Algebraische Probleme der mathematischen und theoretischen Physik

von G. A. Saizew, in deutscher Sprache herausgegeben von H. J. Hoehnke

1979. X, 154 Seiten, 32,– M, Bestell-Nr. 7622456 (2174/2)

Band 3

Kategorielle Algebra

von G. Richter

1979. VIII, 299 Seiten, 58,– M, Bestell-Nr. 7623773 (2174/3)

Band 4

Lectures on Lie Algebras

von J. A. Bahturin (in englischer Sprache)

1978. VIII, 126 Seiten, 28,– M, Bestell-Nr. 7624929 (2174/4)

Band 5

The Concept of Fuzziness in Automata and Language Theory

von W. Wechler (in englischer Sprache)

1978. VIII, 141 Seiten, 27,– M, Bestell-Nr. 7625171 (2174/5)

Band 6

Discriminator-Algebras

von H. Werner (in englischer Sprache)

1978. VI, 92 Seiten, 18,– M, Bestell-Nr. 7625198 (2174/6)

Die Beitragssammlung enthält thematisch geordnet, aber unabhängig von einander lesbar, Beiträge zu folgenden Problembereichen:

A. Allgemeine Kategorientheorie
B. Spezielle Kategorien
C. Gruppen und Gruppoide
D. Algebren über Monaden
E. Realisierungstheorie
F. Modelltheorie
G. Kompliziertheitstheorie

Die Autoren sind international anerkannte Algebraiker, sowohl aus der DDR als auch aus dem Ausland.

This collection of papers contained topically ordered, yet readable independent of one another, contributions to the following problems.

A. General category theory
B. Special categories
C. Groups and groupoids
D. Algebras over monades
E. Realisation theory
F. Model theory
G. Complexity theory

The authors are internationally known mathematicians dealing with algebra from the GDR as well as from abroad.